T0178031

MIGRATION

While the subject of migration has received enormous attention in academic journals and books across the social sciences, introductory texts on the matter are few and far between. Even fewer books have explored migration through a critical and explicit engagement with spatial concepts.

Now in its second edition, *Migration* remains the only text in more than a decade that emphasizes how geographical or spatial concepts can be used critically to understand migration. The multi-disciplinary text draws on insights from human geography, political science, social anthropology, sociology, and to a lesser extent economics. All of the chapters focus on key terms, theories, concepts, and issues concerning migration and immigration. The book argues that in the context of migration, two opposing 'spatial positions' have emerged in the wake of the critique of 'methodological nationalism'. On one hand is the significance of 'transnationalism', and on the other, the importance of 'sub-national' or local processes. Both require more nuance and integration, while many of the concepts and theories which have thus far neglected space or have not been 'treated' spatially, need to be re-written with space in mind. Pedagogically the text combines a carefully defined structure, accessible language, boxes that explore case studies of migrant-related experiences in particular places, annotated suggestions for further reading, useful websites and relevant films and summary questions for student learning at the end of each chapter.

Migration provides a critical, multi-disciplinary, advanced, and theoretically informed introduction to migration and immigration. Revised and updated with new material, new maps and illustrations and an accompanying website (https://migration2ndedition.wordpress.com/), it continues to be aimed at advanced undergraduates and Masters-level graduate students undertaking courses on migration and immigration.

Michael Samers is Associate Professor of Geography at the University of Kentucky, USA. His research interests include the urban and economic dimensions of migration, as well as international finance. He is a co-author of *Spaces of Work: Global Capitalism and Geographies of Labour*.

Michael Collyer is Professor of Geography at the University of Sussex, UK and has held visiting positions at universities in Egypt, Morocco, New Zealand, Sri Lanka, and the USA. He is primarily a political geographer with research interests in the politics of mobility. He is editor of *Emigration Nations: Policies and Ideologies of Emigrant Engagement*.

Key Ideas in Geography

SERIES EDITORS: SARAH HOLLOWAY, LOUGHBOROUGH UNIVERSITY AND GILL VALENTINE, SHEFFIELD UNIVERSITY

The *Key Ideas in Geography* series will provide strong, original, and accessible texts on important spatial concepts for academics and students working in the fields of geography, sociology and anthropology, as well as the interdisciplinary fields of urban and rural studies, development and cultural studies. Each text will locate a key idea within its traditions of thought, provide grounds for understanding its various usages and meanings, and offer critical discussion of the contribution of relevant authors and thinkers.

For a full list of titles in this series, please visit www.routledge.com/series/KIG

MIGRATION

Second Edition

Michael Samers and Michael Collyer

Routledge
Taylor & Francis Group

LONDON AND NEW YORK

Second edition published 2017
by Routledge
2 Park Square, Milton Park, Abingdon, Oxon, OX14 4RN

and by Routledge
711 Third Avenue, New York, NY 10017

Routledge is an imprint of the Taylor & Francis Group, an informa business

First edition published by Routledge 2010

British Library Cataloguing in Publication Data
A catalogue record for this book is available from the British Library

Library of Congress Cataloging in Publication Data
Names: Samers, Michael.
Title: Migration / Michael Samers and Michael Collyer.
Description: Second edition. | New York : Routledge, 2017. |
    Series: Key ideas in geography | "[First edition published by Routledge
    2010]"—T.p. verso. | Includes bibliographical references and index.
Identifiers: LCCN 2016030488| ISBN 9781138924468 (hardback : alk.
    paper) | ISBN 9781138924475 (paperback : alk. paper) |
    ISBN 9781315684307 (eBook)
Subjects: LCSH: Emigration and immigration. | Emigration and
    immigration—Social aspects. | Spatial behavior. | Human geography.
Classification: LCC JV6091 .S36 2017 | DDC 304.8—dc23
LC record available at https://lccn.loc.gov/2016030488

ISBN: 978-1-138-92446-8 (hbk)
ISBN: 978-1-138-92447-5 (pbk)
ISBN: 978-1-315-68430-7 (ebk)

Typeset in Joanna MT
by Swales & Willis Ltd, Exeter, Devon, UK

CONTENTS

FIGURES

MAPS

TABLES

Boxes

PREFACE FOR STUDENTS

Welcome to the second edition of *Migration*. We hope you will find it helpful for your studies or your research, depending on your university level, or even if you are starting out at post-graduate studies (graduate school in North America). This second edition of *Migration* is, like the first edition, designed as a critical, multi-disciplinary, and advanced introduction to migration and immigration. It is generally aimed at advanced undergraduates and Masters-level graduate students (in Canada and the US) or post-graduate students in the UK and elsewhere, who are undertaking courses on migration and immigration. You will probably find that there are thousands of books, academic journal articles, reports, and popular media articles written on migration. At times, it can be hard to know how to make sense of the news, or if you are undertaking research for a paper, where to start, and it all might become rather overwhelming. This book will hopefully be a guide for you since it is designed to be a systematic overview (although at times with quite a bit of detail) of migration and immigration. It includes the perspectives of many disciplines, including anthropology, geography, political science and sociology so it may provide material which is completely fresh or new to you. Now since the authors of this book are geographers, we are especially interested in matters of 'space'. However, no need to worry if you are not a geographer or if your course is in say, sociology; 'space' has become the concern of many social science disciplines, not least the ones mentioned above. So much of the book is devoted to trying to understand migration and immigration from a geographical perspective, which – as we outline in the Introduction – centres on four related issues: 1) The causes and consequences of migration; 2) The conflicted task of governing migration; 3) The question of employment and settlement for migrants; and 4) The question of citizenship and belonging for and among migrants.

We hope you find the language more or less accessible. No doubt there will be a lot of unfamiliar concepts, terms, and words, but if you can work through these, then the rewards should be worth it. We tried to make it at least a lit bit more enjoyable with boxes that explore, for example, case studies of migrant-related experiences in particular places, annotated suggestions for further reading, useful websites and relevant films, as well as summary questions for student learning at the end of each chapter. As with the first edition, we hope that we have offered a readable, enlightening, and lasting volume that will help you understand migration and immigration.

Michael Samers and Michael Collyer
Lexington, KY, USA, and Brighton, UK
30 April 2016

PREFACE FOR INSTRUCTORS

As with the first edition of Migration, this second edition is designed as a critical, multi-disciplinary, and advanced introduction to migration and immigration. It is generally aimed at advanced undergraduates and Masters-level (post-)graduate students undertaking courses on migration and immigration. While the subject of migration has received enormous attention in academic journals and books across the social sciences, introductory texts on the matter are few and far between. Even fewer books have explored migration through a critical and explicit engagement with spatial concepts. In fact, geographers, who might have otherwise placed spatial concepts at the forefront of their discussion, have produced only a handful of books uniquely about migration or immigration.[1]

In this respect, the second edition of Migration has five distinctive features. First, besides the first edition of the book, the second edition remains the only text in more than 15 years to show how geographical or spatial concepts can be used critically to understand migration. Indeed, many of the concepts and theories used in the explanation of migration or immigration are either not explicit about space, or need to be re-written with space in mind. This is not simply a matter of academic concern or disciplinary flag-waving; a critical appreciation of space will enable us to intervene more appropriately in policy debates in order to address some of the terrible practices that compel disadvantaged people to migrate and the difficult lives they lead in the countries of immigration. Second, it is designed to be an avowedly multi-disciplinary text, borrowing as it does from human geography, political science, social anthropology, sociology, and to a less extent economics. Third, all of the chapters focus on key terms, theories, concepts, and issues concerning migration and immigration. Fourth, *Migration* is not simply an encyclopaedic overview of migration 'facts', trends, and migration systems based on world regions or nation-states, as is often the case in other volumes. Rather, it is designed to have lasting intellectual value through developing a *theoretical and conceptual analysis* of migration and immigration, while maintaining its accessibility for an undergraduate audience. A fifth distinctive feature of the book is once again its pedagogical objective. In this regard, we have sought to combine a carefully defined structure, accessible language, boxes that explore, for example, case studies of migrant-related experiences in particular places, annotated suggestions for further reading, useful websites and relevant films, as well as summary questions for student learning at the end of each chapter. While many students might find some sections of this volume difficult in terms of its concepts and language, we thought it important to challenge students to work through theoretical, complex and contradictory debates which illustrate that scholars actually disagree on how they view the social world. This book then is partly an exercise in theoretical and conceptual humility, but with, once again, pedagogical aims.

The first edition of *Migration*, which followed the aims of the *Key Ideas in Geography* series, sought to combine a highly readable, advanced introductory text, with a core argument or set of arguments. This version has dropped some of the constraints of the first edition, namely that the text should have a singular, core argument. In its place, we tried to be a bit more comprehensive, and Michael Samers brought in Michael Collyer's expertise on the relationship between 'development' and asylum and refugee migration for example. Yet, the subject of migration is itself so diverse and wide-ranging that encapsulating it into a single volume has forced us to narrow considerably the range of theoretical and substantive material. Indeed, there are literally thousands of papers and hundreds of books published on various dimensions of migration and immigration every year. We have certainly left out many studies that we would have liked to include. In this respect, the process of narrowing the book involved three choices. First, the text is meant as a portrait of the present, and the majority of the substantive material is gleaned from the twenty-first century. We do however draw on some older comparisons in a few places in the book, but it was simply not possible to engage with the longer-term historical development of the issues in this text, as requested by some reviewers of the first edition and reviewers of the proposal for the second edition. Second, while we ignore a longer historical outlook, we do seek to provide readers with a sense of the intellectual history of migration ideas, since it is our contention that much research on migration suffers from an intellectual amnesia. In doing so, we actually do illustrate some key historical themes, at least for migration from poorer to richer countries. Third, as we touched on above, we have relied on our own strengths in terms of migration to provide further direction to the book. That is, Michael Samers has over the last 20 years or so, limited his research to two dimensions of migration: first, the relationship between work and 'low-skilled', 'low income' or 'low-paid' migration from poorer countries to richer ones, and second, the geopolitics or political economy of migration, particularly with respect to France and the EU but also in relation to NAFTA. In contrast, Michael Collyer has focused on both the relationship between 'development' and (asylum/refugee) migration, particularly from North Africa to the UK. So these two forms and patterns of migration are covered disproportionately in this volume, although the book seeks to move beyond this exclusive thematic and geographical focus by emphasizing 'south-to-south' migration and settlement issues, as well as a special focus on East Asian countries, such as Malaysia, Korea, and Singapore. In fact, we have not only added new themes to this edition (such as the relationship between environmental change and migration, but also 'marriage migration' in Asia), as well as new concepts theories, several new or revised boxes, and new data.

As with the first edition, we hope that we offer a readable, enlightening, and lasting volume that will help us understand migration and immigration, not just in the present, but in the past and future.

Michael Samers and Michael Collyer
Lexington, KY, USA, and Brighton, UK
30 April 2016

NOTE

1   Perhaps the most significant of these Anglophone contributions by geographers include the textbook *Exploring Contemporary Migration* by P. Boyle, K. Halfacree, and V. Robinson, 1998; the introductory text *International Migration: A very short introduction* by K. Koser, 2007; the edited collection *A New Geography of European Migrations*, by R. King, 1993; more specialized monographs including *Labor Movement: How migration shapes labour markets*, by Harald Bauder, 2005; *Writing Across Worlds*, by R. King and P. White, 1995; *The Lie of the Land: Migrant Workers and the California Landscape*, by D. Mitchell, 1996; 6 Billion

Plus, by K. Bruce Newbold, 2007; *Making Population Geography* by Adrian Bailey, 2005; *Managing Displacement: Refugees and the Politics of Humanitarianism*, 2000, by Jennifer Hyndman; *The California Cauldron: Immigration and the Fortunes of Local Communities*, 1998, and *Immigrants and the American Dream: Remaking the middle class*, 2003, both by W.A.V. Clark. Furthermore, all of these were written more than a decade ago. Beyond the literature by geographers and the myriad edited collections, a more recent and popular text is Castles and Miller's *The Age of Migration* (2014, 5th ed.), but this volume does not place the critical use of spatial concepts at the heart of its enquiry.

ACKNOWLEDGEMENTS

We could not have written this book without the assistance of a number of friends and colleagues. In particular, we would like to thank Andrew Mould and Egle Zigaite at Routledge/Taylor & Francis who pushed us to finish the project, and to Katharine Bartlett for copy-editing. We also express our sincere gratitude to Dick Gilbreath and Jeff Levy at the Gyula Pauer Center for Cartography & GIS, University of Kentucky for work on many of the figures and maps. Several anonymous reviewers offered excellent comments and suggestions on the proposal for the second edition, and ensured that this project did not consist solely of our own interests and musings.

Michael Collyer would like to thank colleagues in the Geographies of Migration research cluster at the University of Sussex.

Michael Samers is grateful to the following individuals: Laura Stanganini, who translated the first edition of *Migration* into Italian; the graduate students at the University of Kentucky, and in particular Jonghee Lee (who helped translate the first edition of *Migration* into Korean), Mitchell Snider who offered insights and suggestions for readings and editorial assistance, and Charles Heller, research fellow at the Center for Research Architecture, Goldsmiths College, who pointed to references on issues related in particular to seaward migration. Michael Samers is especially grateful to the Department of Geography at Ehwa Womans University, and in particular Professor Youngmin Lee, Hyunuk Lee, and Hwayong Lee for their hospitality in Seoul, and for pointing him to migration issues in Asia. Michael Samers is also grateful to the professors, staff and graduate students at CERAPS, Université de Lille 2 where he spent a Fulbright year in 2013–14, especially Professors Thomas Alam,

François Benchendikh, Isabelle Bruno, Jean-Gabriel Contamin, Guillaume Courty, Annie Laurent, Remi Lefebvre, Manuel Schotte, Alexis Spire, Julien Talpin, and Karel Yon. Younnes Haddadi and Thomas Soubiran provided friendship and wonderful administrative support; graduate students Julie Aubertin, Gabriela Condurache, Rafaël COS, Alexandre Fauquette, and Alina Stan, among many others at CERAPS, helped me through the year and made it that much more enjoyable. In Paris, Claire Hancock (Université Paris-Est Créteil) and Mustafa Dikeç (LATTS, l'Université Paris-Est Marne-la-Vallée) offered hospitality and intellectual camaraderie.

Michael Samers would also like to thank the professors and graduate students in the Department of Sociology at the University of Amsterdam, especially Jan Rath and Sebastien Chauvin, where he was a Visiting Professor as part of MISOCO; Prof. Dr. Andreas Pott in IMIS at the University of Osnabrück where he gave a visiting lecture as an additional part of MISOCO; Professor Thomas de Vroome, University of Utrecht, Professor Felicitas Hillmann, then at the University of Cologne, and Pieter Bevelander at the Malmö Institute of Migration, Diversity and Welfare, Sweden, for inviting him to speak and for the feedback from faculty and students.

Finally, Michael Samers is especially grateful to Debra Gold who provided indispensable companionship, support, and most of all patience.

PERMISSIONS

1. Figures 1.2 and 1.3 rely on data from the OECD's annual SOPEMI reports; used with permission.
2. We are grateful to the UNHCR, UNRWA and UNDESA for the use of publicly available data that has informed Tables 1.3, 1.4 and 1.5.
3. For the discussion of 'Lillian' in Chapter 1: Reprinted by permission of the publisher from SUBURBAN SWEATSHOPS: THE FIGHT FOR IMMIGRANT RIGHTS by Jennifer Gordon, Cambridge, Mass.: The Belknap Press of Harvard University Press, Copyright © 2005 by Jennifer Gordon.
4. We would like to thank Professor Gavin Jones for Tables 3.1, 3.2, and 3.3 from Doo-Sub Kim (ed), 2012, *Cross-Border Marriage: Global Trends and Diversity*, Seoul: Korean Institute for Health and Social Affairs (KIHASA).
5. For Figure 3.1, we are grateful to Elsevier to reproduce 'Figure 2' from Black, R., Adger, N., Arnell, N.W., Dercon, S., Geddes, A., and

Thomas, D. (2011) The effect of environmental change on human migration, *Global Environmental Change*. 21, Supp. 1: S3–S11.

6. For the discussion of Bangladeshi migrants in Malaysia in 4.8, we are grateful to Elsevier to reproduce 4 excerpts from 'Transnational migration and the transformation of gender relations: The case of Bangladeshi labour migrants', by Petra Dannecker, in *Current Sociology*. 53: 655–74.

7. We would like to thank the International Organization for Migration, for permission to reproduce Figure 5.1

8. For Table 6.1, we are grateful to Marco Martiniello and the University of Amsterdam Press to adapt Table 2 by Harald Waldrauch, Voting rights of third country nationals in western Europe (25 EU states, Norway and Switzerland) pp. 110–11 from Martiniello, M. (2006) Political participation, mobilization and representation of immigrants and their offspring in Europe: *Migration and Citizenship: Legal Status, rights and political participation*, edited by Rainer Bauböck.

1

INTRODUCTION

On 3 October 2013, around 2 a.m., the diesel engine of a boat originating in Libya and carrying 518 migrants, stalled just a quarter mile off the island of Lampedusa, an Italian island in the Mediterranean Sea between the coast of Tunisia and the island of Sicily, Italy (see Map 1.1). Lampedusa is sought-after, not so much as a destination for migrants and the smugglers that bring them there, but as a stepping stone for migration to elsewhere in the European Union. According to witnesses, with the engine silent, the bilge pump ceased to function, sea water began to pour in, and the captain of the boat frantically tried to re-start the engine. When the engine refused, he lit a blanket with diesel fuel to attract the attention of the Coast Guard and other boats and persons near the shore. As he held the lit blanket, the captain burnt his hand on the flames and dropped it, which ignited residual diesel fuel on the upper deck. The flames spread, passengers awakened to the commotion, panicked, and hurried to one side of the boat. The boat leaned heavily to one side and eventually capsized. Those in the hot and crowded deck below stood little chance of surviving. Those on the deck above were plunged into the water, and as many could not swim, instantly drowned. Some survived by holding onto floating corpses. Of the 518 migrants, 366 would perish in the Mediterranean. The survivors of the sunken boat – covered in diesel fuel – were eventually rescued by the Coast

Guard and other boats. The bodies of those who did not survive were transported to the Italian island of Sicily and buried in cemeteries there. A state funeral held by the Italian government had extended an invitation to Eritrean officials, officials of the very regime which had propelled the migrants to leave Eritrea in the first place. Yet Italian officials did not extend an invitation to the survivors of the tragedy and they were even prevented from attending the funeral while still being held in detention camps for asylum-seekers in Lampedusa and Sicily. While those who lost their lives were eventually given posthumous citizenship (Isin 2014), this tragedy added to a long list of other similar ones, not just in the Mediterranean, but in the waters between Indonesia and Australia; in the Andaman Sea between India, Thailand, and Malaysia; in the waters near the Canary Islands off the coast of northwest Africa; in the Aegean Sea between Turkey and Greece; in the southwestern deserts of the US near the border with Mexico; and in the mountains between Iran and Turkey (*Guardian* 2014; Heller and Pezzani 2016; IOM 2014).

What lies at the roots of the Lampedusa tragedy, and why do such tragedies continually happen? Why did the migrants involved agree to undertake such an arduous and dangerous journey to an uncertain destination and future? Or to put it more broadly, why do people migrate, despite significant obstacles, and what sort of reception will they find in

Map 1.1 The location of Lampedusa and countries bordering the Mediterranean

their new destination? Unlike other general (text) books you might read on migration, this book tries to answer these questions in general and many others related to migration through a *geographical* or *spatial* perspective. This perspective involves an attention to such spatial concepts or metaphors as 'space' itself, but also 'place', 'node', 'friction of distance', 'territory', and 'scale'. We adopt such an approach because while so many volumes address the subject of migration, so few involve a *critical* and *explicit* engagement with spatial concepts.

Let us return to the tragedy above. From what is known from survivors' testimony, many of the passengers in this case were not actually from Libya at all, but from the country of Eritrea (see Map 3.2). How are such regions, countries or places connected, and how can an explicit attention to geographical concepts help to shed light on these series of events? To begin with, social networks (see the glossary at the end of the book and the discussion later in this chapter) allow migrants to communicate the value of particular destinations to other would-be migrants, and places such as Lampedusa *may* have both an imagined and potentially real promise for migrants. Second, these harrowing incidents point to different scales (territories) of regulation, among them the continuing ability of territories such as national states to decide who can enter and who cannot, and how supra-national territories such as the European Union step in to shape migration control. Yet the enforcement of migration regulation happens in particular places, and the interaction between migrants and Italian authorities in Lampedusa produce a particular local geography of enforcement. These more local spaces of regulation and enforcement seem to shape an entire migratory system which extends as far south as Eritrea. Above all, the events described above show the desperation of migrants in covering vast distances, often by the cheapest available option to reach the European Union and other wealthier states. What this discussion does not show perhaps is how the relationship between wealthier and poorer countries actually creates these migrations, and the reception that migrants might have once they do succeed in settling in the richer countries, but we will explore these issues later in the book.

This critical engagement with geographical concepts seems equally vital insofar as it enables us to assess what Sheller and Urry (2006) call 'the new mobilities paradigm' in the social sciences; in other words, the idea that the social sciences can be renewed again by exploring ideas of mobility rather than taking stability and stasis as the world's natural state of affairs.[1] In the same vein, Favell (2008) argues that migration should be a subset of mobility studies, and mobility and migration accepted as the norm. Once we accept this, nation-states will no longer be the benchmark against which such migration and mobility is gauged negatively. Rather migration and mobility would be seen as natural, and territory as aberrant. While we applaud much of this emphasis on mobility rather than stability, stasis, or territorial/nation-state centred analyses, and we recognize that mobility is part and parcel of the lives of millions of people across the world, it also questions whether international migration – particularly of asylum-seekers, refugees and low-income migrants – can be uncritically subsumed within this mobility approach. We say this since territorially defined borders and immigration regulations do much to impede mobility, though they also serve to create it and to shape it.[2]

This book draws unashamedly from across the social sciences, including works in anthropology, economics, human geography, political science, and sociology. Moving beyond the tidy world of disciplines that are typical of university departments is essential since migration is multi-faceted, having cultural, economic, political, and social dimensions. Yet its complexity also ensures that encapsulating these various dimensions in a single volume will forever be a challenge. In order to rise to this challenge, we maintain a focus on *international* low-income migration and immigration (including asylum-seekers and refugees), the causes and consequences of such migration, as well as the experiences of migrants and immigrants.

We choose this focus because we seek to provide a purposefully *critical* treatment of migration and immigration, not an arid and detached commentary on the geographical dimensions of migration, or a repository for a barrage of statistics, nor a synoptic review of every type or dimension of migration. By critical, we do not simply mean that this book is an attempt to think 'long and hard' about (spatial) concepts and ideas, but that this text is concerned with those migrants who are on the whole *disadvantaged*.

The volume has four additional foci. First, it places an emphasis on migration from the so-called 'global south' (or broadly speaking, poorer countries) to the 'global north' (in large measure, the richer countries). Nonetheless, we do not neglect internal migration (that is, migration within countries), especially *within* poorer ones. While this distinction between south and north may seem crude given the enormous diversity within these two hemispheres, we show in this book that making this distinction is important for explaining *why people migrate*, but may be less so for other issues related to immigration. Second, we focus on the experiences of migrants and immigrants within the global north, though experiences in poorer countries are also not neglected. Third, although this book focuses on disadvantaged migrants, it does provide some discussion of 'highly skilled' or 'high-income' migrants. It has often been asked by more critical observers to what extent highly skilled or high-income migrants should be the object of academic scrutiny since they are a comparatively privileged group of migrants. We share these concerns, although many of those who are considered highly skilled in their own countries end up performing menial jobs in the country of immigration, and they too are subject to racism and other exclusionary processes. By the same token, our interest in highly skilled migrants also stems from their role in constructing the sorts of economic activity and the kinds of jobs people perform in richer countries, often (but not always) at the expense of many people in poorer ones. Fourth, we also devote some attention to the migration of students, or what is now called international student mobility. Like highly skilled migrants, many student migrants may be relatively privileged compared to other low-income immigrants and asylum-seekers, but they too are subject to security fears and their manifestation in tighter visa controls, racist violence, discrimination and various forms of exclusion from jobs, public services, cultural spaces, and so forth. And like highly skilled migrants, international students serve the interests of governments in terms of economic development, and universities in terms of their search for greater financial resources, a diverse student body, and intellectual prestige.

As the aims of this book suggest, migration involves different people in a variety of situations, some more desperate than others. Most academic discussions of these situations are often expressed in quite abstract terms. In this section then, we begin by discussing four vignettes of migrant lives in order to place a 'human face' on the discussion of migration categories that follow.

MIGRANT STORIES AND KEY TERMS AND CATEGORIES IN THE STUDY OF MIGRATION AND IMMIGRATION

Laika (Jacqueline), the 'illegal' immigrant in Malaysia

Hilsdon (2006: 4–5) recounts the story of Laika, 22 years old, who came to Sabah (the Malaysian part of the island of Borneo), in the 1990s from the island of Mindanao in the Philippines. She arrived as a teenager in Pulau Jaya, Sabah. Her passport and visa were 'fixed' (that is to say, illegally arranged) by a relative before she left the Philippines. In Sabah, she began working at a local restaurant and eventually met a man, Salim, whom she later married, in part because she could not live on her meagre wages of some 300 ringgit (about US$94) per month. In addition, her illegal visa had expired, and for that reason it could not be renewed and she could not obtain the necessary documents to return home legally. Jacqueline was hardly the only person to face this situation as many women (but not just women) have faced questions about whether their visas were 'good enough' to continually stand up to official scrutiny. Thus Jacqueline, like so many others, have avoided public places such as shopping centres, markets, hospitals, government offices and public transport where police and other immigration officials often concentrate their policing activity.

The story of Asha, the asylum-seeker in Finland

Asylum-seeking – especially from the Middle East – is a significant dimension of migration to Scandinavia. Around 1990, many Somali asylum-seekers began migrating to Finland. Tiilikainen (2003) offers us the story of Asha. Born in Mogadishu, Asha got married and had two daughters and a son, while studying at the university for two years. Civil war erupted in Somalia, and Asha fled city and country. As a 24-year-old woman, Asha joined her brother in Finland where she sought refugee status. Asha was never joined by her husband and they eventually divorced. After three years, she was finally given approval to bring her three children from Somalia, but she also adopted another two children from her brother who had passed away. Taking care of her children and other household responsibilities as a single mother, Asha heroically studied to become an assistant nurse and began working at a hospital. As the years passed, trouble frequently found her son, and he eventually ran away. Unable to reconcile the ways of her son with her new-found Islamic religiosity, Asha took her five children and left them with their grandmother in England. She returned to Finland alone, but eventually went back to England, enrolling at the local university and settling in the same city as her mother and five children.

The story of Lilliam, the 'low-income' immigrant in New York

In the suburbs of New York City, many Latino immigrants eke out a meagre living in the wealthy towns on the north shore of Long Island. Gordon (2005) documents the life of Lilliam:

Not far away, night has fallen in the kitchen of a middle class Long Island family. Lilliam Araujo, who lives with the family, cleans their home, and cares for their daughter, sits in the dark at the breakfast table, the phone line wound tightly around her shoulders like an umbilical cord. Her voice, hushed to avoid waking up the sleeping people on the second floor, is warm but firm: "Did you do your homework, *papi*? Is your brother home? No, you can't wait up for him to come from work. It's late already . . ." [. . .] Seven years old and seventeen, they live alone in a small apartment she rents two towns away.

She left El Salvador for them: to keep the older one from being recruited by the military or the guerillas in their increasingly conflict-ridden town, to save the younger one from being caught in the ever more frequent crossfires of the Salvadoran civil war. But the best work she could find on arrival was a live-in domestic job at the rate of $160 a week for 65 hours of work, less than $2.50 an hour. A single mother, she took the job and got the nearest apartment she could afford for her sons. Bathing and dressing and hugging a child not her own, she is plagued by the question of whether she is doing better by her boys here, or worse. As for herself . . . as she moves out of the kitchen and towards the stairs, her mind drifts toward her days in El Salvador, the coffee-growers' cooperative where she served as secretary, the degrees she had been earning at night and on weekend in psychology, social work, and teaching, the courses she taught at the local business college, the house she owned. It feels like another life. (Gordon 2005: 11–12)

The above stories represent only a very small fraction of the kinds of migration that exist in the world, and the sorts of issues that migrants face. What they suggest at the very least is that migration is a complicated, challenging, and diverse phenomenon involving changing statuses and multiple geographical trajectories. Some authors refer to this as a condition of 'migrancy', or "the movement and process rather than stability and fixity across both space and time" (Harney and Baldassar 2007: 192). Whatever the degree to which migrants live fluid lives, governments, citizens, the media, authors of policy reports and authors of books like this on migration routinely use such categories to discuss the lives of migrants. Thus, even if we find these categories unhelpful or downright oppressive, we would find it difficult to make sense of national immigration policies as well the vast literature, academic or otherwise on the subject of migration. It is for this reason that we turn to a careful analysis of some basic terminological and categorical issues that will figure centrally in the chapters that follow.

Migrants seem to fit both into and across different types or categories of migration that imply a certain citizenship or residence status (e.g. internal or international, temporary and permanent, legal and undocumented), and different modes of entry (e.g. as asylum-seekers, refugees, low-income and highly skilled workers, students, and so on). Concerning their mode of entry, they may also be classified by academics, policy-makers, or statisticians as 'forced' or 'voluntary'. Given the apparent fluidity of migrant lives across these different categories and modes of entry, it has now become common in migration studies to reject the significance of these categorizations in some cases (e.g. Faist 2008; Richmond 2002). As Faist argues, if the aim is to "map trajectories of mobile populations" (2008: 36), then distinctions between 'origin' and 'destination', between 'emigration' and 'immigration', between 'temporary and permanent', or between labour migrant and refugee' are no longer tenable. He points to Turkey as an example of a country that is considered to be both a country of emigration and a country of transit (see Box 3.2), and now we would argue of return or re-migration as well.

Ironically, collapsing together or at least confusing categories is also a feature of the popular press. For example, at one point during the early 2000s when the issue of asylum had flared in political debates in the United Kingdom (UK), the more salacious newspaper tabloids concocted nonsensical terms like 'illegal asylum-seeker'. Quite simply, someone cannot be an asylum-seeker and an illegal migrant. If an asylum-seeker's claim is rejected and that person chooses to stay in the UK without the knowledge of UK authorities, then and only then could they be considered 'illegal', but then they cease to be an asylum-seeker. The point of this anecdote is to assert that categories around legal status and modes of entry should still be seen as meaningful, even if this book acknowledges the complexity of migrants' trajectories; that migration is a process rather than an event (King 2012), and that migration categories may reinforce an 'us' and 'them' politics (Anderson 2013). Given our claim above concerning the significance of such migration categories, but also our reservations about their use, below we elaborate critically on some commonly used terms and the different types or categories of migration, reasons for migration, and various modes of entry.[3]

Let us begin with the simple distinction between internal and international migration. Internal migration involves those who move within their own countries, for example from rural to urban areas. Often this assumes the form of 'circular migration' in which migrants move back and forth between rural and urban areas. Although we will touch upon large-scale internal migrations (such as in China), our main concern in this book is again with international migration, even if the two are commonly related (e.g. King and Skeldon 2010). International migration can be defined as the act of moving across international boundaries from a country of origin (or country of emigration) to take up residence in a country of destination (or country of immigration).[4]

Such international migration may involve just one country of origin and destination, but it might also involve different steps or stages between various countries, before a migrant moves on to her or his final destination. This is sometimes referred to as temporary or sojourner migration. We saw this in the story above of Asha. In official terms, temporary migration refers to international migrants whose duration of stay in a given country is greater than three months but less than 12 months (this is the UN Department of Economic and Social Affairs[5] definition of a 'short term migrant'). It is certainly a plausible definition, given how many countries limit migrants to, let us say, a tourist visa of three months before they must leave the country, but international migration might also be 'permanently temporary', entailing frequent returns to the country of origin, and much has been written in the last decade about this common practice of circular migration.

Others will stay in a foreign country for years as permanent residents without 'naturalizing' (that is becoming a citizen of a particular country). It may make sense to call these individuals immigrants rather than migrants. Thus, we will generally use the term migrants and migration to refer to those who find themselves in a condition of more temporary residence in a country of destination. While in practice, the distinction between the two is far from clear, referring simultaneously to migrants and immigrants, and migration and immigration all of the time would become tiresome for the reader. Thus, we will sometimes use the term migration to refer to both conditions, especially when it concerns a very broad or general discussion.[6]

We would highlight one significant caveat to this whole discussion though. While there is a tendency to focus on an individuals' longer-term *legal* residential and work status, such as whether they possess Green Cards or permanent residency stamps in their passports, the same person may be constantly dreaming of returning to the country of origin, or what is often called the 'myth of return'. Thus, the temporality or permanence of a migrant's stay may also be read from a psychological, rather than simply a legal standpoint. As Sayad (1977, 1991) and Bailey et al. (2002) have argued, migrants live with a 'permanent sense of the temporary' and a 'temporary sense of the permanent'.

Another crucial distinction is between *legal* and *undocumented* migration. Legal migrants are those individuals who have express authorization of (usually) a national government to enter, reside or work in the country of destination. This covers a large number of different official statuses, each with a distinct set of rights. At one end of the spectrum, migrants may enter a country legally but not be permitted to be resident, since they are obliged to leave after a short time and have no right to bring family or use any public services. At the other end of the spectrum, migrants may have permanent residence and have rights very similar to citizens. In between these two extremes is a wide range of possibilities. In fact, Morris (2002) identified 25 different statuses of legal migrants in the UK, which she refers to as 'civic stratification'. In contrast, undocumented migrants are those individuals who cross international boundaries either without being detected by authorities (often called clandestine entry) or who overstay their visas. 'Legal' migrants may also violate their work conditions stipulated by their visas such as working more hours than is permissible (what Anderson et al. (2006) call 'semi-compliance'). In this case, migrants may also be considered 'illegal' ('semi-compliant') and subject to the terrible reality of deportation. The widely used term 'illegal migrant' is inaccurate since individuals may behave in ways that are technically illegal (such as crossing borders or working without authorization) but they cannot be illegal. Quite apart from their accuracy, the terms 'illegal migrant' or 'illegal migration' are widely considered to be pejorative. Some observers of migration, and migrants themselves, have insisted that 'no-one is illegal'. That is, no-one can be ever outside the law (e.g. Cohen 2003). A range of alternative labels is widely used such as irregular, clandestine, unauthorized, and most recently 'illegalized' migrants (Bauder 2014a; Hannan 2015), which highlights the process of transforming migrants' status. We prefer the term 'illegalized' but this has not yet become a term widely used by either the majority of scholars or immigrants themselves. Thus, throughout this volume, we will generally employ the word 'undocumented' as it appears to be the preferred term of many migrants themselves, at least in the US, Canada and many European countries.[7]

A prominent academic distinction in the migration literature is between forced and voluntary migrants. This also provides some grounds for discussing different modes of entry. It should be emphasized that the reasons people migrate exist on a continuum between forced and voluntary, and thus determining precisely who is forced and who is voluntary is difficult. However, it is common to distinguish between two types of forced migration: the migration of asylum-seekers and refugees, as recognized by international conventions, and those who are forced to migrate for reasons of poverty or low wages – what is commonly called 'economic migration'. A further distinction within forced migration reflects the separation between internal and international, as already discussed. Internally Displaced Persons (IDPs) are those who are forced to move within their own countries, while asylum-seekers and refugees must have crossed an international border. The main international convention that governs, but does not completely determine, who will and who will not receive refugee protection is the United Nations 1951 'Convention relating to the Status of Refugees'. This Convention was modified in 1967 by the Protocol Relating to the Status of Refugees (often called the 'New York' Protocol).[8]

Using the definitions generated by these agreements (we elaborate on these below), asylum-seekers are to be understood as individuals who are seeking asylum or refugee status in another country. They may claim asylum on arrival and so enter a country as 'asylum-seekers' or they may have been living in that country for some time when circumstances change, obliging them to request asylum. They may or may not in turn be granted asylum or refugee status by a particular national government. Seeking asylum in this way is one of two main ways of receiving protection as a refugee. The second way involves the direct resettlement of individuals who have already been recognized as refugees, by a nearby government or the United Nations High Commissioner for Refugees (UNHCR), prior to their arrival in another country. It is possible then that the mode of entry into another country may be very different for an asylum-seeker and a refugee. An asylum-seeker may enter a country clandestinely (and thus illegally) but later claim asylum in that same country. It is a government's decision based on their interpretation of the Geneva Convention and the 1967 protocol whether to grant refugee status or some other status to the migrant. The 1951 Geneva Convention defined refugees as persons who:

> As a result of events occurring before 1 January 1951 and owing to well-founded fear of being persecuted for reasons of race, religion, nationality, membership of a particular social group or political opinion, is outside the country of his nationality and is unable or, owing to such fear, is unwilling to avail himself of the protection of that country; or who, not having a nationality and being outside the country of his former habitual residence as a result of such events, is unable or, owing to such fear, is unwilling to return to it.[9]

In addition to the stipulations set out above, a refugee or an asylum-seeker should not be subject to refoulement; that is no state should return a refugee or asylum-seeker to a country where she or he fears persecution without 'due process' (in other words, the right to a legal hearing) (Goodwin-Gill 2014). The Geneva Convention emerged out of the context of the Second World War and applied only to persons migrating before 1951 in Europe. It was thus geographically exclusive and rested upon "Eurocentric, Orientalist, even racist constructions of African peoples and politics" (Hyndman 2000: 11). The 1967 protocol dissolved this geographic exclusion for persons claiming asylum after 1951, but we could hardly call the effects of the protocol to be liberal and welcoming. As of April 2015, 148 states have ratified one or both of these legal instruments (UNHCR 2015a, see footnotes 7 and 8), and the determination of a 'refugee' is in part determined by how state parties interpret both the Geneva Convention and the 1967 protocol. Much of this determination weighs on the interpretation of 'fear' (Hyndman 2000; Goodwin-Gill 2014). Indeed, the definitions of refugees and asylum-seekers are contested, and actual practices of treatment diverge from any definitional accuracy laid down in international conventions and international law.

Nonetheless, in principle, if an individual is granted refugee status, and depending on the state in question, this usually means that such a person and their family will be granted similar rights to those of legal migrants (perhaps even more rights), and a certain degree of social support. This might include anything from legal assistance to education and housing. However, most of the world's refugees are not part of the requisite networks, nor do they have the resources to reach the richer countries where an adequate level of support can be provided. Thus, it is in the poorest countries (rather than in the wealthiest countries) of the world where by far the greatest number of refugees eke out a living in shanty-towns and similar sites, or rest on the generosity of refugee and other humanitarian agencies such as the UNHCR in refugee camps.

We might categorize 'economic migration' as another form of forced migration. People may be seeking an escape from poverty or unemployment and below-subsistence wages, chronic disease, malnutrition, or some long brewing environmental calamity (which may have decidedly human causes) and which may be either the cause or the effect of the above. For governments of the richer countries, and not just the richer countries, adjudicating between who is or is not a 'genuine' asylum-seeker is far from straightforward. Given the low acceptance rates of asylum-seekers over the last decade in territories such as the EU, as well as the restrictive means by which many richer country governments interpret international refugee conventions, those who are not deemed to be fleeing immediate and serious political danger are unlikely to be accorded much sympathy by these governments at the beginning of the twenty-first century (e.g. Papastergiadis 2006). In short, those who are forced by poverty are imagined by governments as voluntary and so-named 'economic migrants'. The term 'economic migrant' and 'economic migration' may have negative connotations and assume a pejorative tone; that is, economic migrants are often imagined in relation to 'more deserving' refugees. In the UK for example, when the government and the public have perceived certain migrants to be economic migrants rather than 'true' asylum-seekers, they have been labelled as 'bogus' asylum-seekers. A significant question then is whether and how rich country governments should distinguish between those suffering from political persecution and those suffering from severe economic hardship, particularly when the conditions of the latter may be an effect of rich country policies.

This brings us to the fault lines between forced and voluntary migration, and the relationship between highly skilled and low-skilled/low-income migration.[10] It would be tempting to simply map low-skilled and highly skilled on to forced and voluntary migration respectively, but this would be in many instances an erroneous mapping. First, labour sociologists have long admonished that there is no globally accepted definition of 'skilled' and 'less skilled', and certainly no-one is 'unskilled' (e.g. Gallie 1991). In fact, people who migrate have all kinds of skills, but they are not recognized by the governments or firms of the countries of immigration at particular times and even for particular areas within countries. In that sense, governments' and firms' definition of skilled and not-so-skilled varies over space and time. Nonetheless, many states have immigration categories which determine who is considered skilled, less skilled, unskilled, or indeed anywhere in between. Thus, so-called highly skilled migrants (those with secondary or post-secondary education, including doctors, computer engineers, nurses and bankers, among others) can be contrasted with low-skilled migrants, those who generally lack secondary or post-secondary education, or who lack the requisite professional qualifications to obtain highly paid jobs.

Second, there are many highly skilled migrants whose movement may be deemed voluntary. That is, they seek to gain international working experience, open a new business, join their families, and so forth, and millions of low-income or low-skilled migrants who may be forced to one degree or another to migrate, depending on their available economic options. However, there are also many highly skilled migrants who are fleeing political persecution and poverty, and low-skilled and low-income migrants who may be migrating for higher wages, to join families, to seek adventure or any combination of these. In this case then, we should shy away from a neat correspondence between skill and the voluntary or forced motivations for migration.

The use of the above categories whether by governments, the media, academics or anyone else makes it clear that migration is clearly not just a matter for migrants. We therefore turn our attention to some issues and debates that involve the relation between migrants and governments and citizens. Our discussion is not designed to be comprehensive nor does this book address all of these issues head on. Rather, it is a selective exposition, aimed at laying the groundwork for the analysis in the five chapters that follow. The first of these issues concerns the causes and consequences of migration.

KEY ISSUES AND DEBATES CONCERNING MIGRATION

The causes and consequences of migration

Why do people migrate? What encourages or enables specific kinds of migration to continue? How does 'space' matter to these questions? In short, what are the causes of migration? These questions are of far more than academic concern. For example, the policies of governments in the richer countries, manifested for example in an unequal trading system and its detrimental effect on farmers in poorer countries, partly explain international migration. War, environmental stress and chronic unemployment which may also be directly or indirectly related to the policies of richer or poorer countries can shape migration patterns too. The causes of particular kinds of migration, like undocumented migration, are partly a product of national states' restrictive immigration policies. The causes may also be found in the cultural, political, and social marginalization of specific groups of people in a particular nation-state or sub-national regions, which impel or encourage them to migrate. The causes may be found in social networks which connect individuals between different places, whether they be family members, asylum-seekers or students, and whether it involves state policies in the form of emigration and recruitment agencies, or smugglers and traffickers. The causes might be located in gender expectations and oppressions from the impact of domestic violence on women's propensity to migrate, to men's desire to migrate in order to 'be a man', as sometimes seems to be the case, for example, among Albanians migrating to Italy or the UK (King et al. 2006). Given these various causes, we may find that we wish to address any or all of them, not necessarily because they lead to migration, but they may themselves reflect social problems and inappropriate policies that cry out for our attention.

These various causes cannot be divorced from each other and may be mutually reinforcing. What seems clearer now after more than a century of intensive migration research is that the causes of migration are not unrelated to the consequences of migration in the countries of emigration. Indeed, over the last 15 years especially, many scholars of migration have turned their attention to the relationship between migration and development, now dubbed the 'migration-development nexus' (Van Hear and Sørensen 2003). This is not surprising, since for much of the last 40 years in at least the richest countries, low-skilled and low-income migration from poorer countries has been seen generally as a 'bad thing' by governments. Governments have sought ways to stem the movement of low-skilled and low-income migrants. And it was a commonplace assumption that 'development' (in the form of industrialization and perhaps democratization) would stem such a movement of people. This assumption has now given way to a vigorous debate that we will explore further in Chapter 2.

The question of employment for migrants

A second key issue is the relationship between migrants and work. Work is a core dimension of migrants' livelihoods, even if migrants such as asylum-seekers, family members or students may not move simply for the purposes of work and they may never engage in waged labour when they reach their destination. Thus for governments, citizens, migrants, and migrant organizations, the world of employment raises a number of concerns. Chief among them is the character of work that migrants perform. Migrants work across the occupational spectrum but many migrants around the world are relegated to some of the most low-paid and arduous jobs in agriculture, care-work, construction, mining, and other service work such as in hotels and restaurants, regardless of the skills they have. Why are migrants relegated to such work, or why do migrants find the work they do? What obstacles do they face in finding work? A simple response to these questions is that they lack the necessary education, qualifications or skills to compete with citizens in job markets. A more rounded response might turn to the stereotypes and racist assumptions held by employers, which are decidedly geographic in character. Another fuller response might turn to the networks of information among migrants about the availability of certain jobs in neighbourhoods with an already existing immigrant population.

Often the work that migrants perform is informal in character, that is, unregulated by governmental authorities, and migrants may not be always paid for the work they undertake, or as mentioned above, they may be paid so little that they find it hard to pay the rent and eat, let alone have enough to send back to perhaps desperate relatives waiting for these transfers of income. What policies address the widespread use by employers of undocumented immigrants for informal employment, and should further policies tackle this issue? This is a question for the governance of migration.

The conflicted task of governing migration

A third issue is how various levels or 'scales' of government (international, national, regional and municipal) and their respective citizens view and respond to migration. When migration and other social phenomena are regulated at and through various levels of government, this regulation is often referred to as *governance*. Governance is also used to refer to the growing role played by private companies and civil society groups in responding to migration, often referred to as the 'migration industry'. For some governments and citizens, migration is a process to be actively encouraged; for others, it is something to be vigorously resisted, sometimes at great financial and social cost. The encouragement of especially low-income migrants or the acceptance of asylum-seekers and refugees (as opposed to foreign students or highly skilled migrants) seems to spark the ire of many citizens and the more nativist-leaning (that is citizen-centric or nationalist) media in countries across the world, though there are significant exceptions, such as the more *laissez-faire* attitudes of, let us say, employers and their representatives who rely on low-paid immigrants. The reaction to migration is not however simply a matter of the level of government involved nor a product of distinct groups in society with different interests, but related to complex geographies, with some regions, cities or towns more welcoming than others. For example, areas where there is a large concentration of citizens of immigrant origin may be more accepting of migrants than other regions with very few residents born overseas, though the reverse may be true as well (Money 1999; Wright and Ellis 2000b). This points to a sometimes unrecognized dimension of governance, and that is how migration, migrants and pro-migrant non-governmental organizations (NGOs) also serve to shape migration policy. To put it differently, we can say that governments are far from in control of migration.

Governments of the richer countries struggle to find a balance between on one hand, meeting the literal dictates of the Geneva Convention, the 1967 protocol on refugees and other regional directives (e.g. in the EU), and on the other, the desire to severely limit the number of asylum-seekers and refugees. A serious issue is the increasing criminalization of asylum-seekers and refugees, and the impact of security concerns on policies of protecting those fleeing persecution. This criminalization and securitization is now manifested in the proliferation of off-shore island detention centres and other similar spaces which now seem to operate partially outside the purview of national and international law (e.g. Ong 2006; Mountz, 2011). Nonetheless, it would be incorrect to assume that these same governments are not actively seeking immigrants. In fact, before the global financial and economic crisis of 2007–2009, the so-called global search for 'talent' meant that countries were deeply involved in discussions of how to recruit highly skilled migrants, while at the same time attempting to restrict or manage other kinds of migration. Paradoxically, governments also recognize the vital contribution of undocumented migrants to perform work now abandoned by more upwardly mobile citizens. This need to balance complex objectives has given rise to the doctrine of 'migration management', a host of international migration management committees and organizations, and increasing chatter about the possibilities of a 'global mobility regime'.

In many respects, the immigration politics of poorer countries is the mirror image of the rich countries. The former seem to be as preoccupied as much with *emigration* policies as they are with *immigration*. That is, poorer countries are concerned with how to reduce unemployment; how to retain skilled labour and how to export low-skilled labour in order to ensure the continual flow of money sent back by migrants (remittances). Nevertheless, the government agencies in poorer countries that are responsible for managing migration are *also* preoccupied with so-called 'transit migration' (we talk about this later in Chapter 3), the migration of refugees across international boundaries, large-scale rural to urban migration, whether forced or not, and the movement of low-income workers into their own countries.

The question of citizenship and belonging for and among migrants

A final issue concerns citizenship and what is often called 'belonging'. One of the fundamental desires for many (but certainly by no means all) migrants today is obtaining *legal* (or *formal*) citizenship in a country of immigration. Nation-states, administrative regions within nation-states (such as the German *Länder*) and even cities have different laws and construct different rules for especially formal citizenship. While citizenship in most countries is not easily obtained, some countries' citizenship is more easily obtainable than others, depending on one's officially recognized ethnic or national background as well as other characteristics that a migrant may possess (money, skills, time spent in the country, to name just a few). Meanwhile, the different levels of government and publics in the countries of destination continually question the value of migration and thus the ease with which formal citizenship may be obtained. For migrants, obtaining such citizenship is only one part of the process of what might be called 'integration' within the countries of immigration. Migrants are also plagued by problems that impact on their *substantive* citizenship. Substantive citizenship can be understood as the issues that concern the daily lives of immigrants: matters of family, finding an adequate place to live and work, choosing decent and appropriate schools, participating in relevant organizations and events, the problems of finding quality legal advice, and accessing health care. These challenges are exacerbated on a daily basis by racism or expectations of certain kinds of cultural behaviour, often generated by a variety of state-based organizations, or citizens, and even other migrants. Substantive citizenship also concerns a sense of 'belonging'. Though migrant identities constantly change vis-à-vis citizens, other immigrants, and compatriots remaining in the country of origin, they are shaped by influences associated with the country, region or village of origin and other *axes of differentiation* including age, ethnicity, gender, religion, and sexuality, among others. The ability to express this identity or these identities in the country of immigration in which they settle is a particular concern for many migrants. At the same time, many migrants also wish to adopt at least some of the cultural, political and social practices of the majority of citizens in the country of immigration. Yet if the expression of people's identities were only an individual issue then neither governments nor the media would afford it much attention but managing diversity, that is managing people's cultural, political, and religious identities is something that governments feel is necessary to regulate the relationship between migrants, citizens, economic development, and indigenous political institutions and parties.

GLOBAL TENDENCIES AND ESTIMATED PATTERNS OF MIGRATION ACROSS THE GLOBE

In the latest edition of their *Age of Migration*, Castles *et al.* (2014: 16) outline six general trends associated with migration:[11] 1) the globalization of migration, or what might be better called its diversification; that is, an increasingly wide array of countries are involved in migration with a consequential diversity of migrant backgrounds. However, roughly 67 per cent of all international migrants are concentrated in just 20 destination countries, and about two thirds of all international migrants live in Europe and Asia (UN 2016, see also Czaika and de Haas 2014). Furthermore, the US is home to roughly 19 per cent of the world's total (some 47 million migrants), followed by Germany (12 million), Russia (12 million), and Saudi Arabia (10 million); 2) The changing direction of dominant migration flows, including the growing migration to the Gulf countries, such as Kuwait, Saudi Arabia, or the United Arab Emirates; 3) the differentiation of migration, in other words, the diversification of types and modes of entry discussed in the previous section; 4) the proliferation of migration transition, in other words, that some long-time countries of emigration such as the Dominican Republic, Poland, South Korea, and Turkey have themselves either become countries of transit[12] or countries of more permanent immigration; 5) the feminization of labour migration, or the higher percentage of women migrating relative to men; and 6) the growing politicization of migration, meaning that migration has moved to the centre of global and national political debates.

Here we focus only on two of the above trends: the globalization of migration and the changing direction of migration flows in order to provide the reader with some idea of the number of migrants residing in countries outside their country of birth. It is these patterns that theories of migration seek to explain, and which we will explore in some detail in Chapters 2 and 3. Unfortunately, to provide some broad overview of global migration requires that we use existing data which is commonly glued to a 'methodological nationalist' frame, meaning that data is collected on a country basis. The data tends to involve measurements of stock (a horrible metaphor used to describe the number of immigrants in any one country at a particular moment in time) and flow (an equally problematic metaphor which documents – generally speaking – the one-way movement of people, as if migration was in fact only one way). Some of the data for poorer countries especially, can be taken as probably very inaccurate, and a notable problem of data on migration for all countries is its incompatibility. Countries measure migration and who is a migrant and immigrant in often very different ways, but there are also differences within countries from year to year and from government agency to government agency. Nonetheless, while it has become common to critique the problems of data, international organizations such as EUROSTAT, the International Organization of Migration (IOM), the OECD, and the United Nations are aware of the incompatibility of data, and are in fact working towards providing reasonable data on return migration and rectifying the incompatibility of nationally collected data (e.g. UN 2012).

These data issues aside, worldwide migration has increased in volume from about 173–176 million migrants in 2000 to 244 million in 2015.

And depending on what strategies motivate us to collect data, we might include the vast number of asylum-seekers and refugees. In this respect, at the end of 2014 there were approximately 19.5 million refugees, 1.8 million asylum-seekers and an unprecedented 38.2 million IDPs (UNHCR 2015b; UN 2016). In total though, international migrants comprised 'only' 3.2 per cent of the world's population in 2015 (UN 2016), which suggests that migration *as a percentage of the world's total population has remained relatively stable since* 1960 (Czaika and de Haas 2014). Nevertheless, let us explore some of these statistics in a little bit more depth.

Table 1.1 provides rough comparisons of 30 OECD and other countries[13] ranked by absolute and relative number of migrants. Table 1.1 is also graphically represented in Figure 1.1. Table 1.1 and Figure 1.1 do not include asylum-seekers, refugees, and undocumented persons. What is striking as noted above is the *large number* of migrants residing in the Gulf States (in particular the United Arab Emirates and Saudi Arabia), but also neighbouring states such as Jordan. This is equally noticeable in terms of the *percentage of the total population* (notice Qatar, Kuwait, Bahrain, and Oman).

Table 1.1 Migrant destinations: countries hosting the largest absolute/relative numbers of international migrants by number of migrants.

Top 30 countries by absolute numbers of migrants (2015)			Top 30 countries by relative numbers of migrants (2015)		
Country	Migrants (millions)	% of population	Country	% of population	Migrants (millions)
USA	46.6	14.5	UAE	88.4	8.1
Germany	12	14.9	Qatar	75.5	1.7
Russian Federation	11.6	8.1	Kuwait	73.6	2.9
Saudi Arabia	10.2	32.3	Bahrain	51.1	0.7
UK	8.5	13.2	Singapore	45.4	2.5
UAE	8.1	88.4	Oman	41.1	1.9
Canada	7.8	21.8	Jordan	41	3.1
France	7.8	12.1	China, Hong Kong SAR	38.9	2.8

Continued

Top 30 countries by absolute numbers of migrants (2015)			Top 30 countries by relative numbers of migrants (2015)		
Country	Migrants (millions)	% of population	Country	% of population	Migrants (millions)
Australia	6.8	28.2	Lebanon	34.1	2.0
Spain	5.9	12.7	Saudi Arabia	32.3	10.2
Italy	5.8	9.7	Switzerland	29.4	2.4
India	5.2	0.4	Australia	28.2	6.8
Ukraine	4.8	10.8	Israel	24.9	2.01
Thailand	3.9	5.8	New Zealand	23	1.0
Pakistan	3.6	1.9	Canada	21.8	7.8
Kazakhstan	3.5	20.1	Kazakhstan	20.1	3.5
South Africa	3.1	5.8	Austria	17.5	1.5
Jordan	3.1	41.0	Sweden	16.8	1.6
Turkey	3.0	3.8	Cyprus	16.8	0.2
Kuwait	2.9	73.6	Ireland	15.9	0.75
China, Hong Kong SAR	2.8	0.1	Gabon	15.6	0.29
Iran	2.7	3.4	Estonia	15.4	0.2
Malaysia	2.5	8.3	Germany	14.9	12
Singapore	2.5	45.4	USA	14.5	46.6
Switzerland	2.4	29.4	Norway	14.2	0.74
Cote d'Ivoire	2.2	9.6	Croatia	13.6	0.58
Argentina	2.08	4.8	Latvia	13.4	0.26
Japan	2.04	1.6	UK	13.2	8.5
Israel	2.01	24.9	Spain	12.7	5.9
Lebanon	2.0	34.1	Belgium	12.3	1.4
			Libya	12.3	0.77

Source: Authors' selection of data from *International Migration 2015* by UN Population Division.

N.B. Relative numbers list has 31 countries as Belgium and Libya tie for 30th place.

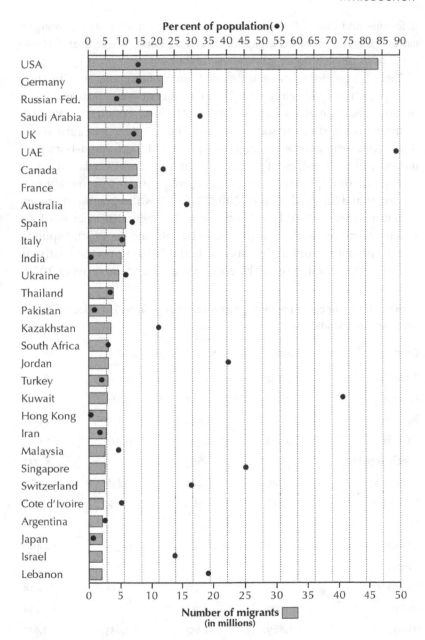

Figure 1.1 A comparison of 30 OECD and non-OECD countries with the largest number of migrants (with percentages)

If Europe and North America are considered to have a high *percentage of migrants*, this is relative to Singapore and Hong Kong, where nearly 40–45 per cent of the people are considered migrants.

Table 1.2 provides data on the *absolute number* of migrants (the 'foreign-born population') in selected OECD countries between 2010 and 2013. In most OECD countries (Spain being a notable exception), the number of migrants has risen substantially since 2007. Table 1.3 in turn shows *changes in the relative* number of migrants (percent of the 'foreign-born population') in the majority of European countries (and Canada) based on averages for three periods, corresponding roughly to before and after the economic and financial crisis of 2008. Table 1.3 is also graphically represented in Figure 1.2. Table 1.3 and Figure 1.2 show that in most of the selected countries, with the exception of Estonia, and the Czech Republic, that the proportion of immigrants rose in the 2008–12 period relative to the 2003–07 period, despite the economic and financial crisis of 2008.

Table 1.2 Total number of migrants ('foreign-born population') in selected OECD countries (thousands)

Country	2010	2011	2012	2013
Australia	5,994	6,029	6,209	6,392
Austria	1,316	1,349	1,365	1,415
Belgium	-	1,644	1,690	1,725
Canada	6,778	6,933	6,914	7,029
Czech Republic	661	669	-	745
Denmark	429	442	456	476
Finland	248	266	285	304
France	-	7,358	-	-
Germany	10,591	10,689	10,918	10,490
Greece	-	751	730	-
Hungary	451	473	424	448
Ireland	773	752	-	754
Israel	1,869	1,855	1,835	1,821
Italy	-	5,458	5,696	-
Luxembourg	189	215	226	238

Mexico	961	-	974	991
Netherlands	1,869	1,906	1,928	1,953
New Zealand	1,013	1,041	1,066	1,261
Norway	569	616	664	705
Portugal	669	872	-	-
Slovak Republic	-	-	158	175
Slovenia	229	272	300	331
Spain	6,660	6,738	6,618	6,264
Sweden	1,385	1,427	1,473	1,533
Switzerland	2,075	2,158	2,218	2,290
UK	7,056	7,430	7,588	7,860
USA	39,917	40,382	40,738	41,348

Source: Authors' selection from OECD (2012, 2013, 2014, 2015)
The following countries appear in the OECD's SOPEMI reports but annual data on foreign-born population: Bulgaria, Chile, Estonia, Japan, Korea, Lithuania, Poland, Romania, Russian Federation, Turkey

Table 1.3 The number of migrants ('foreign born population') as a percentage of the total population in selected OECD countries comparing average from 2003–07 to average for 2008–12

Country	2003–07 average	2008–12 average	2015
Australia	24.2	26.5	28.2
Austria	14.5	15.7	17.5
Belgium	12.1	14.5	12.3
Canada	18.7	19.7	21.8
Czech Republic	5.3	6.7	3.8
Denmark	6.5	7.7	10.1
Estonia	17.1	14.9	15.4
Finland	3.4	4.7	5.7
France	11.2	11.7	12.1

Continued

Country	2003–07 average	2008–12 average	2015
Germany	No data	12.6	14.9
Hungary	3.3	4.3	4.6
Ireland	12.8	16.7	15.9
Luxembourg	36.2	40.8	44
Netherlands	10.7	11.2	11.7
New Zealand	20.3	23.2	23
Norway	8.4	11.7	14.2
Spain	11.1	14.4	12.7
Sweden	12.6	14.7	16.8
Switzerland	23.9	26.8	29.4
UK	9.4	11.5	13.2

Source: 2003–07 and 2008–12 are from OECD (2015); 2015 figures from UN (2016)

Table 1.4 shows net migration (that is the difference between immigration and emigration) based on available and comparable UN data. We note that Germany, Turkey, and the United States have experienced the most net migration between 2010 and 2015, while more people left China, India, and Spain than arrived in those countries.

Table 1.5 shows 22 countries hosting the largest population of refugees. The world's refugee population is very unevenly distributed. At the end of 2014, a third of the world's refugees were hosted by just three countries (Turkey, Pakistan, and Lebanon). The ten countries in Table 1.5 accounted for 57 per cent of the total refugee population. Some countries have remained on this list for a long time; both Pakistan and Iran have hosted large numbers of Afghan refugees for decades. UNHCR refers to any situation in which more than 25,000 refugees have been in exile for more than five years as a 'protracted refugee situation'. Approximately 6.4 million refugees lived in such situations of long-term exile. Other countries have only appeared very recently; although Lebanon has hosted Palestinian refugees for decades the refugee population has increased dramatically since the arrival of large numbers of Syrians. More than a quarter of Lebanon's population are now refugees.

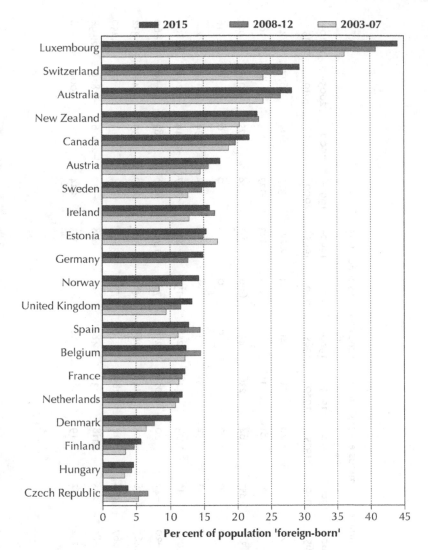

Figure 1.2 The number of migrants ('foreign-born population') in selected OECD countries in 2010–13

Finally, Table 1.6 provides some figures for student migration, or international student mobility as it is now also called. From 2001 to 2014, the number of overseas students more than doubled from about 2.1 million to about 4.5 million (IIE 2015b). The geographic concentration

Table 1.4 Net migration (immigration minus emigration) in selected countries 1960–2015

Country	1960–1965	1965–1970	1970–1975	1975–1980	1980–1985	1985–1990	1990–1995	1995–2000	2000–2005	2005–2010	2010–2015
Afghanistan	−20	−20	−20	−1,117	−3,418	−1,484	2,227	−379	804	−677	473
Angola	−135	−201	−81	11	234	−150	143	−127	173	85	102
Australia	383	853	264	236	456	666	351	378	575	1,133	1,023
Brazil	0	0	0	0	0	0	0	0	0	0	16
Chile	−34	−43	−52	−51	−37	8	63	72	94	139	201
China	−1,059	−13	−1,113	−428	−258	−154	−824	−507	−2,144	−2,202	−1,800
Ethiopia	−20	−42	−60	−2,025	250	780	1,445	−156	−83	−50	−60
France	1,467	416	805	147	268	482	303	189	739	481	332
Germany	667	918	714	289	0	1,711	3,233	744	1	32	1,250
India	−85	−245	2,118	1,132	483	45	−678	−717	−2,206	−2,829	−2,598
Italy	−232	−232	19	165	266	−10	153	224	1,624	1,006	528

Jordan	36	254	-30	-81	82	118	401	-188	-94	450	230
Libya	46	48	59	58	113	5	4	9	-12	-82	-502
Mexico	-411	-620	-900	-1,212	-1,519	-1,536	-1,303	-1,845	-2,841	-410	-524
Poland	-10	-143	-171	-107	-82	-459	-241	-244	38	5	-74
Republic of Korea	-281	-389	-450	-165	317	435	-633	-293	229	405	300
Russian Federation	-1,394	-551	-290	640	1,107	907	2,520	2,308	1,735	2,157	1,118
South Africa	115	201	232	73	175	-133	805	159	1,072	1,403	600
Spain	-194	-145	97	77	-43	-68	319	896	2,829	2,250	-593
Thailand	0	0	403	287	342	514	-808	696	1,103	-859	100
Turkey	-259	-320	-315	-368	-78	-150	-200	-150	-100	-50	2,000
UAE	38	54	245	391	176	261	327	484	1,180	3,493	405
UK	143	-85	106	39	-97	99	205	499	968	1,524	900
USA	959	1,495	2,840	3,873	3,279	3,686	4,569	8,692	5,149	5,070	5,008

Authors' selection from UN (2016)

Table 1.5 Most significant refugee hosting countries as of 31 December 2014, by refugees' proportion of the total population

Country	Refugees (n)	As % of total population
Palestine (West Bank and Gaza)	2,051,096	45.2
Jordan	2,771,502	37.4
Lebanon	1,606,709	28.6
Chad	452,897	3.3
South Sudan	248,152	2.1
Turkey	1,587,354	2.0
Iran	982,027	1.3
Kenya	551,352	1.2
Cameroon	226,489	1.0
Uganda	385,513	1.0
Yemen	257,645	1.0
Afghanistan	280,267	0.9
Iraq	271,143	0.8
Pakistan	1,505,525	0.8
Ethiopia	659,524	0.7
Sudan	244,430	0.6
France	252,264	0.4
Egypt	236,090	0.3
Germany	216,973	0.3
Russian Federation	235,750	0.2
USA	267,222	0.1
China	301,052	0.0

Source: Data on refugees is based on authors' selection of data from UNHCR (2015) *Statistical Yearbook 2014* and UNRWA (2015) *About UNRWA*.
Data on total population is for 2014 from UN (2016) World Population Prospects 2015.

of students in the wealthier countries (approximately 62 per cent of all foreign students, down from 85 per cent in 2001) is also marked by an increase in the number of students studying in China (8 per cent, up from less than 1 per cent in 2001), as well as a concentration in the Anglophone countries (the US, UK, Australia, Canada, and New Zealand in that order): 45 per cent of the worldwide total, up from 42 per cent in 2001. This may be explained largely by the possibility of learning English combined with the perceived quality of higher education institutions in these countries, and the possible employment and settlement prospects that one might be afforded afterwards. The US hosts by far the greatest number of foreign students (some 974,000) but they account for only 5 per cent of the total student population in the US, compared to the UK (roughly 22 per cent) or Australia (about 21 per cent). Countries in East Asia and the Pacific now have an international student

Table 1.6 Selection of countries with the highest percentage of international students in tertiary education in 2001 and 2014

2001	% of World Total1	Number of international students hosted2	2014	% of World Total1	Number of international students hosted1
US	28	475,169	US	20	974,926
UK	11	225,722	UK	11	493,570
Germany	9	199,132	China	8	377,054
France	7	147,402	Germany	7	301,350
Australia	4	120,987	France	6	298,902
Japan	3	63,637	Australia	6	269,752
Spain	2	39,944	Canada	5	268,659
Belgium	2	38,150	Japan	3	139,185
Netherlands	n.a	16,589	Netherlands	n.a.	90,389
New Zealand	n.a	11,069	New Zealand	n.a.	46,659
Denmark	n.a	12,547	Denmark	n.a.	32,076
Norway	n.a	8,834	Norway	n.a.	25,660
Mexico	n.a.	1,943	Mexico	n.a.	12,789

Sources: IIE, Project Atlas *Project Atlas* Trends and Global Data Fact Sheet (2015); OECD. *stat* (2016) Foreign/international students enrolled
1 Data from IIE, Project Atlas
2 Data from OECD.*stat*

population of some 750,000 while sending out more than a million students. Approximately two thirds of international students in richer countries are from poorer countries, and China and India alone are responsible for about 44 per cent of all international students, with Korea and Saudi Arabia sending another 6.5 and 6.1 per cent of the world total, respectively.

We have seen the numerical significance of migration in the section above, but we have hardly spoken of its diversity beyond the vignettes cited in the previous section. In the late nineteenth and early twentieth centuries, there were certainly some less documented and perhaps surprising migrations, such as Welsh and Lebanese-Syrians to Argentina, Indians to Fiji, or of Japanese to Peru. Yet with the exception of maybe Australia, there are grounds for making the case that migrants' countries of origin have diversified, while their 'destinations' have narrowed, especially since the post-WWII period in the wealthier countries (on this, compare Faist 2008 and Czaika and de Haas 2014). From the 1950s to about the 1980s, most migrations to the wealthier countries could be labelled as 'post-colonial' or 'neo-colonial' (Samers 1997b) in character, consisting of people from former European colonies migrating to the former colonial powers in northern and Western Europe in particular, or from countries that were dominated by the prevailing interests of the US government, such as the Dominican Republic, Mexico, and Puerto Rico. At the end of the twentieth century and the beginning of the twenty-first, migrants to the wealthier countries hailed from a remarkable number of countries. For example, as Smith (2005) notes, some 24 nationalities existed in Los Angeles and New York in the 1920s, whereas now there are some 150. There are Filipinos in Italy and Lebanon, Somalis in Liverpool and Minneapolis, Algerians in London, Sri Lankans in Paris, and South Koreans in the United Arab Emirates, to name just a few examples of both post-colonial migrations and migrations unrelated to past colonial linkages. In some cases though, this diversity is exaggerated or misunderstood, such as in Los Angeles. It may be taken as the paradigmatic city in terms of poly-ethnicity, but it appears to be far less diverse than London or New York, given the high percentage of Mexican migrants relative to all migrants in Los Angeles (Benton-Short et al. 2005; Price and Benton-Short 2008). In this sense, Figure 1.3 provides a depiction of the 25 cities with the largest number of 'foreign-born residents' around the world in approximately 2015.

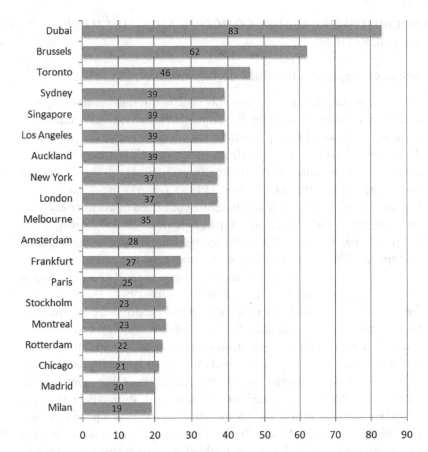

Figure 1.3 Twenty-five cities with the largest number of 'foreign-born residents' around the world

Source: International Organisation for Migration (2016) *World Migration Report 2015: Migrants and Cities. New Partnerships to Manage Mobility.*

At this point, we should have a fairly comprehensive picture of the different types of migration, the migration patterns based on national territorial statistics, and a hint of the diversity of migrations. We will now devote our time to a necessary discussion of some basic building blocks of society which are commonly used in human geography and the social sciences more broadly. These will serve as the foundation for the arguments in this book and ultimately for explaining the many facets of migration.

SOCIAL THEORY, SPATIAL CONCEPTS AND THE STUDY OF MIGRATION

Social concepts and the study of migration

For many social scientists, the concepts of structures, institutions, agents, and social networks provide the fundamental, if conventional elements of what is called 'social theory'. Perhaps the most contested are structures. There are at least two understandings of 'structure'. For so-called 'structuralist' or 'functionalist' thinkers, structure(s) are akin to the skeleton of buildings; 'things' that as Giddens (1984) remarks are "'external' to human action" (p. 16). They are therefore conceived of as either enabling or constraining certain kinds of human behaviour. Thus, in a way, structure might be viewed as what keeps humans in their places, both socially and spatially. This understanding of 'structure' is witnessed in some of the more 'crude' Marxist and neo-Marxist writings of the 1970s, but also in some of the earlier work on 'globalization' in the 1990s. There was (and for some, still is) a tendency to see the 'global economy' or 'global capitalism' (the more critical term for the global economy) as a structure which bears down upon or constrains nation-states, institutions, groups and individuals in society. Many observers agreed however, that this 'reified' the global economy or globalization; in other words, that structuralist thinkers attributed causal power to abstractions (e.g. the global economy) that in fact neither existed nor had causal power (see e.g. Gibson-Graham 1996; Massey 2005). Certainly, other corners of human geography and the social sciences wielded their own structures, such as the feminist literature and its conception of patriarchy.

In any case, in joining this critique of reification, the sociologist Anthony Giddens (1984) proposed a second and different understanding of structure, namely that it consisted of a combination of 'rules and resources' that shapes social action (p. 25), and that structures did not just constrain social action, but "enabled involvement in this action" (Cloke et al. 1991: 98). Hence, in his duality of structure, Giddens maintained that "the structural properties of social systems are both medium and outcome of the practices they recursively organize" (1984: 25). What he means is that structures constrain and enable social practices, but social practices also shape structures (i.e. rules and resources). This 'duality of structure' formed part of what he calls 'structuration theory' and we will explore this theory in more depth in Chapter 2 as a potentially useful way of understanding at least the causes, if not the consequences and experiences of migration and immigration.

Returning to the idea of the 'global economy' or 'global capitalism' above, we will tread a fine line in this book between on the one hand, viewing global capitalism as a set of forces that has a *relatively* fixed architecture and which, we argue, *does* impel people to migrate from poorer to richer countries; and on the other, the idea that the global economy is itself composed of institutions, individuals, and social networks that make the rules and control resources.[14]

A second frequently used concept in the social sciences is that of 'institution'. Institutions can certainly be organizations, which might include the UK Home Office, the US Department of Homeland Security, the United Nations High Commission on Refugees, an immigrant-run 'hometown association', a school, a refugee help centre, employer recruitment agencies, or any number of similar organizations. Such institutions deploy a whole range of discourses and procedures (what the social philosopher Michel Foucault calls 'technologies' or 'techniques of power' — Foucault 1977) in order to exercise power and control over policy domains and people. Yet 'institutions' might also include less tangible entities such as marriage, the family, household, and other forms of social arrangements. This brings us to the third social concept or category that concerns us, namely 'agents'. By 'agents', we mean human groups (e.g. Moroccan immigrants in a small town in Spain) and individuals (a particular migrant). This differs from some other notions of agents in the social sciences (see e.g. Fuller 1994; Murdoch 1997), but it is one that we will adhere to in this volume. In this respect, agents do not come 'pre-formed' but are shaped by structures and institutions, and develop in interaction with other agents. Here, agents are not to be understood as the passive "dupes of structural determination" (Cloke *et al.* 1991: 97); that is, they have 'agency', or the capabilities, behaviours, and practices of individuals or groups. They can and do exercise power to shape structures, institutions, other agents, and social networks. However, the degree to which different agents (that is migrants and migrant groups) can in fact exercise different forms of power over, through, or against these structures, institutions, and other agents is partly one that we will explore in the rest of this book.

The idea of social networks has become extremely popular in the social sciences and particularly in migration studies as a way of overcoming 'dualisms' such as global/local, and macro/micro, to connect structures, institutions, and agents, as well as quite simply, to explain migration and understand immigration. For Goss and Lindquist (1995) social networks "are generally defined as webs of interpersonal interactions, commonly comprised of relatives, friends, or other associations forged through social and economic activities that act as conduits through which information, influence, and resources flow" (p. 329). Such networks assume many different forms, from relationships of kin to relations between institutions and other institutions, between institutions and individuals; between individuals who are distant from each other and whose knowledge of each is restricted to the function or functioning of the network. Furthermore, these networks may be official or unofficial, visible or secretive, long-distance or local, and they may involve relatively 'weak ties' (Granovetter 1973) or strong ones. They may involve symmetrical relations of power or asymmetrical relations (where everyone in the network does not exercise power equally). They may be beneficial to migrants' well-being insofar as they strengthen 'social capital'[15] or they may be detrimental insofar as they impede access to other networks, institutions, markets, and so forth. Nonetheless, the concept of social networks and more specifically *migrant* networks (Boyd 1989; Massey *et al.* 1987, and Portes and Sensenbrenner 1993) has been criticized as a 'chaotic conception' (Sayer 1984), that is, a loose concept which obscures more than it illuminates. At the same time, the structure and operation of such networks have been viewed as inadequately theorized for at least two reasons. First, they miss significant agents and institutions such as employers and employment recruitment agencies (Goss and Lindquist 1995; Krissman 2005). Second, since social or migrant networks have become quite dominant in sociological understandings of migration, this has the effect of eclipsing matters of space. In other words, one might get the impression that migrants are un-problematically connected across the globe without the impediment of distance or borders. It is for this reason that we divert our attention now to some *spatial* concepts.

Spatial concepts and the study of migration

As we discussed earlier in the Introduction, the reason for undertaking such a task is to address the paucity of careful and critical 'spatial thinking' in migration studies. In particular, we are motivated by a spatial dilemma created by two seemingly opposed concepts. On one hand is the problem of taking the nation-state as the starting point or basic lens for analysing migration issues. This has been variously called 'embedded nation-statism' (Taylor 1996), the 'territorial trap' (Agnew 1994) or now more commonly 'methodological nationalism' (Beck 2000b; Wimmer and Glick-Schiller 2003; Favell 2015). Methodological nationalism refers to the ongoing academic practice of viewing social processes through the lens of the 'nation-state' despite more than a decade's critique of this particular perspective on the world. On the other hand are the limitations of what might be labelled 'methodological transnationalism' in the context of migration (see e.g. Harney and Baldassar 2007), that is, the emphasis is on the "multiple ties and interactions linking people or institutions across the border of nation-states" (Vertovec 1999: 447). Though both 'isms' are far from opposites, there is no doubt that the turn towards transnational perspectives reflected in part dissatisfaction with 'methodological nationalism'. Unfortunately, the literature which mushroomed in the 1990s around the concept of transnationalism (see Vertovec 1999 and Portes *et al.* 1999 for reviews) failed to adequately conceptualize what 'national' means, and how 'supra-nationalism' or 'localism' figured in shaping migration and immigration.[16] Clearly, what is needed is a more sophisticated treatment of space than that which is offered by either methodological nationalism or methodological transnationalism. One solution might be to think in terms of both territories *and* networks *together* (Jones 2009; McCann and Ward 2010). The problem of resolving these debates surely stems in part from the desire among many social scientists to find *the* spatial concept to understand migration (Leitner *et al.* 2008). It is our argument in this book that such a pursuit is not fruitful. Rather, we need a plurality of spatial concepts to understand migration. Below then, we first explore some meanings of 'space' itself, and then outline five spatial concepts that are essential to understanding migration. These are:

- Place
- Nodes
- Friction of distance
- Territory and territoriality
- Scale and Scalar

Some meanings of space

Like other geographical concepts, the idea of 'space' is slippery and contested (Massey 2005). Perhaps one of the most simple, but significant, arguments about the nature of space is that it cannot be understood without reference to society. In short, space has no meaning by itself, and we should therefore speak of 'socio-spatial' relations (Soja 1989), in other words, how space and society interact. The celebrated theorist of space Henri Lefebvre (1974 [1991]), famously developed a triadic understanding of space, involving 'spatial practice', 'representations of space' and 'spaces of representation' (p. 33). It is only with the second and third of these with which we will be concerned here. Thus, representations of space refer to the 'spaces' conceived by architects, government officials, urban planners, and well, authors of books on migration! In contrast, the spaces of representation refer to people's lived understanding of space, the vernacular spaces if you will of migrants (Delaney 2005). This vernacular understanding of space brings us to the way in which 'space' is viewed in relation to 'place'.

Place and migration

For humanistic geographers, space is more abstract and more 'empty' than place (Cresswell 2004). For example, if we speak of 'regulatory spaces of immigration', this would presumably not invoke strong emotional feelings among readers. In contrast, if for instance we asked an asylum-seeker about her stay in a migration detention centre, let us say the Yarl's Wood Immigration Removals Centre near the British town of Bedford, this particular detention centre becomes a 'place' of lived experience and a site of meaning. But place may also be simultaneously the site (or not) of the provision of basic necessities, such as food, clothing, and shelter. Cities and towns, neighbourhoods and workplaces, and cafes and parks are all examples of places that involve at the same time lived, meaningful experiences and the provision of basic necessities. In that sense, it is common to view place as something more localized – or on a smaller scale than space (Cresswell 2015). In this volume, we accept this definition for place (and by contrast the definition of space) even if it is not accepted by everyone, and even if it remains problematic (Massey 2005). Indeed, we must tread cautiously with this conception of place because there is a danger in thinking that places always involve desirable experiences, fond memories, homogeneity, safety, stability, or supportive and inclusive relationships (Harvey 1996; Massey 2005). The example above of the Yarl's Wood detention centre in the UK may be a place of supportive relationships among co-asylum-seekers but it is unlikely to be a place of desirable experiences or fond memories. Nonetheless, as Massey (1994) has pointed out, places may also be diverse, cosmopolitan, and open to outsiders such as immigrants, or they may be exclusionary and characterize some people *as* outsiders (Sibley 1995; Cresswell 2004).

Nodes and migration

A second concept associated with space (and place too) is the idea of nodes. As a vast and long-standing literature on networks now argues, nodes form part of networks in a 'space of flows' (Castells 1996: 410–18). The idea that networks and nodes form basic elements of a flow-based global society has proved enormously popular. For example, Voigt-Graf (2004) explores and maps the lives of Indian transnational communities in Australia and elsewhere, and insists that one cannot understand 'transnational spaces' without referring to networks and nodes. For Voigt-Graf, nodes refer to "Countries, regions or places that are linked by flows" (p. 29). A node can be a 'cultural hearth', that is, an area where "the culture of migrants originally developed" (p. 29) such as in the Punjab in north-western India.[17] Nodes can also refer to the 'new centres' of a transnational community, let us say Sydney, where roughly half of the Punjabis in Australia live. The important point here is that such nodes are part of networks or the trajectories of certain migrant groups that can span the globe, but that these nodes are also real and complex places (cities, towns, neighbourhoods, etc.), in which migrants grow up, work, find housing, face ethnic and racial discrimination, raise sons and daughters and build communities.

Despite the popularity of Castells' vision of society as composed of an apparently seamless combination of networks and nodes, we as authors remain critical of this flow or fluvial metaphor in the case of low-income migrants especially. Low-income migrants may be as networked as their wealthier compatriots but as we argued earlier in this introduction, national governments and the territories they govern make some forms of mobility and migrancy difficult, though they also enable or force it.

Friction of distance and migration

'Friction of distance' refers to 'the time and cost of overcoming distance' (Knox and Marston 2007: 25), and it is not entirely surprising that it seems to have been forgotten in studies of migration. This is probably the case because innovations in transport (air travel, etc.) and communication technologies (the internet, email, phone) have apparently lowered the cost and reduced the time of overcoming distance and because migrants seem to transcend vast distances regularly. There is no reason however, to assume that this is uniformly so for all migrants in all parts of the world, or that transport and communications will be ever cheaper relative to migrants' income and other expenditures. Fluctuations in energy costs and airline prices over the last 15 years suggest that the problem of overcoming distance will not disappear. Besides, technology does not determine migration (e.g. Czaika and de Haas 2014).

Indeed, distance *alone* may be a poor explanation for understanding patterns of migration (think for example of the settlement in the 1960s of the Moluccan Indonesians half a world away in the Netherlands, because of Dutch colonialism in Indonesia). But it also might help to explain the large number of Indonesian women now working as domestics in Malaysian and Singaporean middle-class households only a few hundred miles away across the Strait of Malacca, or the large number of Ukrainians working across the border in Poland, or Polish migrants working across the border in Germany. Thus, the tired refrain that 'distance no longer matters' needs to be assessed through careful comparative analysis and by differentiating among different kinds of migrants. Contrary perhaps to popular opinion, geography is not history (see e.g. Graham 2002).

Territory, territoriality, and migration

Let us now move to our third set of spatial concepts: territory and the idea of 'human territoriality' (Sack 1986). At its simplest, according to Storey (2001), 'territory' refers to "a portion of geographic space which is claimed or occupied by an institution or person or group of persons. It is thus an area of 'bounded space'"(p. 1), and human territoriality is "a human strategy to affect, influence, and control" (Sack 1986: 2). Territory and territoriality entail therefore an 'inside' and an 'outside' (Delaney 2005). These 'insides' and 'outsides' are however more stark and policed in some cases than in others, and with different implications (Collyer 2014a).

As Delaney (2005) puts it, "Unauthorized incursions into a co-worker's cubicle may be grounds for disciplinary action but not grounds for a military reprisal" (p. 14). He argues further that, "not every enclosed space is a territory. What makes an enclosed space a territory is first, that it signifies . . . [something] . . . and second, that the meanings it carries or conveys refer to or implicate social power. But meaning and power are not independent of each other" (p. 17). According to Delaney, the existence of territory 'works' or is acceptable because the relationship between power and meaning are seen as natural and obvious. Indeed, for many non-migrants and perhaps some migrants as well who benefit from a certain kind of territoriality, it might seem odd to question the sanctity of territories (especially national territories) and the right for governments to exercise power over such bounded spaces. In contrast, for migrants who resent and protest the existence of territorial borders or certain territorial regulations such as national immigration policies, territories and territoriality are not seen as 'natural', but malleable. Migrants live in hope for the construction of new forms of territoriality. We can say then that territories and territoriality are not permanently fixed and territories do not simply act upon migrants; rather, territories are porous and migrants themselves shape the nature of territory and territoriality (Collyer and King 2015).

Human territoriality can involve anything from the territory of a national state to the territory of a firm or organization. It can concern the United Kingdom, a state in the interior of Brazil (a sub-national region), the European Union (a macro-region), disputed zones such as the 'Western Sahara', a Canadian province, a section of the city of Istanbul, the 'space' of a textile plant in India, a neighbourhood, and someone's work cubicle. In short, territories should not be thought of as restricted to national states, and 'micro-territories' can be as significant as 'macro-territories'. While human territoriality is enormously varied (Delaney 2005), territoriality is a significant dimension of national states in particular, and they exercise enormous control over migration and immigration. It is for this reason that a certain version of territoriality will figure centrally in our spatial toolbox.

'Scale', 'scalar' and their relationship to understanding migration

Territory is also associated with 'scale' and 'scalar', the final set of spatial concepts to be discussed. The use of 'scale' as a term or concept in critical human geography and related disciplines is nothing short of ambiguous and vague, and it has rarely been employed in studies of migration (e.g. Glick-Schiller and Caglar 2010). Never really defined outright by critical human geographers, 'scale' has become a difficult and contested term (Marston et al. 2005). During most of the 1990s, it seemed to refer mainly to spaces (or territories?) of political-economic processes (e.g. the 'local', 'national', 'macro-regional', and 'global'). Hence, 'scale' seemed to be a substitute for different 'levels' of governance. It was common in the 1990s for instance, to speak of the 're-scaling' of economics, politics, or culture from the 'national scale' upwards to the 'global scale' (the process of globalization) or downwards to the local scale (the process of localization) (e.g. Jessop 1997). But others criticized this idea of fixed scales (or territories?) and argued that scale should be viewed as fluid and relational (in other words, scales only exist insofar as they relate and interact with each other). They are therefore socially constructed over time, and one should not necessarily privilege any scale (let's say the global or national) over another beforehand (e.g. Brenner 2001; Leitner and Miller 2007; Mansfield 2005; Swyngedouw 1997). In the same vein, Mansfield argues that the notion of 're-scaling', that is when for example, global processes become more important than national ones, should be thrown out in favour of an analysis of the 'scalar dimensions of practices' (Mansfield 2005). What Mansfield means is that no social process fits neatly or should be associated with any one scale. Instead, we should focus on the process first and then explore its spatial (or scalar) dimensions. Note here that the word 'scalar' appears to have eclipsed 'scale' in twenty-first-century geographical writings. In any case, the important point is that 'scalar' now seems to imply the 'spatiality' of certain social processes, but what is spatiality? We might say that it refers to the overlapping or complex spaces of creation, interaction, resistance, and other aspects of the exercise of power by governments, institutions, citizens, and migrants alike. Certainly, different social processes have different spaces, spatial dimensions, or 'spatialities', but it is contradictory, confusing and inconsistent to refer to scalar as the spatiality of a process, particularly when in many studies scale itself is either never defined, is not considered a fixed container for social processes, or both rejected and accepted as a territory or the spatial extent of some process.

Since this unclear and/or confusing usage of scale and scalar has become central to the literature in critical human geography, how can we resolve this problem? One way might be to better define how we use the term scale and scalar in this book, and then explain what is meant by spatiality or even 'scalar spatialities'. In this book, we define scale as 'territory' (a 'container' so to speak, such as a national state, a sub-national entity such as an American state, a macro-region such as the European Union, but also a human body, and so on). However, we do not see these 'containers' as forever fixed, non-porous, or so stable that they cannot evolve. We also define scale in terms of the spatial extent of any given process expressed in territorial terms. Thus, the migration of people can cross or encompass many scales (territories). When the migration of people involves multiple scales and when multiple scales (territories, bodies, cubicles, etc.) are involved and/or are interrelated in regulating this migration, we will use the term 'scalar' or 'scalar spatialities' inter-changeably, because 'scalar' is an adjective that describes when a process is subject to, but also transcends particular scales.

A brief example can better illustrate our point here. At the beginning of the twenty-first century, migrant workers in Germany were regulated by a 'double regulation system'. If a German firm wished to employ an 'unskilled' non-EU worker, it had to first search at the (Federal) Central Placement Agency. Once a provisional job offer had been made, the firm had to ensure that no EU worker was available from the local employment office associated with a specific municipality (city, town, etc.). Once a certificate had been issued from the Local Employment Office, the for-eigner could obtain a visa to enter Germany, and subsequently the employment contract. In turn the employment contract generated a resi-dence permit (OECD 2000). Thus, we can say that migrant workers were subject to different scales of regulation, or that the scalar regulation of migrants is both federal and local. Thus, we can talk both of the scale of governance but also the scale of migration control and its territorial extent (Germany). However, this scalar regulation can also be referred to in terms of the 'scalar spatiality' of labour market regulation, because this process of regulation in particular involves many scales. In sum, our argu-ment in this book is that scale should be used in two ways, as a synonym for territory, and as the spatial extent of a process or 'something' in terms of scale (the national or global scale of a process for example). Scalar or scalar spatiality refers to the way in which a process is subject to or tran-scends multiple scales.

SUMMARY OF THE INTRODUCTION AND STRUCTURE
OF THE BOOK

We opened this Introduction with the argument that a critical appreciation of spatial concepts or metaphors should be central to our
understanding of migration. We initiated the remainder of the discussion
in this chapter with first, a tragedy of migration in the Mediterranean Sea,
and second, with some vignettes of migrant lives to humanize migration
and show the challenges that migrants face. The stories suggest a set of key
issues and debates around migration that are difficult to understand without a fuller analysis of some of the terms and categories that are used in
debates about migration. These terms and categories of migration are
both useful and problematic because migrants are subject to these categories by citizens, governments, NGOs, private companies and organizations,
including the media, but they also transgress them. We then explored
some data on the 'globalization' (i.e. diversification) of migration and the
change in its direction. Migration involves a diverse range of people:
highly paid and low-paid workers, asylum-seekers and refugees, students
and family members, those who emigrate voluntarily and those who are
forced to one degree or another. Some may migrate for a couple of
months, others for several years, and still others permanently.

We then outlined four key issues relating to migration including first,
the causes and consequences of migration, especially how migration is
related to 'development' in poorer countries; second, the question of
employment for migrants; third, the governance of migration, including
some differences between the richer and poorer countries, and finally
some of the issues that governments, citizens, and migrants face with
respect to citizenship and belonging. This discussion was not meant to be
exhaustive, but rather to highlight some important cultural, economic,
political and social fault lines across the globe with respect to migration,
as well as to 'set the scene' for the chapters that follow. However, in order
to think about or understand these issues, we argued that we first require
a set of social concepts (structures, institutions, agents, and networks),
and we suggested a more 'structurationist' perspective which acknowledges the 'rules and resources' associated with 'global capitalism'.
However, consistent with a more structurationist approach, understanding migration through global capitalism is necessary but insufficient since
agents (such as migrants and citizens), institutions, and migrant networks

also shape global capitalism and migration. Finally, we discussed some spatial concepts (the meaning of space itself, place, node, friction of distance, territory, and scale) and the necessity of deploying them with more critical reflection in order to better frame migration. The aim of this book once again is to build a more explicitly spatial approach to migration, one that does not rely *solely* on world regions, nation-states, or transnationalism, and one that seeks to understand in particular the movement of the less advantaged.

Structure of the book

The chapters of this book are designed so that each chapter clearly follows the other. Chapters 2 and 3 seek to explain migration across international boundaries, drawing upon some major theories of migration. Chapter 2 focuses in on what are labelled 'determinist theories' to migration, while Chapter 3 diverts attention to 'integrative theories'. Both of these chapters will review very briefly the various theories of why people migrate, follow with a short set of criticisms and finish by explaining how to inject space into these various theories in order to develop a more robust understanding of migration. If Chapters 2 and 3 provide a study of the diversity of migrant imperatives and decisions, Chapter 4 shows how these imperatives and decisions are shaped by state and state-related forms of control and regulation.

Chapter 4 is divided into two parts. The first part reviews a range of theories of migration politics and policies, and in the process we explore actual practices in relation to these theories. We focus in particular on how these various approaches or theories fail to make explicit issues of space, especially the importance of sub-national territories such as towns. While there are common processes in the production of migration control across the globe, we maintain that wealthier countries are facing a different set of migration issues from poorer countries. The response to this has been the general doctrine of migration management, and we show how migration policies frequently encounter political opposition from a variety of groups and institutions. Yet the chapter also highlights how these oppositional movements are neither simply international nor national in character, but are often sub-national and highly localized.

The second and considerably briefer part of the chapter focuses on poorer countries. We elaborate on how southern governments pursue a tense balance of opening borders to refugees and labour migrants, consistent with a loose regionalism on the one hand, and on the other strictly control the movement of such persons because of desperately high unemployment levels and the risks of political stability. We provide examples of this restrictionism, and we summarize this section of the chapter by highlighting this mixture of cooperation and often militarized control that marks the migration regimes of southern governments.

Chapter 5's main purpose is two-fold: first, to explain how labour markets operate beyond the often simplistic representations peddled by the media, and to elucidate some of the working conditions under which migrants must labour. This entails relating the operation of labour markets and their labour market outcomes to theories of labour market segmentation and to complicate such theory through a discussion of spatial metaphors and settlement geographies such as global cities, ethnic and immigrant enclaves, 'hetero-localism', dispersal, and 'ethnoburbs'. This chapter then suggests that the nature of the work that migrants perform (and the conditions and social relations bound up with it) may shape their sense of 'belonging' in the countries of destination, as discussed in Chapter 6.

Chapter 6 (on migration, citizenship and belonging) claims that social scientists' reading of citizenship and belonging employ under-developed notions of 'socio-*spatial* relations' with respect to migration. Following the general argument of the book, we demonstrate in this chapter how using a *geographic* lens can shed light on the nature of, and relationship between migration, citizenship, and belonging. The emphasis is on the wealthier countries of the global north, though we use examples from other countries in the global south to illustrate our arguments. After an explanation of the chapter's geographical premise, we begin by drawing an initial distinction between legal forms of citizenship; citizenship as economic, political, and social rights; citizenship as political participation and citizenship as belonging; and we use this four-fold distinction to divide the chapter's discussion into four separate, but related sections.

In the Conclusions (Chapter 7), we bring together the summaries of each chapter, and emphasize how a *geographic* lens on migration issues should not be tacked on as an afterthought to what seem to be somehow non-geographic issues, but that the issues themselves are constituted through a geography of social relations (including the borders, institutions, regulations, laws, and cultural hopes, fears, and expectations that spring from them). It concludes finally that the question of migration is likely to remain at the heart of economic, political, social, and cultural discussion in the twenty-first century, and that 'spatial thinking' has much to contribute to debates about making a world somehow less terrible for the disadvantaged among us.

FOR FURTHER READING

For overviews of the history of global migration prior to the twentieth century, see Goldin et al.'s *Exceptional People* (2011). For the twentieth century, see Boyle, Halfacree, and Robinson's *Exploring Human Migration* (1998), *The Age of Migration* by Castles, de Haas and Miller (2014), the annual reports of the IOM (*World Migration Reports*) and the annual OECD's *International Migration Outlook*. These all present widely received synopses of migration across the globe, and discuss many of the terms and issues described in this chapter for various regions of the world. For other useful surveys, students and professional scholars can also consult the websites listed further below in a separate section. For the UK, Chapter 41 by Koser in Cloke et al.'s *Introducing Human Geographies* (2014) represents a simple and very useful exposition of some chief issues concerning migrants and refugees. Chapter 6 by Susan Hardwick in Brettell and Hollifield (2015) entitled 'Place, Space, and Pattern' offers a clear, if American-centric introduction as to how geographers might look at some migration themes.

Other more advanced discussions can be found in *The Migration Reader*, edited by Messina and Lahav (2005); *Rethinking Migration: New Theoretical and Empirical Perspectives*, edited by Portes and Dewind (2007) and *Migration Theory: Talking Across Disciplines*, edited by Brettell and Hollifield (2015). All three offer insightful surveys of different disciplinary and theoretical approaches to migration, what makes them distinctive, and some of the convergence between them. The journals *Asia and Pacific Migration Journal*, *Ethnic and Racial Studies*, *International Migration*, *International Migration Review* (more oriented towards the United States Canada, and to a less extent Australia and New Zealand), *Journal of Ethnic and Migration Studies* (for the UK and Europe in particular), *Journal of Refugee Studies*, *Migration Studies* and *Population, Space, and Place* all provide a bounty of articles at least on immigration in the richer countries.

On the social and spatial concepts discussed in this book, see Cloke, Philo, and Sadler (1991) *Approaching Human Geography*, for a comprehensive foundational discussion. Delaney (2005) has written a great introduction to 'territory'. The first half of the article by Marston, Jones III, and Woodward (2005), the first half of the chapter by Samers (2010b) and Herod (2010) all provide critical summaries of the problems of using scale.

In terms of statistics, the *World Migration Report* (listed above) published by the International Organization for Migration (IOM) provides global and national level statistics for most countries. The *International Migration Outlook* published annually by the OECD (Organization of Economic Cooperation and Development) provides a wealth of information on the OECD countries (essentially countries in the EU, North America, as well as Japan, Korea, Mexico and Turkey).

USEFUL WEBSITES

The following are the principal, official government websites for a selection of predominantly English-speaking countries:

- Australia: Department of Immigration and Citizenship (http://www.border.gov.au)
- Canada: Citizenship and Immigration Canada (http://www.cic.gc.ca/english/index.asp)
- New Zealand: Immigration New Zealand (http://www.immigration.govt.nz/)
- Singapore: Immigration and Checkpoints Authority (http://www.ica.gov.sg/index.aspx)
- United Kingdom: The Home Office (https://www.gov.uk/government/organisations/home-office) and within it, especially UK Visas and Immigration (https://www.gov.uk/government/organisations/uk-visas-and-immigration)
- United States: Homeland Security (http://www.dhs.gov), and within it, especially US Citizenship and Immigration Services (http://www.uscis.gov) and US Immigration and Customs Enforcement (http://www.ice.gov/). The Migration Policy Institute (http://www.migrationpolicy.org/) offers a useful information and data bank.
- From an international perspective, the International Organization for Migration: http://www.iom.int provides a varied range of publications and statistics, as does the United Nations (http://www.un.org), particularly their Population Division, but also the Department of Economic and Social Affairs, International Migration and Development, and the Global Migration Group.

Some websites that provide data and related resources (these sites also contain numerous links themselves):

- Age of Migration website (the authors of The Age of Migration have established a very useful website with its own list of links, photographs, and additional case studies): http://www.age-of-migration.com
- Asia Pacific Migration Research Network: http://www.forcedmigration.org/organisations/asia-pacific-migration-research-network-apmrn
- IOM (International Organization for Migration): http://www.iom.int
- Institute of International Education: http://www.oecd.org/migration/
- IMISCOE (International Migration, Integration and Social Cohesion) a network of 38 European migration research centres: http://www.imiscoe.org
- Forced Migration Online: http://www.forcedmigration.org/
- Missing Migrants Project (part of IOM): http://missingmigrants.iom.int
- Metropolis International: http://international.metropolis.net/index_e.html
- Migration News: http://migration.ucdavis.edu/mn
- Migration Policy Institute: http://www.migrationpolicy.org/
- Organization for Economic Co-operation and Development (OECD): http://www.oecd.org/migration/
- United Nations, Department of Economic and Social Affairs, Population Division: http://www.un.org/en/development/desa/population/theme/international-migration/

SOME SUGGESTED DOCUMENTARY AND FICTIONAL FILMS CONCERNING MIGRATION AND IMMIGRATION ON DVD*

*The country indicated after the date indicates where the film mainly takes place

A Day without a Mexican (2004), US, 100 min.

America's New Religious Landscape (2001), US, 60 min. Documentary available at www.insight-media.com

April Children (Aprilkinder) (1999), Germany, 85 min.

Chain of Love (2001), US, 50 min.

Bread and Roses (2000), US, 110 min.

Beautiful People (1999), UK, 107 min.

Brick Lane (2007), UK, 102 min.

Brothers in Trouble (1996), UK, 102 min.

Crossing Arizona (2006), US, running time unknown

Dirty Pretty Things (2003), UK, 97 min.

El Norte (1983) US, Mexico, 141 min.

Escape from Greece (2014), Greece, 15 min.

Exiles (2004) France/Algeria, 104 min.

From the Other Side (2002), 99 min.

Go Back Where You Came From (2015), Australia, 60 min.

God Grew Tired of Us (2006), US, 89 min.

Golden Venture (2006), US, 90 min.

Hostage (Omiros) (2005), Greece, 105 min.

Inch'Alla Dimanche (2001), Algeria and France, 98 min.

In this World (2002), UK and various countries, 88 min.

La Haine (1995), France, 95 min.

La Misma Luna (Under the Same Moon), 2008, Mexico and US, 106 min.

Le Havre (2011), France, 93 min.

Little Jerusalem (La Petite Jerusalem) (2005), France, 94 min.

Living in Paradise (Vivre au paradis) (1998), France, 105 min.

Lost Boys of Sudan (2004), US, 87 min.

Maria Full of Grace (2004), US and Mexico.

Mayan Voices: American Lives (1994), 56 min.

My Son the Fanatic (1999), France and UK, 87 min.

Night of Henna (2005), 92 min., USA

Permanently Temporary: The Truth About Temp Labor (2014), USA, 33 min.

Perspectives: 'Temporary' Migration in Canada (2011), Canada, 54 mins.

Romántico (2004), USA, 80 min.

Saimir (2004) Italy, 88 min.

Saving Face (2004), USA, 96 min.

Sentenced Home (2007), US, 87 min, Documentary available from http://www.pbs.org/independentlens/sentencedhome/

Taxi to Timbuktu (1994), 51 min.

The Gatekeeper (2002), US, 103 min.

The Invisible Mexicans of Deer Canyon (2006), US, 73 min, Documentary.

The Namesake (2007), US, 122 min.

The Other Europe (2006), Various European countries, 58 min. Available from www.newsreel.org

The Refugee Show: The Plight of the Padaung Long-Necked People (2007). Available from www.films.com

Welcome (2009), France, running time unknown.

Tea in the Harem (Le Thé au Harem d'Archimède) (1985), France, 110 min.

SUMMARY QUESTIONS

1. What are some of the pitfalls of collapsing migration into mobility studies?
2. Why is 'skill' a difficult term to use in the context of migration?
3. In what ways is migration 'diverse'?
4. Why is migration not simply a matter for the countries of immigration?
5. This book suggests that the 'global economy' or 'global capitalism' is an example of which social concept? What are the problems of using this concept to explain migration?
6. What are the difficulties of using the term 'scale' and how does the book deploy the idea of scale?

NOTES

1 For a review of cultural geographical studies along these lines, see for example Adey (2010), Blunt (2007) and Cresswell (2006).

2 King (2012) offers both a defense and criticism of conflating migration and mobility studies, which we discuss later in the text.

3 A more detailed listing of various types of migration can be found in King (2012: 137)

4 There are some exceptions to the general use of this definition, such as the case of France's DOM-TOM (overseas Départements and territories). In French statistics, those who migrate to the French 'mainland' from the DOM-TOM are considered immigrants or 'international migrants', but are not considered 'foreigners'.

5 See http://unstats.un.org/unsd/demographic/sconcerns/migration/migrmethods. htm

6 At the same time, the literature on the United States tends to employ the term 'immigrant' and 'immigration' more frequently. When discussing the American context in particular then, we have retained the use of the term immigrant and immigration.

7 To muddy the waters further however, the term 'undocumented' in official European Union policy reports literally means migrants who have lost or had their documents stolen. This is in contrast to those immigrants who simply over-stay their visas. The term 'sans papiers' – the equivalent of 'undocumented' has also become popular in France and with many NGOs in the EU.

8 UNHCR (2008) 'Convention and Protocol Relating to the Status of Refugees', http://www.unhcr.org/cgi-bin/texis/vtx/home, accessed 15 September 2008.

9 Article 1, pt. 2 of the Convention and Protocol Relating to the Status of Refugees, http://www.unhcr.org/cgi-bin/texis/vtx/home/opendocPDFViewer.html?docid= 4ec262df9&query=1951%20Convention, accessed 1 July 2015.

10 We should not lump together highly skilled and 'high-income' migrants (e.g. wealthy entrepreneurs) since the latter possess resources which the former may not, including the ability to literally buy citizenship.

11 We would note that for some of these trends, Castles et al. (2014) do not indicate a time frame, and so any reader should ask precisely when such trends began.

12 A 'country of transit' is a country that migrants go through to access another country that they may have difficulty reaching in more direct ways. It can only be understood in relation to migration policy and so has been widely critiqued as an analytical category (for example, Collyer et al. 2012).

13 The OECD or Organization for Economic Co-operation and Development is a set of wealthier countries, including eastern European countries and a number of so-called 'middle-income' countries, including Mexico and Turkey. We recognize the problem of defining or categorizing countries under this rubric. As we hinted at in the beginning of this section, perhaps the whole exercise of recording national data and classifying countries as wealthier or poorer should be avoided? We will leave this open to readers to decide.

14 Perhaps less critical or less radical scholars who are more uncomfortable with the idea or concept of (critiquing) global capitalism, might refer instead to 'asymmetric relations' between the wealthier and poorer countries.
15 Social capital may be understood as the resources that may or may not be provided by durable social networks between individuals and institutions. We will discuss this further in Chapter 2.
16 One notable exception, among others, is Smith (2001).
17 Voigt-Graf warns that this does not mean that 'cultures' are fixed or frozen in time and space.

2

EXPLAINING MIGRATION ACROSS INTERNATIONAL BORDERS: DETERMINIST THEORIES

INTRODUCTION

Migration is a bewildering set of processes to understand, and there is no shortage of theories to explain why and where people migrate. Exploring theories should be more than just an academic exercise however. Understanding why people migrate may point to processes of global structural inequalities and disadvantage that warrant our attention. A key theoretical issue then is that the explanation of migration may be different for different (groups of) people over time and space. This in turn suggests that an overarching theory of migration is an impossibility, or at least too abstract as a lens on the variety of migrations that have occurred across the world and throughout history (e.g. Brettell and Hollifield 2008b, 2015). One helpful way through the range of explanations is to distinguish, as Massey et al. (1998) do, between theories that explain the creation or the initial phase of a particular migration, and theories that explain subsequent phases, that is, the 'continuation' or the 'path dependency of migration systems' (Collyer 2005: 700). No doubt, this distinction has value, but in practice there is an overlap between that which initiates and that which continues different forms of migration. Boyle et al. (1998)

offer another distinction, between *determinist* theories (theories that on their own determine migration behaviour and patterns) and *integrative* theories (theories that bring together different theoretical and conceptual propositions). This distinction has merit too, but often so-called determinist arguments integrate a number of different political, cultural, economic, environmental and social processes, and integrative theories can be remarkably determinist. We would add to this that we can delineate between *explanatory* and *critical* theories, though what exactly constitutes 'critical' is not always clear either. As a consequence of the limitations of these various distinctions, we are faced with a dilemma: how to make migration comprehensible? While we discuss whether certain theories seem to be primarily concerned with the initiation or continuation of migration, and whether they are explanatory or critical, we have organized our discussion of the various theoretical approaches around Boyle *et al.*'s framework. We do this because they focus on the foundations of the theory involved,[1] rather than on often thorny assumptions about when and how a particular migration began and/or continued, or indeed whether the theory is in fact critical.

With the above in mind, this chapter is the first of two that explores various theories of, or approaches to, international migration. This chapter only reviews a set of more *determinist* theories,[2] namely 1) Ravenstein's laws and push-pull approaches, 2) neo-classical economic analyses, 3) behaviouralist approaches, 4) new economics approaches, 5) dual labour market and labour market segmentation approaches, and 6) structuralist and related understandings. Massey *et al.* (1993) remind us that these different theories or approaches have such different 'levels of analysis' and that they are not 'inherently incompatible' (p. 433), but

> the various models reflect different research objectives, focuses, interests, and ways of decomposing an enormously complex subject into analytically manageable parts; and a firm basis for judging their consistency requires that the inner logic, propositions, assumptions, and hypotheses of each theory be clearly specified and well-understood.

Nonetheless, some of these approaches have overlapping premises or units of analysis (let us say households and networks, for example), which are not mutually exclusive. In the pages that follow, we will explore these connections by systematically reviewing the theories above. At end of this theoretical review, we briefly discuss some of their limitations, including two glaring problems. The first is that the state tends to disappear in 'pure' versions of these theories. Some earlier theories, namely push-pull, neo-classical, and behavioural suffer egregiously from this problem. Yet strikingly, it also appears in some more recent arguments about globalization and transnationalism. So, while the role of states in migration is crucial, we will have to wait until Chapter 4 for a more in-depth discussion of this.

A second, but perhaps less obvious problem is the continual failure to integrate sophisticated understandings of 'space' in these approaches, despite all of the attempts in the cavernous literature on 'transnationalism' (we discuss this later in Chapter 3) to overturn the methodological nationalism of previous approaches and studies. This may be as true of accounts of migration by geographers, as it is of other social scientists. In light of the problem of space then, we draw on the various spatial concepts discussed in the Introduction to this book, along with an eclectic theoretical approach to migration as a means of understanding migration. The emphasis will be on low-income migration (including asylum-seekers and refugees) from poorer countries to richer ones (the subject – for better or worse – of most international migration theories), though some attention will be paid to explaining other types of migration.

THE BEGINNINGS OF MIGRATION THEORY: RAVENSTEIN'S 'LAWS'

Let us begin by revisiting the work of the nineteenth-century geographer Ravenstein. While it may seem odd to return to a geographer's lengthy and painstaking studies in the *Journal of the Statistical (or Royal Statistical) Society* in 1885 and 1889 concerning internal migration in the United Kingdom, Ravenstein's ideas still haunt neo-classical approaches to international migration. Troublesome as it may seem for researchers on migration today, there seems to be some return to the 'methodological individualism' of his analysis. Methodological individualism simply means that individuals serve as the unit of analysis. Importantly, Ravenstein's undertaking involves a study of migration using mainly sub-national census data from counties, cities, towns and villages. Yet he also categorizes migrants into 'short-distance', 'stage-migrants', 'long-journey' and 'temporary migrants', thus avoiding a homogenization of different forms of migration. For Ravenstein, the primary reasons for migration are higher wages or better work ("work of a more remunerative or attractive kind" – 1885: 181), and in that sense, an economic determinism prevails. At the same time, and as noted above, Ravenstein's approach can be characterized as 'methodologically individualist'.

Ravenstein is most noted for his 'laws of migration'. Although there is some question over how many 'laws' he actually stated (Tobler 1995), below we provide an abbreviated selection of seven of his laws from his 1885 paper (see Ravenstein 1885: 198–9, and Lee 1969). These supposed laws are as much about empirical patterns (that is a description of migration) as they are about the *causes* of migration.[3]

1. He argues from the point of view of the demand for migrants that groups of migrants travel relatively short distances and develop into a 'current' of migration to the 'great centres of commerce and industry', and this current is a reflection of the number of people in the area of origin as well as the number of people in the area of destination.
2. His second law relates to the 'natural outcome' of the first; namely that the residents of a rural area will move to a surrounding and rapidly growing town or city. As migrants move from the surrounding rural area to the neighbouring town or city (so-called absorption), rural depopulation occurs and these rural areas are then attractive to migrants from even further afield. The town or city then eventually becomes a pole for migration from throughout the country, in this case the United Kingdom.
3. Thirdly, he claims that a process of 'absorption' occurs (the in-migration and reception of migrants from certain areas) at the expense of 'dispersion' (out-migration from certain areas).
4. From 'law' number 3 immediately above then, every 'current of migration' produces a counter-current.
5. Long-distance migrants generally migrate to the 'great centres of commerce and industry'.
6. Those in rural areas migrate more than the natives of towns or cities.
7. Women migrate more frequently than men.

What should be gained from Ravenstein's seven 'laws'? At first sight, they may seem dreadfully antiquated, if nothing else than for their apparent determinacy and lack of any explicit theoretical foundation.

The former is especially noteworthy, given that the social sciences (or at least its critical corners) have abandoned such attempts at law-making for more complexity and indeterminacy. Yet this is to dismiss some of the value of his proclamations too easily. Indeed, he seems to point to some persistent patterns and processes associated with international migrations that are witnessed today by many scholars of migration. To begin with, whereas global migration is not simply the result of differences in wages between countries and the prospects of a better job, there is no doubt that below-subsistence wage levels, unemployment, and dangerous or demeaning work, do drive people to migrate (e.g. Castles and Miller 2003). In short, migration may partly be determined by economic concerns, without us having to necessarily subscribe to economic determinism. Second, his categories of migrants and migration seem to anticipate many of the categories used in the analysis of migration today. In fact, what he calls 'stage migration' was initially recuperated by dependency theory in the late 1960s and 1970s (we discuss this theory later in the chapter) and has remarkable resonance for the increasing interest lately in stages of internal and international movements. This sort of stage migration is sometimes referred to as 'circulation' or a matter of 'diverse trajectories', rather than migration as a single movement from point A to point B. Third, his claim that 'long distance' migrants generally move to the 'great centres of commerce and industry' is reflected in much (but certainly not all) of the migration of the second half of the twentieth century and the beginning of the twenty-first, at least from poorer countries to richer countries. We only have to think of Frey's (1998) notion of 'immigrant gateway cities' and Sassen's 'global city hypothesis' (1991)[4] to illustrate this relationship between immigration and large, economically dynamic cities. Fourth, Ravenstein observed that one should not ignore the role of women as migrants. Consider for example, the following passage:

> Woman is a greater migrant than man. This may surprise those who associate women with domestic life, but the figures of the census clearly prove it. Nor do women migrate merely from the rural districts into the towns in search of domestic service, for they migrate quite as frequently into certain manufacturing districts, and the workshop is a formidable rival of the kitchen and scullery. (Ravenstein 1885: 196)

His emphasis on women as migrants is notable, given that over the next hundred years, most theories of migration remained remarkably silent about the role and experience of women in migration (Morakvasic 1984).

Ravenstein's analysis suggested a set of push-pull factors that drove migration – an idea later elaborated upon by Lee (1969), that is, a set of factors that 'pushed' migrants from one region (or country), and a set of factors that 'pulled' them to another region (or country). Such push factors might include rapid demographic growth (but see Box 2.1), ethnic cleansing, environmental crises, political repression, poverty, and war or other forms of violence. Pull factors might include job opportunities, a better standard of living, medical treatment, freedom from political repression, and even the ability to just buy and sell goods (cross-border migrants from Lesotho, Mozambique, and Zimbabwe, to South Africa during the late 1990s come to mind here – McDonald et al. 2000). In some cases, it might not be clear whether something can be considered a push or pull factor, such as the search for 'adventure' among younger migrants (Goss and Lindquist 1995).

Box 2.1 IS LABOUR MIGRATION A RESULT OF DEMOGRAPHIC DISPARITIES?

A now widespread view is that rapid demographic growth in many poorer countries of the global south combined with the demographic slow down ('below population replacement levels') in countries of the global north (especially, Germany, Italy, eastern Europe, Russia and Japan) is driving labour migration. This seems to make sense. Employers in richer countries are unable to find workers so they either send for them or hire those that are already resident in the country. The logic is simple and compelling: there is too much work to be done in the richer countries and not enough hands to do it. In contrast, there are too many hands in poorer countries and not enough work to make use of them. But let us take a moment to think more critically about this.

First, some Eastern European countries with very slow demographic growth, such as Moldova and the Ukraine, experience significant emigration.

Continued

Second, some *sub-national* regions in both the richer and poorer countries are growing rapidly and the demand for migrant workers is high, while others are growing very slowly, even declining, and demand is low, so this methodologically nationalist approach is problematic from the very start (e.g. England 2015).

Third, states or firms may recruit migrant workers regardless of whether there are citizen workers available, in part to demand company loyalty and lower labour costs. Firms in California's Silicon Valley have been accused of this.[5]

Fourth, national and more local economies in the rich countries *can* adjust to the absence of migrant workers, though this may be extremely painful for everyone. How do they adjust? Substituting technological innovation for workers is one way; another way is by making existing employees work harder, ceasing certain forms of production or services, and/or moving them elsewhere, when and where this is possible.

Fifth, state policies stimulate or connect particular regions and migrants, often regardless of the conditions of the number of workers in each region, and states even encourage some countries to have children (pro-natalist policies). In fact, many EU countries (France and Germany to name just two) have responded to slow demographic growth through such pro-natalist policies. For example, in 2007, the German government implemented a programme called 'Elterngeld' to encourage *professional* mothers to exit the workforce and have children by paying these would-be mothers (Carle 2007). Notice that the programme was designed for 'professional' women, meaning German and not generally speaking, Turkish or other immigrant women.

Lastly, perhaps the most significant problem with the demographic disparity argument is that it deploys a 'neo-Malthusian' view of population growth in poorer countries. In other words, it sees population as a problem relative to economic (or environmental) opportunities and resources, rather than seeing a problem with the socio-economic organization of the societies that have rapid demographic growth. In other words, rather than thinking of labour migration as the result of a demographic disparity, an alternative way of viewing labour migration is to see it in relation to the type of development that occurs in poorer and richer countries (and how policies of the rich world and institutions such as the International Monetary Fund and the World Bank shape development in the poorer countries) and not a problem of too many people in one part of the world and too few in another.

A review of a number of studies in the 1990s and 2000s revealed that the idea of 'push' and 'pull' persisted in studies that seek to explain migration, mainly but not exclusively from poorer countries or regions to richer ones (see e.g. Hugo 1996; Hamilton et al. 2004; Li and Bray 2007; Perrin et al. 2007; Van Wijk 2010; Wilson and Habecker 2008). Lee's understanding of push and pull (see above) left conceptual room for intervening opportunities or "factors which come into play between the origin and destination" (Schapendonk 2011: 4), but these factors simply enter into a rational decision-making process in which a migrant chooses between origin and destination. Many see this framework therefore as involving the dubious economic rationality of the neo-classical framework (described in the next section), yet it may be extended to all kinds of rationalities, which only reinforces the ambiguity of such a metaphor (e.g. Van Wijk, 2010). The consequence is that we miss an understanding of the social networks, the state processes, and serendipitous events that connect the many 'heres' and 'theres'.

An alternative to employing the language of push and pull factors is witnessed in Mahler's (1995) wonderfully evocative but also sobering book *American Dreaming*. In this work, she emphasizes that most Salvadoran migrants to Long Island in the suburbs of New York arrived during the period immediately after the 1980s civil war in El Salvador, and many because of events relating to that civil war, including declining opportunities for employment. But she also shows how often plentiful (but very poorly paid jobs) existed on Long Island; how the two countries were linked through foreign policies, how US employers actively recruited Salvadorans in the 1960s, and how Salvadoran migrant networks figured in bringing migrants to Long Island specifically (see also Bailey et al. 2002). That is, she connects a range of simultaneous processes in the country of emigration and immigration without discussing them as a set of push and pull factors. More than halfway around the world, we could consider another example that calls into question the many assumptions of the push-pull metaphor, that of the Rohingya from Myanmar (see Box 2.2), or the migration of refugees from Syria to the European Union (see Box 2.3). Nonetheless, the notion of push and pull factors continue to flourish in the neo-classical economic approach to migration, and it is to this literature which we now turn.

Box 2.2 EXPLAINING THE EMIGRATION OF THE ROHINGYA FROM MYANMAR (BURMA): PUSH AND PULL?

In 2012, 140,000 Rohingya people of Myanmar were forced into camps after an attack by Buddhist extremists, which seriously injured many and killed dozens. It was then that their plight became a matter for the global press, and their horrible conditions re-surfaced again in 2015, when some 25,000 crossed the Bay of Bengal and the Andaman Sea and began landing in Thailand and Malaysia on rickety boats (see Map 2.1.) The Rohingya are said to originate from Rakhine State (otherwise known as Arakan) in eastern Myanmar (Burma) and have emigrated especially over the last decade to mainly Bangladesh, Malaysia and Thailand, although there are significant communities in countries such as India, Indonesia, and Pakistan, among others. One might immediately turn to push factors such as stigmatization, marginalization, ethnic cleansing or even genocide in Burma, in order to explain their emigration. Indeed, they have been victims of Islamophobia; barred from local schools and paid employment, subject to limits on how many children they can have, and constantly the targets of forced labour, physical violence and even death. They are frequently denied citizenship in Burma, and many are stateless refugees, caught between deportation from Bangladesh, Malaysia, and Thailand, and refused official re-entry in Burma. On the so-called pull side, Bangladesh, Malaysia or Thailand appear to many Rohingya as offering *relative* shelter from the misery of life in Burma (*The Economist* 2015a; Palmgren 2014; Parnini et al. 2013).

We could therefore conclude that they are 'pushed' from Burma and 'pulled' to Bangladesh, Malaysia or Thailand. Yet the history and dynamics of emigration of the Rohingya suggest that this push-pull metaphor is inadequate, at least if we want to understand why they emigrate to *where they do*. Indeed, as we noted in the Introduction, and as we will see in Chapters 3 and 4, part of the limitations of a push and pull view is first, the methodological nationalism that often pervades such arguments; second, the existence of state-based institutional linkages between Burma and the countries of destination, and third, the presence of social (migrant) networks – often 'illegal' – between Burma and countries such as Bangladesh, Thailand and Malaysia that also calls into the rationality and methodological individualism of the push-pull metaphor.

In terms of methodological nationalism, yes, many Rohingya may look favourably towards Malaysia as a country, given both its prosperity and Islamic heritage, and this may in turn affect their destination decisions (*The Economist* 2015b), yet the movement of Rohingyas to Bangladesh, Malaysia and Thailand occurs through, and is concentrated within, particular sub-national sites (the neighbouring town and region of Cox's Bazar and southern Bangladesh, the eastern border district of Mae Sot, Thailand, and the capital cities of Bangkok, and Kuala Lumpur, as well as particular port cities in Thailand and Malaysia as smuggling decisions shift on the Andaman Sea). This concentration can itself be explained partly through the location of general pro-migrant NGOs such as the UNHCR or even Rohingya-run NGOS. For example, some Rohingyas are trying to leave Thailand for Malaysia, not strictly because of significant differences in national immigration policies, but because the UNHCR seems to have better access to detention centres in Kuala Lumpur than it does in Bangkok, and in the former they can apparently do a better job of preventing deportation (Palmgren 2014).

In terms of state institutions, there are a variety of institutions and policies that connect the four countries including Bilateral Memorandums of Understanding with respect to legal low-income labour recruitment, although most Rohingyas are actually undocumented, asylum-seekers, or recognized refugees. There is also illegal institutional involvement in the form of state officials

Continued

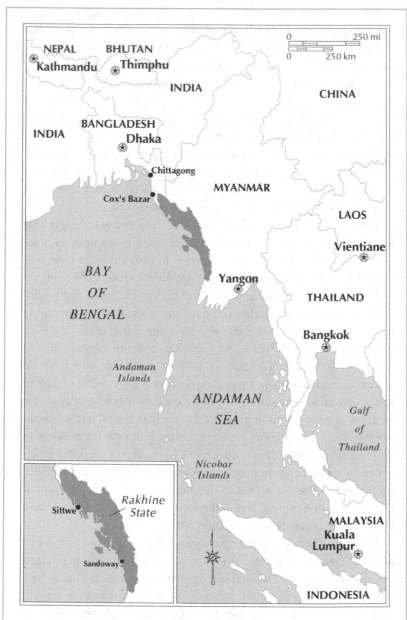

Map 2.1 The emigration of Rohingya from Burma/Myanmar, to south-eastern Bangladesh, Malaysia, and Thailand

who are often bribed by Rohingyas or smugglers to cross the Thai or Malaysian border.

Migrant (social networks) also shape *how, when, and where* Rohingyas migrate. In this process, most Rohingya asylum-seekers rely on the information of family, friends, NGOs, and smugglers to cross the Andaman Sea to reach Thailand, and failing that, to Malaysia or even Indonesia.

Malaysia and Thailand may indeed be appealing 'pull' countries, and there is some rationality in the choosing of destinations, but life is far from easy in both countries. Most Rohingyas will eke out only a meagre living (as in Burma, they are barred from formal paid employment in Bangladesh, Malaysia and Thailand) and frequently sell *roti* (fried South Asian bread) illegally on the streets of Bangkok or Kuala Lumpur, and bribe local police to not arrest them for selling food. If caught, they may be deported across the Thai border into Burma, only to bribe authorities to travel back across the border into Thailand. Thus, Rohingyas, like other asylum-seekers, refugees, and labour migrants, leave Burma to seek out a better life elsewhere, but their emigration is always mediated through particular routes, nodes/places, and territories, along with both formal and informal state institutions, their respective actors, and migrant networks. The methodological individualism and rationalism of the push-pull metaphor may therefore seem obvious unless we carefully examine the various theories that are laid out in this and the next chapter (for further reading, see Aziz 2014; *New Straits Times* 2015; Palmgren 2014; Parnini *et al.* 2013).

Box 2.3 RE-THINKING THE SYRIAN REFUGEE CRISIS

On the face of it, the 2015 'Syrian refugee crisis' seems to be easily explained. In short, there has been a devastating civil war in Syria (or rather a series of complex 'conflicts' as a number of observers prefer to call it) that began sometime in 2011, which in turn led to an ever growing number of migrants or refugees in especially the European

Continued

Union (EU) by 2014–2015. On one hand then, violence and poverty pushes individuals from Syria while the possibilities of a 'better' life pull them to the EU, for example. Yet let us again think about this in another light. Refer back to Figure 1.4 in the Introduction, and notice the different elements of social theory that can be used to explore an issue. Thus, we can begin with an 'event' (such as the refugee crisis), and then think in terms of *structures*. So, we could ask then what generated the movement of refugees in the first place, such as conflict, unemployment, poverty, environmental degradation, and so forth. We could even dig deeper to suggest that rich country governments might be partially responsible for either producing, reproducing, or escalating the conflict in Syria, and if so does that mean that the western powers have actually inadvertently created the migration of refugees? Second, what *institutions* are involved in regulating and producing the crisis and its subsequent migrations (such as the UNHCR, or EU migration institutions of immigration)? In other words, institutions figure centrally in – among other things – naming migrants *as* refugees (or not), and thus in need (or not) of protection and certain basic services. At the same time, migrants are not simply buffeted by the winds of structures and institutions. Rather, migrants are *actors* or *agents* insofar as they make conscious decisions to hire a smuggler in order to cross the Aegean Sea between Turkey and, let us say, the Greek island of Lesbos. This is a very complex and fraught decision, and may not involve the same sort of rationalities envisioned by Lee (1969). In any case, the decision to travel with a smuggler may be more collectively decided upon by a whole *network* of families from villages in Syria, some of whom will already have relatives in the country of destination, such as Germany or Sweden. But if smugglers are involved, then smugglers as *intermediaries* are significant actors in the process of migration, and thus migration is not simply a weighing of 'pushes' and 'pulls'. In sum, such language should by now begin to look a little thin and problematic.

THE NEO-CLASSICAL ECONOMIC APPROACH

Neo-classical economic writings on migration began with studies of mainly internal migration *within* either poorer or richer countries, rather than international migration *between* poorer and richer ones (Massey *et al.*

1998). Sometimes called the 'functionalist' or 'traditional' approach to migration, the neo-classical economic literature on migration is considerable and diverse, but here we consider only a select set of widely rehearsed studies.[6] According to Massey et al. (1998), the neo-classical approach can be divided into macro and micro theory[7] (see also Radu 2008). From a macro neo-classical perspective, the work of Lewis (1954) and Ranis and Fei (1961) stand out as initial contributions, albeit in the context of explaining economic development. More specifically, these were attempts to show the relationship between the demand for labour in urban areas and the supply of labour in rural areas, and how internal rural-to-urban migration shaped the economic development of both. Broadly speaking, they argued that urban labour markets would eventually absorb supplies of cheap labour from rural areas which would decrease the supply of labour in rural areas, thus raising wages. However, in urban areas, the labour supply would increase, thus causing wages to fall. The result would be an equalization of wages between rural and urban areas, and when wages are equalized, migration ceases (Enke 1962). While these studies were based on internal migration, they informed the neo-classical debate on explaining international migration, and are labelled 'macro' because they did not investigate the individual reasons for migrating.

The neo-classical perspective also pointed to another migration which mirrored (Massey et al. 1993) the movement of migrant workers from poorer countries to richer ones. This other migration is initiated by the relative lack of capital in poorer countries which also attracts international capital (foreign investment) and along with it individuals with higher levels of human capital (in other words managers, professionals, and other skilled workers from richer countries) who can reap substantial awards in poorer countries which are relatively bereft of both capital and higher levels of human capital. The temporary migration of European and North American engineers to countries as distinct as the Philippines and Saudi Arabia illustrates this quite well. One of the implications of this is that migration causes wages to rise in the poorer country, and wages to decrease in the richer country, ultimately leading to wage equalization across the world. The end result, according to the basic neo-classical framework, is that this migration will lead to equilibrium as people migrate from job-poor regions to job-rich regions, thus balancing people with available economic opportunities.

In the micro-perspective, following a general neo-classical model, migrants are 'rational' individuals, responding to 'perfect', or 'various

pieces of information' (Borjas 1989: 461) about their economic opportunities in both the country of emigration and the country of immigration.
They are also 'utility maximizers', usually by maximizing income, the
possibility of employment, and/or other employment conditions, and
who seek better opportunities in the region or country of immigration.
The neo-classical micro-approach to migration therefore involves a 'methodological individualism' (Boswell 2008a). In the earliest contributions
to a micro-theory of migration, Todaro (1969) and later Harris and Todaro
(1970) and Todaro (1976) were responding to the limitations of Lewis's
(1954) and Ranis and Fei's (1961) arguments about wage differentials,
and provided notable revisions to their model. To be more specific, Harris
and Todaro argued that the presence of a large group of unemployed or
underemployed people in cities in the poorer countries would affect the
probability of finding employment. Migration is therefore only partly a
response to real income differentials; it is also a response to the expected or
perceived wage differentials based on the probability of eventually securing a job in the modern industrial sector, rather than in the traditional
sector (or informal employment). Here, time is important and migrants
may move with their longer-term total income in mind; the larger the
expected income, the more likely are individuals to migrate. Their
approach is more behavioural than the strictly wage-differential theory,
and suggests other socio-psychological explanations for migration
(Kearney 1986; Molho 1986; Radu 2008). In fact, Radu insists that the
assumption that the analysis of migration by economists only involves an
analysis of wage differentials is a caricature. We will return to the question
of socio-psychological explanations very shortly, but for the moment, let
us move to the human capital version of neo-classical theory.

In the human capital version of neo-classical theory, Sjaastad (1962)
for example, viewed migration as a means of investing in one's human capital
(though Sjaastad did not use that term), but also as a function of one's
human capital. One can then discern the likelihood of migrating from the
amount of human capital a person possesses. In order to maximize
migrants' human capital, migrants must weigh the costs (travel costs, the
time spent searching for work and being unemployed, immigration regulations, the problems of learning a new job and language, as well as
other less tangible psychological costs associated with migration) with
the benefits (higher wages, better employment conditions, and so forth)
(Borjas 1989).

THE BEHAVIOURAL APPROACH

Like neo-classical analyses, the behaviouralists focused on individuals as the unit of analysis, but they remained wary of neo-classical explanations, in part because they seemed to be guilty of the so-called ecological fallacy, that is, making inferences about individuals from the study of groups of people, or to be more specific, because it appeared that particular nation-ally defined groups of people moved from low-wage countries to high-wage countries; that did not mean that *all* individuals migrated for reasons relating to wage differentials (Boyle *et al.* 1998).

Scholars working in the behaviouralist vein, such as Mueller (1981) Clark (1986), and Wolpert (1965) were concerned to understand a migrant's cognition and decision-making (or the psychological reasons) for choosing a particular place as a destination (Boyle *et al.* 1998). And like many neo-classical analysts, they were also concerned primarily with internal, rather than international migration. However, what distinguishes the behaviouralists from the neo-classicists is that the former were as much interested in the apparent irrationality (rather than the rationality) of indi-vidual decision-making, especially in terms of why migrants chose certain destinations. Wolpert (1965) in particular sought to understand what he called the 'place utility' of particular-migrants, and he argued that migrants chose particular destinations because they offered the highest place utility (or satisfaction). This utility did not necessarily concern expected wages, nor did it necessarily involve an individual cost-benefit analysis around improving one's human capital for example. Migrants may choose certain locations to be closer to their relatives for emotional reasons, or because they have simply heard of a place before migrating. Yet, the behaviouralists also focused on why people migrate altogether, not just why they migrate *where they do*. In this sense, Wolpert (1965) focused on the question of tol-erating a certain degree of stress (or discomfort) in the country of emigration. However, stress in the country of emigration did not lead migrants to compare this stress with the possibilities of stress in the coun-try (or place) of destination. Rather it was the degree of stress in the country of emigration, and whether it passed a certain threshold which would impel people to migrate. From Wolpert's perspective, migrants are therefore 'satisficers' (seeking satisfaction), rather than 'maximizers' (seek-ing optimization) (Boyle *et al.* 1998; Conway 2007). While this earlier literature – especially by demographers and geographers – could be char-acterized as more analytical, quantitative and pre-occupied with generating laws, or at least patterns or theories of migration, scholars across the social sciences are incorporating increasingly a more ethnographic and interpretative approach to answer why and where people migrate.

Without acknowledging the idea of 'place utility', Wilson and Habecker (2008) asked in their ethnographic fieldwork why African migrants chose Washington DC as a destination instead of other cities. What they found is that African migrants *perceived* all capital cities to be centres for business, cultural and educational opportunities, because the migrants insisted, capital cities in African countries are also the primary locus for such opportunities. They also found that many Africans perceived Washington to be more accepting of black and 'international' people, full of international institutions, quieter and less expensive than New York for example, and a better place to raise their children. Without also referring to place utility, Beech (2014) too speaks of the significance of the *imaginations* of places by international students – in this case about cities in the UK or the US – before they choose to migrate. Imaginations of particular places – often shaped by post-colonial relationships,[8] influence students' decision of where to study. These imaginations of places (or countries) may even manifest themselves in 'destination hierarchies' (Paul 2011) or differential rankings of places or countries by men and women, as Hofmann (2015) finds in her study of a diverse range of migrants from the country of Georgia. Clearly then, place utility is complicated.

THE NEW ECONOMICS APPROACH

A fourth approach to understanding international migration is the so-called 'new economics paradigm'. This term is really an invention of Massey *et al.* (1993, 1998) and not a term to which economists would necessarily subscribe. At any rate, the emphasis here is not so much on individuals, but on families, households, and other units larger than individuals. The work of Stark (1991) in particular, stands out. From this perspective, families and other units allow the maximization of income but they also allow for the minimization of risk associated with the problem of labour and other markets. In the language of the new economics approach, households can address risk by diversifying the allocation of scarce resources (Goss and Lindquist 1995; Massey *et al.* 1993). That is to say, the labour of some family members (or 'scarce resources') may stay behind in the country of origin while others may migrate. If economic conditions deteriorate in the country of origin, then family members overseas can compensate through the use of remittances (the money that is sent home by migrants). Massey *et al.* (1993) identify four examples where risk minimization occurs: crop insurance markets, futures markets, unemployment insurance and capital markets. We briefly summarize these issues below. In terms of the first, most families in poorer countries do not have crop insurance in the case of crop failure; where this is the case, remittances play a role in providing funds. Similarly, in poorer countries where there is usually no futures market (a commodity market in which a certain price is guaranteed for the future delivery of some commodity) and usually no large investors to absorb any losses if the future guaranteed price falls below the cost of producing the good, migration once again serves as a means of addressing income risks from fluctuations in crop prices. In terms of unemployment, unemployment insurance is once again either non-existent or weak in poorer countries, and migration (and any potential remittances) plays a role in mitigating the risk of unemployment. With respect to capital markets, farm families might need capital in order to buy certain kinds of seeds, fertilizers, irrigation and other equipment. Non-farm families may wish to finance the education of family members, or purchase capital goods for re-sale in local markets. The problem is that banking institutions in poorer countries can be unreliable in terms of savings and loans; loan capital is scarce, and in any case, it is difficult for many families to qualify for a loan. The remittances generated by migration serve as a means of providing capital and savings for investment purposes. The point here is that it is not so much the total family income which is important, but the source of income. In other words, families wish to diversify the sources of their income to ensure against risks. Furthermore, families are not simply interested in maximizing absolute income as in neo-classical theory, but maximizing income (or minimizing their deprivation) relative to other families, in let us say, the village of origin, by using migration as a strategy (Massey *et al.* 1993, 1998).

DUAL LABOUR MARKET AND LABOUR MARKET SEGMENTATION APPROACHES

This approach is most associated with its pioneer, the economist Michael Piore. In his widely regarded book, *Birds of Passage*, Piore (1979) argues that it is not so much push factors in the countries of emigration but pull factors and namely the presence of a dual labour market in the richer countries (or in the language of the time – modern industrial societies) that drives migration. For Piore, this dual labour market consists of a primary and a secondary sector. The first is dominated by 'native' (i.e. citizen) workers, and contains more highly paid and stable jobs with better working conditions and greater possibilities for promotion. Employers may invest considerable training in these jobs, and firing workers may be expensive and/or difficult. In contrast, the secondary sector dominated by migrants contains poorly paid and unstable jobs with poor working conditions and few possibilities of promotion, employers invest very little training in this sector, and firing workers may be easier. It is difficult to attract native workers into secondary jobs because of their low pay and poor working conditions, and thus employers look to migrant workers to fill these positions. The creation of these jobs precedes the migrants who fill them. While Piore's overall argument figured prominently in studies of migration and labour markets in both Europe and the United States, it rarely appeared as a central argument for the initiation or continuation of migration. Some subsequent literatures do approximate his approach however.

The first is Sassen's 'global city hypothesis', but again we will discuss this later in the chapter. The second is a substantial body of work that connects the growth or existence of informal secondary-sector-type jobs[9] with the demand for undocumented immigrants in richer and poorer countries. The work in this vein goes something like this: there is a reduction in the number of more formal employment opportunities in poorer countries, leaving informal jobs as the only option for citizens and migrants alike. At the same time, there is a proliferation in small sub-contracted firms that offer only informal jobs in the richer countries. Given that a large proportion of migrants in both poorer and richer countries are undocumented, and given their undocumented status, researchers argue that their illegality means they have little choice but to accept very low-wage informal jobs. The result is that undocumented immigration facilitates (but does not determine) the expansion of informal employment which in turn creates a demand for undocumented immigration, and the two processes are argued to be mutually reinforcing.[10]

Unfortunately, the dual labour market hypothesis and related studies employ very simplistic categories. Observers of labour markets in the US fathomed that there were not just two sectors in the labour market, and

Reich *et al.* (1973) used the term 'labour market segmentation' to refer to the innumerable 'cells' within American job structures. Distinct operational rules are said to regulate each cell (a cell can be considered a group of similar jobs within a firm), with some workers in certain cells having more chances of promotion than others, some having higher pay than others, and some having better working conditions than others. While the study of labour market segmentation itself has evolved enormously (Peck 1996; Samers 2010a), few studies developed a more sophisticated conception of how labour market segmentation might stimulate migration (Bauder 2005).

This became clear in a third and related literature, which is the large body of work that exists on 'immigrant' or 'ethnic' entrepreneurship and its spatial manifestation: 'ethnic enclaves' (Light (2005) provides a review). This literature concerns both immigrant *entrepreneurs* and the migrant *wage earners* they employ, but once again it tends to rely on a very crude distinction between the two. Nevertheless, Massey *et al.* (1998) summarize the consequences of the development of such enclaves, which tend to demonstrate the 'continuation' of migration rather than its initiation. If and when these enclaves form, they begin with a small number of immigrant entrepreneurs who have the cultural, financial, human, and social capital to begin a business in an urban area, under the necessary social and economic circumstances. They call on other migrant workers of the same ethnic background to work in these firms.[11] Employers benefit because they have a large pool of potential workers which allows the firms to be competitive vis-à-vis firms outside the enclave who may not be able to draw on such a large pool of workers. Employers in the enclave also benefit because they profit from a spatially concentrated demand for their ethnic or immigrant-specific products and services. Through the ethnic solidarities between employers and employees, workers balance the possibility (or even promise) of employment and future advancement with the long hours, arduous work, and loyalty to their current employer (e.g. Ahmad 2008a; Ram *et al.* 2003a). These same employees may themselves become entrepreneurs, and through norms of ethnic solidarity, are eventually also expected to hire their co-ethnics. As the enclave expands, so does the need for migrant workers, and as Massey *et al.* (1998) put it "immigration can, quite literally, generate its own demand" (p. 32).

STRUCTURALIST APPROACHES

Structuralist explanations for migration – sometimes labelled 'macro' or 'political economy' approaches – generally have their foundations in various Marxist, neo-Marxist, and historical-sociological readings of capitalism. In this sense, it remains unclear whether they are designed as theories of migration specifically, or theories of capitalism, (neo-colonialism), imperialism or neoliberalism, in which migrations are viewed as pivotal. In any case, there are various strands within this literature, including dependency theory, articulation theory, world systems theory, globalization arguments, global city arguments, and neoliberalism.

Dependency theory, articulation of modes of production theory, world systems theory

During the 1970s and 1980s, scholars employed the above approaches to focus on the considerable labour migrations that were occurring during the second half of the twentieth century, namely from the Caribbean and Latin America to the United States, from Europe's former colonies to the countries of northern Europe (e.g. France, Germany, the Benelux countries, and the UK), or from a number of southern African countries to South Africa. In that respect, they tended to emphasize the problems of political-economic inequality, and the 'development of underdevelopment' through international capitalism. Other perspectives within this broad approach focused on trade inequality manifested in unequal exchange (wage differentials) between richer and poorer countries linked through a history of class inequality, colonialism, imperialism, and the racism and xenophobia that both justified and facilitated their persistence (e.g. Burawoy 1976; Castells 1975; Castles and Kosack 1973, Cohen 1987; Miles 1982; Portes and Walton 1981).

In the Latin American-inspired dependency theory, migration within Latin America and between Latin America and the United States in particular, was argued to be the result of the inequalities generated by the penetration of US agribusiness operations in Latin America, and Mexico especially. With the cooperation of the Mexican government's support for more commercialized agriculture, these operations led to the gradual impoverishment of Mexican peasants (small subsistence-oriented land-owners) by dispossessing them from their land and reducing the demand for agricultural labour in Mexico. This in turn drove Mexican peasants to seek work in US agriculture (Wilson 1993). Yet US corporate investment also contributed to this relation of dependency by siphoning off the most highly productive workers from Latin America, and contributing to what many call a 'brain drain' (Goss and Lindquist 1995; Kearney 1986; Castles and Miller 2003).

The 'articulation of modes of production' theorists claimed that capitalism (or capitalist social relations) had not flattened or eliminated so-called pre-capitalist modes of production in an even manner, and countries and regions were unevenly integrated into the capitalist system (Cohen 1987; Portes and Walton 1981).[12] The penetration and development of capitalism in so-called traditional societies had the effect of disrupting existing agricultural, household, and other social relationships, thereby throwing people off the land which led to their incorporation into capitalist sectors and eventually to their international migration to the richer countries (e.g. Portes and Walton 1981; Samers 1997b). At the same time, this penetration and development of capitalism relied on the social reproduction (see Box 2.4) performed under pre-capitalist relations, often by women (e.g. Wolpe 1980; Kearney 1986; Meillassoux 1992).

Box 2.4 WHAT IS 'SOCIAL REPRODUCTION'?

Social reproduction is a Marxist-Feminist inspired term which has come to mean the process by which people are housed, fed, clothed, educated and generally raised to become workers and/or citizens under capitalism. In short, people need to be reproduced in particular ways to make them 'ready' for capitalism. In the so-called domestic labour debate of the early 1970s, feminist scholars began to criticize Marxist theories of capitalism because they neglected the contribution of women's work to reproducing capitalism. In other words, household work by women, feminist scholars argued, not only created economic value in itself, but was integral to the reproduction of men and their work. The assumption then was that there was a functional relationship between processes of social reproduction and the continuation of capitalism. In other words, capitalism *relied on* this social reproduction in order to exist, and

Continued

social reproduction existed to further the development of capitalism (see e.g. Malos (1980) on early theorizing in general, and in the
context of immigration, Sassen-Koob (1984). Cravey (2003) and
Smith and Winders (2008) provide more recent discussions of
social reproduction and immigration in the US South).

When migrants worked in capitalist sectors but were supported by
pre-capitalist social reproduction or when workers' wages were so low
that they could not even adequately house, clothe or feed themselves, it
was commonly labelled 'super-exploitation'. The point is this: since workers in the poorer countries were being uprooted and migrating overseas,
capitalists could now rely on the so-called internationalization of
super-exploitation. Since capitalists in the rich countries now relied on
the internationalization of super-exploitation, this allowed for, or facilitated, the movement of workers from the global south to the north
(Kearney 1986).

A third and related, but also more influential approach emanates from
the writings of the sociologist Immanuel Wallerstein (1974, 1979) and
his world systems theory. Wallerstein saw the world as a single capitalist
system, comprised of nation-states within regions that were progressively
incorporated into this world-system, and which were part of either the
'core' (at the time he was writing, this meant North America, Europe,
Japan, Australia and New Zealand), the 'semi-periphery' (for example,
Argentina, Brazil, Hong Kong, Mexico, Singapore, South Korea, and
Taiwan), and the 'periphery' (most of the remaining countries in the
world). These different zones of production and consumption were linked
through an international division of labour. For Wallerstein, the gradual
penetration of capitalist relations into these peripheral non-capitalist societies since the sixteenth century had created a single world capitalist
system that impelled the migration of workers to the 'core' or the
'semi-periphery'. In this respect, world systems approaches to understanding migration differ somewhat from dependency theories, insofar as
the former did not so much stress a relationship of inequality and dependency between a 'developed' and an 'underdeveloped' set of countries,
but rather how states (and relations within states) were incorporated into

a constantly transformed capitalist world-system. However, as we will see immediately below, there were noticeable similarities in the focus of both these theories (Wilson 1993).

Writings that reflected Wallerstein's general framework focused increasingly on multinational corporate investment and other dimensions of neo-colonialism in more peripheral countries since the 1960s. Such studies concentrated on the myriad disruptions to non-capitalist social relations, especially in agriculture. This might include the break-down of traditional land holdings by capitalist agribusiness and other interests, the mechanization of agriculture, the development of cash crops, and the application of synthetic fertilizers and pesticides. But it also might include raw material extraction by either multinational corporations or the *comprador* governments[13] of the global south, for sale on world markets. Together, world systems proponents argued (like articulation theorists) that these would reduce the demand for traditional farming labour, disrupt existing patterns of employment, and in general destabilize non-capitalist forms of social and economic organization. This would in turn encourage internal and eventually international migration. At the same time, the development of export processing zones (EPZs) and the industrialization of other regions of the (semi-)periphery led to at least two migration-related processes: first EPZs increasingly involved the employment of women, and with it, the growing unemployment of men, which also produced greater internal – essentially rural to urban – migration. By the same token, where EPZs were established, a concomitant increase in traded goods and services through improved transport and communications took place, and where such links became both more intensive and extensive, migration followed (Fröbel *et al.* 1977; Sassen-Koob 1984). This could not have happened, these authors argue, without the accompaniment of ideologies or discourses of cultural domination which are mediated and reinforced through mass communications (Massey *et al.* 1993, 1998).

Over time, a variety of macro-sociological analyses built on Wallerstein's fundamental argument, among the most notable perhaps being Cohen's (1987) sympathetic critique of world systems theory. Cohen maintained that capital in the richer countries had created what Marx called a relative surplus population (a 'reserve army of labour' ready and willing to work in difficult conditions for low pay). Against the trade-obsessed world systems theorists then, Cohen argued that trade relations were only a surface manifestation of the ability of richer countries to "unlock a giant migratory stream" (1987: 42) that served two purposes. First, the populations of the poor countries were held in reserve (literally) by capital, to be 'called up' when the business cycle expanded. Second, that the existence of this reserve drove down the value of labour power,[14] as the working classes of the rich world were constantly threatened by other immigrants who could or would work for less. While these arguments have considerable resonance today, and we will reassert their importance later in this chapter, their popularity has waned in the migration literature.

Globalization, global cities, and neoliberalism

A second set of structuralist arguments relating to migration, namely globalization, global cities, and neoliberalism intersect with each other but we can nonetheless discern differences in their emphasis. The first of these – globalization – is not so much a theory as a concept that entails a combination of processes and forces that operate in the world together. Observers on both sides of the political spectrum argued that by the 1990s we were living in a globalized world. Global capitalism had truly extended its reach everywhere (or nearly everywhere), so that few subnational regions were untouched by the complex networks and flows of capital, goods, services, information, and people (Held et al. 1999; Mittelman 2000). According to many scholars of globalization, the world was experiencing unprecedented interconnectedness in terms of the speed of such connections, their extensity (the reach of these connections), and their intensity (the density and strength of those connections) (Held et al. 1999).

Very soon however, this depiction of globalization as novel, omnipotent, and all-encompassing gave way to more sober, spatially sensitive, feminist and historically nuanced treatments of the subject (Amin 2002; Cox 1997; Nagar et al. 2002). To highlight but one example of this re-thinking, Held et al. (1999) identified three perspectives on globalization: the hyperglobalist perspective, which saw globalization as a fundamental transformation of the global economy, including the worldwide liberalization of capital, commodities, services and people, the end of the nation-state, and perhaps the 'Americanization' of culture. This hyperglobalist perspective was mobilized by both the left and the right although in a different terminology and tone. More right-leaning observers saw globalization as a form of liberalization that would favour US companies and profits, as well as spread western prosperity and western values (e.g. democracy). Scholars on the left of the political spectrum offered at least two related arguments. First, they viewed globalization as the unfortunate manifestation of a 'neoliberal' ideology and neoliberal policies. In brief, neoliberalism means favouring markets and trade liberalization as a solution to a range of social ills, rather than emphasizing social policies. The result is the obliteration of worker-favourable regulations, social protection, and the privatization of public goods in richer countries.[15] Simultaneously, it extended structural adjustment measures in poorer countries,[16] including the further opening of these economies to foreign investment, further promoting export-led industrialization, commercializing agriculture, reducing welfare and other state budgets, limiting overseas development aid and encouraging remittances instead (e.g. Harvey 2005).

Somewhat on a different track, the sceptic perspective denied that globalization was a particularly new phenomenon, claiming that the world was actually more globalized during the period of the 'gold standard' at the end of the nineteenth century than it was at the end of the twentieth. These scholars compared the levels of migration from poorer to richer countries during these two periods, and found little justification for the claim that the more contemporary period of globalization (from the 1960s to the 1990s) signalled an age of vastly increased numbers of migrants (Hirst and Thompson 1996; Zlotnick 1998). The transformationalist perspective emphasized that there were both continuities with earlier periods, but also unprecedented changes and contradictory tendencies, such as the formation of regional trading blocs instead of truly global trading.

Another set of arguments questioned whether globalization should be seen as some sort of unstoppable juggernaut or a structure 'out there' pressing on people to immobilize them. To view globalization in this way would have the effect of actually dis-empowering people, especially women (e.g. Gibson-Graham 2002; Nagar et al. 2002). Finally, many geographers dismissed the argument that globalization was eradicating the uniqueness of 'place' or that globalization simply determines what happens at the so-called 'local scale'. Rather, the global is in the local, and the local is in the global, and we should instead speak of what Doreen Massey (1994) called a 'global sense of place' or 'glocalization' (Swyngedouw 1997).

Despite these divergent arguments, the globalization approach (including the transformationalist perspective) points to the role of transport and communications in facilitating and promoting migration. Specifically, scholars of globalization argue that the cost, time, and difficulty of longer-distance travel and communication fell dramatically over the course of the twentieth century. These changes allow for the construction of social (or migrant) networks between the countries of origin and des-tination, for – among other things – the sending of remittances, the ease of circular (rather than just one-way) migration, visits of family mem-bers, encouraging the movement of asylum-seekers, and smuggling and trafficking through media-generated images of material wealth in the countries of destination (Massey et al. 1998; Richmond 2002). For those who subscribe to the above, the world had entered an 'Age of Migration' in the twentieth century (Castles and Miller 2003; Brettell and Hollifield, 2008a).

Since globalization is a concept that encompasses so many processes or variables, it may be difficult for researchers to use it to analyse migra-tion. Nonetheless, this has not stopped some from using the banner of globalization to focus on the ability of individuals, groups, and their respective networks to span the globe in unprecedented ways (e.g. Richmond 2002; Ong 1999). Ong (1999) for one, shows how wealthy Hong Kong businessmen use flexible forms of citizenship in the form of dual citizenships and a variety of visas to respond to the global political economy (read 'globalization'). So-named 'astronauts', because of the amount of time they spend on airplanes, they move back and forth routinely between Hong Kong and the west coast of the United States. Others, such as Samers (1999) have used the concept of globalization reservedly to discuss the relationship between geo-political economic restructuring in France, and the lives of North African immigrant work-ers in the Paris automobile industry. Global competitive pressures in the automobile industry shape the lives of migrant workers, who in turn shape the fortunes of the industry. Still others have examined the relationship between globalization, migration, and the impact of remit-tances on the country, region, or village of origin (e.g. Orozco 2002). Though without any comprehensive engagement with the concept (or discourse) of globalization, Hyndman's (2003) study of the relationship between Canada and Sri Lanka is a useful illustration of the links between processes of globalization (such as aid and remittances) and asylum-seeking (see Box 2.5). It also serves to illustrate a certain 'migration-development nexus', which we address in the next chapter.

Box 2.5 GLOBALIZATION AND SRI LANKAN MIGRATION TO CANADA

Canada has had rather loose links with Sri Lanka owing to the British Empire and the two countries' membership of the British Commonwealth. The first Sri Lankans arrived in Canada in the early 1950s. Given its export-oriented turn during the late 1970s and 1980s, Sri Lanka became a favourable target for development aid and one of the largest recipients of foreign aid by the 1980s, including from Canada. The escalation of the civil conflict between the Sri Lankan government dominated by Sinhalese and the ethnically persecuted Tamils in 2001 forced Tamils to seek asylum in a number of countries, including Canada. Tamils chose Canada in particular for at least three reasons. First, there was an already existing Tamil community there, especially in Toronto. Second, Canada had already allocated additional allowances for Sri Lankan asylum-seekers during the 1980s, because it had "comparatively generous refugee legislation and adjudication policies" (Hyndman 2003: 265). Third, there was a sizeable immigration programme, and the generally strong educational qualifications of Sri Lankans met the immigration and human capital criteria of the Canadian government. Over time, this only further cemented the strength of the already existing Tamil diaspora in Canada which by the early 2000s was mainly living in Toronto and numbering anywhere from between 110,000 and 200,000. Many Tamils read and listen to Tamil-oriented newspapers and radio shows in the Toronto area, and many, *but not all*, Tamils in Canada maintain close cultural, financial, political, and social links with Tamils in Sri Lanka, including funding for the Tamil Liberation Tigers (which have waged armed conflict against the Sri Lankan government and army). Such connections point to another approach to migration: transnationalism, which I explore later in this chapter (for further reading, see Hyndman 2003).

Global city arguments

In *The Global City: New York, London, Tokyo* (1991), Sassen develops yet another line of thought in the neo-Marxist/globalization vein, but this time she focuses on labour markets in the countries of immigration in which the arguments are redolent of Piore's (1979) dual labour market hypothesis. Sassen (1991) offers a considerable contribution to the 'world city' debate (Friedmann and Wolff 1982) by claiming that international migration from poorer to richer countries would not have happened without the related development of 'global cities', and that immigration has itself contributed to the development of these cities. Among others, she argues that global or world cities emerged during the 1970s to become the centres of multinational corporate headquarters, and related producer services (such as accountancy, legal services, management consulting, and finance). Producer services create a demand for highly skilled (migrant and non-migrant) labour, but also for low-income (migrant) labour to service the very demands of high income labour. To be more specific, a seeming army of low-income migrants is said to increasingly work in the restaurants where the wealthy dine, clean the homes and offices where the wealthy live and work, and care for their children or their elderly parents. These are only examples, and migrants fill vacancies in countless other positions in the burgeoning service industries of these putatively global cities because employers either choose not to, or cannot hire citizen labour. In this respect, Sassen (1991) develops a 'demand side' argument in which changes in jobs creates a demand for migrants. In a later piece, however her argument is more supply-side-oriented (see Sassen 1996b). That is, she claims that the presence of a large pool of immigrants shapes the structure of the labour markets in these cities, thus reinforcing the duality of these labour markets, and thus the demand for low-income migrants (Samers 2002).[17]

Neoliberalism

In the discussion of globalization in an earlier section, we spoke of the imposition of structural adjustment policies (see note 16) in the late 1980s. Structural adjustment can be viewed as a policy extension of what many critical social scientists call neoliberalism. Neoliberalism as a concept is contested and observers of it interpret it in different ways. For the purposes of our discussion here, neoliberalism will refer to a combination of policies, programmes, and discourses (sometimes the term ideology is also used). Anthropologists, geographers, and sociologists have all stressed that neoliberalism has different variants in different national and sub-national contexts – what Brenner *et al.* (2010) call 'variegated neoliberalization'.[18] Yet, despite the innumerable misgivings about the term neoliberalism, we investigate and outline Peck and Tickell's (2006) idea of 'roll back' and 'roll forward' neoliberalism.

1) *Roll-back* neoliberalism in the richer countries entails cutting or severely reducing social programmes, especially those associated with public housing, food provision, unemployment insurance, medical care and public health. Social reproduction is no longer viewed as the sole responsibility of the state. This is accompanied by the elimination or severe curtailing of trade union power and the de-regulation (or 'flexibilization') of labour markets. In the *poorer* countries, many of the same processes occur, in large measure because of a deepening of structural adjustment policies.

2) *Roll-forward* neoliberalism in both richer and poorer countries concerns a privileging of markets and the logic of competition over the logic of government-managed and government funded social reproduction as the most efficient and effective means of solving social problems and allocating goods and services in societies. This impels people to be responsible for their own individual welfare (what might be called the 'individualization of responsibility'). In the richer countries, roll-forward neoliberalism involves the tacit or explicit support (financial or otherwise) of corporate business (capital welfare) as well as regressive taxation (cutting the taxes on the wealthiest citizens) at the expense of progressive taxation (reducing taxes on the poorest). With regard to immigration, the recruitment of highly skilled immigrants (including students) by governments is privileged and liberalized by visa policies and other procedures. To complement highly skilled migration, and despite any anti-low-income immigrant rhetoric or policies by national and other governments, there is also a 'malign neglect' by governments of undocumented migrants to ensure a cheap supply of labour. The above policies and practices, whether they involve immigration or not, are encouraged by a pervasive set of discourses circulated by governments, think tanks, institutes, organizations, and the popular media who espouse neoliberal ideas (see e.g. Bauder (2008) on

immigration and the neoliberal media in Germany). In the *poorer* countries, the process of structural adjustment is deepened through the ongoing liberalization and encouragement of foreign direct investment, remittances, and trade (e.g. the reduction in tariffs and quotas, and the development of cash crops for export). These are imposed, willingly or not, on the governments of poorer countries by such international bodies as the International Monetary Fund, the World Bank and the World Trade Organisation.

Neoliberalism encompasses other elements, but those are the basic components that should concern us here. Most geographers' accounts of neoliberalism agree that it assumes different forms around the world, with some national states or sub-national regions more neoliberal than others. What should be gained here is that the neoliberal policies of international institutions that are largely dominated by the United States and the richer countries, such as the IMF and the World Bank, *seem* to have resulted in greater poverty and deprivation among many of the poorest countries from the 1980s onwards. As a consequence, this seems to have created a greater need for migration, but it should not be seen as simply evidence of a push factor. Rather, it should be viewed as an indication of how the policies and practices of richer country governments, firms, and international bodies are intertwined with migration. However, what is striking about the migration studies literature is that, despite all the chatter about neoliberalism, much of the literature tends to focus on the country of destination (e.g. Varsanyi 2008) and there is a paucity of studies which actually break down neoliberalism into a set of principles or variables that can be analysed in the countries of origin to gauge the effect on migration. One of the few studies in this vein is by Massey and Capoferro (2006)[19] (see Box 2.6).

Box 2.6 NEOLIBERALISM AND EMIGRATION FROM THE CITY OF LIMA, PERU

Massey and Capoferro (2006) analyse the effects of structural adjustment policies along neoliberal lines that were imposed upon Peru in 1987. This led to rising unemployment and underemployment, declining real wages and a growth in informal employment relative to formal employment. Inflation rose from an annual rate of about 89 per cent from 1980 to 1987, but jumped to 4,000 per cent in 1990 and peaked at 7,000 per cent in 2000. As inflation soared, real incomes (i.e. incomes adjusted to prices) fell sharply, by about 40 per cent. To explore the effects of this on emigration from Peru, Massey and Capoferro use data from ethnographic research and their 'ethnosurveys' as part of their Latin American Migration Project (see http://lamp.opr.princeton.edu/). They collected data on the life histories of family members in approximately 500 families in three middle-/lower-middle class neighbourhoods of Lima. What they found was that structural adjustment/neoliberal policies had a real effect on international migration between 1988 and 2000. Massey and Capoferro show that *before* structural adjustment policies were implemented in 1987, families with higher levels of education, who had migrated before, and who had children over the age of 18, were more likely to migrate. *After* 1987, the level of 'human capital' of migrants diminishes in importance and instead, knowing someone who migrated previously in the neighbourhood ('social capital' generated from social networks) seems to matter more in explaining who migrates. However, what they also found is that the number of migrants going to other Latin American and Caribbean countries fell from 36 to 24 per cent, since those countries were also suffering from structural adjustment. New destinations such as Spain and other countries in Europe became popular, while the US became slightly less popular. Why did Spain for example, become more attractive as a destination? Perhaps it is partly because the Spanish government actively privileged Spanish-speaking migrants over others in the early 1990s (Cornelius 2004).

Neoliberalism and international student mobility?

Besides labour migration, we might also view international student mobility (ISM) through the lens of neoliberalism. ISM has grown significantly (by 7 per cent on average per annum between 2000 and 2012 (OECD 2014)) and it has diversified geographically. It is possible that neoliberalism is at the root of the growth of this phenomenon, although it may be easier to explain *why* student mobility has expanded rapidly through neoliberalism, than to explain *where* it has expanded. In terms of the former, one might view the growth of student mobility from at least three perspectives: from the perspective of migrants themselves; from the strategic sights of national governments; and from the outlook of universities. For students, the desire to improve one's career prospects either in their home country or in other national spaces, as a means of securing emigration in the future, for language acquisition, or simply for adventure and an opportunity to be away from their parents, are all possible reasons. While some of these reasons may not be related directly to neoliberalism, that job opportunities have declined or are limited in certain poorer countries, or that English has become a dominant world language may in fact be related to neoliberalism. For governments, visa policies are used to favour the migration of students (barring the couple of years after 11 September 2001) in what Faist (2008) calls a move from a 'red card' to a 'red carpet' strategy (p. 33). Indeed, governments recognize that international students in engineering and the sciences in particular can provide the seeds of innovation and patent creation, which are central to the perception, reality, or rhetoric of a global neoliberal landscape of economic competition. There are certainly other reasons that governments encourage ISM, such as the desire for cultural exchanges, but this can only be very indirectly attributed to neoliberalism. For universities, who actively recruit foreign students, they are once again required to fill positions in science and engineering programmes since these programmes often struggle to find able and willing domestic students. This brings universities both considerable financial resources but also prestige and international diversity in their student bodies (IOM 2008a). That the pressure of national and international rankings might drive the desire to enhance certain academic programmes and hence the search for foreign students, is not new. However, these pressures may have intensified over the last twenty years, and they may be partly attributable to (as well as a part of) neoliberalism.

One of the trends in education which may actually *stem* ISM is the construction of western universities overseas, such as the broad range of American universities with campuses in the United Arab Emirates (UAE) and Qatar for example. The same might be said for British universities, such as Nottingham University's campuses in Ningbo, China and Kuala

Lumpur, Malaysia. The question remains however, whether the opening of foreign branches of universities as a competitive strategy will tend to lessen or further encourage migration. From what we know about other forms of migration, investments overseas may in fact increase rather than deter international student mobility, especially if it is linked to highly skilled migration. Nonetheless, many foreign students have chosen to follow distance-based courses online, (a product one might say of the growth of technologies associated with globalization) rather than to migrate overseas or even attend the foreign branch (IOM 2008a). Since there appears to be no comprehensive study to discern this relationship between international student mobility, and this form or stage of the internationalization of education, its implications remain to be seen. Perhaps more pertinent is tighter support for students in the form of reduced scholarships in the richer countries, and if this has any significance, it may be showing up in the *relative* slowing of international student mobility worldwide from 7 per cent on average per annum during the 2000s to only 2.5 per cent from 2013 to 2014 (OECD 2014).

CONCLUSIONS

So far, we have explored what have been labelled 'determinist' theories of migration. We want to emphasize that, in general, explanations of migration may be better compared if we recognize them as originating from different political standpoints (what critical social scientists call the 'situatedness' of theory or particular explanations); that they are devised of different philosophical foundations, different substantive foci (in other words, they are designed in part to address only certain kinds of migration), different units of analysis (individuals, households, economies, etc.) and spatial assumptions. It is with these latter three elements (substantive focus, unit of analysis, and spatial assumptions) with which we are most concerned. In Table 3.4 (see the next chapter), we summarize the differences in the approaches we have discussed in this chapter and the next one, based on these three elements. In their wake, we propose a more spatially explicit approach to migration. But before we get there, we are going to have to review a whole range of other theories, and that is where we are headed now. Let us turn to Chapter 3 then.

NOTES

1   When we say 'foundations', we are referring to assumptions about how society is organized and the relations between actors, institutions, and wider structural forces, as discussed in Chapter 1.

2   In the lengthy discussion of migration approaches in this chapter, we draw heavily on the analytical discussion of Massey et al. (1993, 1998) but also from the similar, extensive and impressive reviews of migration theories by Arango (2000), Boyle et al. (1998), Castles and Miller (2003), Goss and Lindquist (1995), Jennissen (2007), Molho (1986), and Wilson (1993). Yet we also hope that we have made these theories more accessible for a student audience. We have eliminated some of the approaches that the authors discuss, but we have also expanded on their reviews to cover more recent approaches to, and studies of migration.

3   Ravenstein was himself a little bit hesitant to call them laws (see 1885: 198).

4   Rest assured, we will also discuss these two ideas later in this Chapter and in Chapter 5.

5   See, for example, the report 'Silicon Valley tech workers locked in jobs with dubious legal tactics' by the Center for Investigative Reporting (https://www.revealnews. org/article-legacy/silicon-valley-tech-workers-locked-in-jobs-with-dubious-legal-tactics).

6   The summaries of the neo-classical migration literature tend to homogenize this very diverse literature, if not caricature it. Some studies are more complicated and sophisticated than others and cover vastly different kinds of migration, but the main thrust of most studies, not surprisingly, is that an economic rationality is the determinant of behaviour.

7   There is also now a large literature on the 'economic impacts of immigration', particularly in Europe and North America, but we do not cover these issues here.

8   By 'post-colonial, we mean the enduring cultural, economic, and political relations between former colonial powers and their ex-colonies.

9   When we say informal, secondary sector jobs, we are referring to employment which is "unregistered by or hidden from the state and/or tax, social security and/or labour law purposes, but which are legal in all other respects" (Williams and Windebank 1998: 4).

10 There is a vast range of studies on the relationship between immigration and informal employment (the latter otherwise called 'undeclared employment', and more vaguely 'the informal economy', 'the shadow economy', or the 'underground economy'). For overviews of this relationship during the twentieth century in at least the richer countries, see for example, Portes et al. (1989); Quassoli (1998); Samers (2005); Williams and Windebank (1998); in the twenty first, see e.g. Marcelli et al. (2010); and Williams and Nadin, 2014).

11 Employers in an ethnic enclave do not necessarily hire migrant workers of the same nationality or ethnicity (co-nationals and co-ethnics). For example, many Mexican and Ecuadorians work in Korean-owned businesses in Los Angeles and New York City, and Spanish may be the first language of Korean entrepreneurs, rather than English (Davis 1999; Light et al. 1999; Logan et al. 2000).

12 Capitalism is generally understood to be a combination of the widespread use of wage labour, the generalization of private property and the extraction of surplus value (e.g. Harvey 1982). Pre-capitalist modes of production are usually understood as a set of relations between groups and individuals involved in the transformation of nature (production) that combine *some* of these capitalist elements to one degree or another with let us say barter and the collective ownership of land. We use the term 'so-called', as some argued that there is nothing inevitable about such relations becoming capitalist over time, and therefore it was not appropriate to put the 'pre-' before them.

13 *Comprador* governments are generally understood as governments of poorer countries whose policies and practices coincide with more dominant, 'western' interests.

14 Very roughly speaking, this is the Marxist equivalent of what is understood as 'labour costs' in conventional terms.

15 We elaborate on the idea of neoliberalism in a subsequent section.

16 Structural adjustment refers to a general policy instituted by the World Bank and the IMF which sought to re-structure the characteristics of the economies of poorer countries, especially in Africa. This involved a 'carrot and stick' approach that encouraged or really demanded an increase in export-led development by poorer countries, opening up poorer country economies to foreign imports, reducing state subsidies to firms and industries, and reducing more social and welfare-oriented spending. In this sense, 'structural adjustment' may be considered part of neoliberalism.

17 There are innumerable critiques of Sassen's global city ideas. See for example McCann (2002) and Robinson (2002).

18 There is a large literature on what constitutes neoliberalism. Some prefer to speak of 'neoliberalization' to emphasize its ongoing, politically contested, geographically contingent, and incomplete nature, even its exhaustion after the 2008 economic and financial crisis in the form of post-neoliberalism in North and South America (e.g. Ward and England 2007; Brenner et al. 2010; Leitner et al. 2007; Peck et al. 2010). For challenging discussions far more critical of the concept altogether, see for example Barnett (2006) and Collier (2012), and note 19 further below.

19 See also Canales' (2003) study of liberalization and industrial restructuring in both Mexico and the United States and its implications for migration. However, there is now a growing literature on post-neoliberalism in Latin America (see e.g. Macdonald and Ruckert 2009; Grugel and Riggirozzi 2012), although we do not address this emerging literature here.

3

EXPLAINING MIGRATION ACROSS INTERNATIONAL BORDERS: INTEGRATIVE THEORIES

In the previous chapter we discussed determinist approaches to migration. In this chapter, we turn our attention to seven integrative (or mixed) approaches, including: (1) social network (or migrant network) analysis; (2) transnational arguments; (3) gender-aware analyses; (4) structurationist and agency-centric approaches; (5) the migration-development nexus; (6) forced-migration studies; and (7) arguments concerning migration that incorporate the environment. We conclude with an explicitly non-theoretical approach, namely Castles' 'social transformation perspective'. So let us begin with social network explanations.

SOCIAL NETWORK EXPLANATIONS FOR MIGRATION

Research on the importance of social networks in migration increasingly became a subject of analysis in the 1980s owing to, among many others, the work of Douglas Massey and his colleagues on Mexican migration to the United States (e.g. Massey et al. 1987) and later Singer and Massey (1998), but there were certainly many earlier studies in anthropology and sociology that contributed to this approach to migration (see the reviews in Boyd (1989) and Brettell (2008)). Research on networks might be encapsulated within what is broadly called a 'migration systems' paradigm (Mabogunje 1970; Massey et al. 1987; Fawcett 1989; Gurak and Caces 1992), a way of looking at migration through the historically rooted and network-based cultural, economic, political, and social linkages between the country of origin and destination, often in regional terms, for example between the oil-exporting countries of the Middle East and South Asia, between Europe and its former colonies, between southern African countries and South Africa, between the US and Latin America, between Southeast Asian countries, and so forth. It is important to recognize that the analysis of networks emanated from studies of labour migration and family reunification, and did not directly involve the study of asylum-seekers or refugees. In this latter case, we will look at the role especially of smugglers as intermediaries in networks that involve asylum-seekers and refugees across different spaces.

For authors concerned with social networks associated especially with labour migration and family reunification, networks are more than just 'migration chains' (e.g. MacDonald and MacDonald 1964);[1] they are defined as the ties that bind migrants, previous migrants, and non-migrants within and between the countries of origin and destination. They are viewed as mediating between structural forces and the individual agency of migrants (Massey et al. 1993, 1998) or to put it differently, they connect the social and individual reasons for migrating (Goss and Lindquist 1995). Commonly called 'migrant networks' (Massey et al. 1987) or 'network-mediated migration' (Wilson 1993), such networks might include kin and friendship ties through villages (sometimes referred to as 'strong ties'), or other networks based around a perception of common cultures or ethnicities (sometimes called 'weak ties', both of which rely on a certain degree of mutual trust – Tilly 2007). Both of these types of networks or 'ties' are sometimes manifested in hometown associations ("an organization of migrants from the same town or parish in a host country who congregate primarily for social and mutual aid purposes" – Caglar 2006: 1–2). Alternatively, Mercer et al. (2009) prefer the term 'home associations', since such associations do not simply involve connections between towns. Social networks, whether or not they involve hometown or simply home associations, can provide food, shelter, information about, and access to jobs, information on health care and services, religious organizations, as well as recreation and emotional support. Often the nature of information, connections, and support are gendered; that is, they may involve either networks of only men or women in specific cases. When these networks entail information about jobs, these gendered networks may be tied to the expectations of jobs in the country of destination as specifically men's or women's work. The consequence is that for example, some women who emigrate from the former Soviet republic of Georgia for example, choose and sustain migration to Turkey because the perception is that only women's work (e.g. working as home help) is available in Turkey, which only encourages other Georgian women to migrate to Turkey (Hofmann 2015). So social networks not only provide the financial and other resources to migrate, but they figure prominently in settlement and the continuation of (specific forms of) migration (Boyd 1989; Levitt 2003; Massey et al. 1993, 1998).

Some proponents of the migrant network idea argue further that unlike the very first migrants from a particular country of origin (or region within that country) who may bear enormous costs and risks in migrating, it is commonly claimed that later migrants can rely on social networks to lower the costs and risks of migration, which only further increases migration. As migration increases, social networks also increase and the process is reinforced. Eventually, broad sections of the societies from which migrants originate also migrate. This process might even be cemented further if immigration policy is more restrictive. In this case, migrants increasingly settle in the country of destination, which enhances the development of migrant communities, which in turn strengthens networks both within the country of destination and between the country of origin and destination. For many authors, these network resources (sometimes called 'social capital') are viewed as positive (read supportive) and are argued to change with the length of settlement. As the period of settlement grows, family reunification is argued to be more likely, increasing the existence of family-based networks in the receiving society. However, over time, the volume and amount of remittances (see Box 3.1) may decline (they may also increase), and membership in ethnic and non-ethnic based voluntary associations in the country of destination may also increase (or decline) (Blue 2004; Boyd 1989; Massey et al. 1993, 1998).

Box 3.1 ON REMITTANCES

Remittances are central to households, communities and countries of origin. Financial flows have received by far the greatest attention, particularly from governments interested in maximizing their access to reserves of foreign capital. For the last 60 years, transfers of money from migrant workers abroad have figured in the official financial accounts systematized by the International Monetary Fund for most countries in the world but the significance of these and related transfers of money were often overlooked until the World Bank began promoting their potential significance for development in the early 2000s. This encouraged much more widespread enthusiasm for what was soon referred to as a 'new development mantra' (Kapur 2004).

This enthusiasm seems justified. Global remittances tripled from $149 billion in 2002 to $581.6 billion in 2015 (OECD 2006; World Bank 2016). Almost three quarters of this financial flow ($431.6 billion) was sent to developing countries. This repeats a picture that has become well established over the last few decades. Financial remittances are argued to have a substantial net impact that, in purely financial terms, is substantially larger than Official Development Assistance (ODA). In 2015, ODA was $131.9 billion, less than a third the total remittances sent to developing countries. Financial remittances are also much more stable than other international capital flows. Whereas Foreign Direct Investment dropped dramatically following the 2008 financial crisis and took years to recover, remittances to poorer countries fell only slightly in 2009 and recovered their aggregate increase by 2010, surprising many observers (Sirkeci *et al.* 2012). Financial remittances are private transfers, often between family members, but they provide an important source of foreign exchange, particularly when they are sent through official financial institutions. They also have a multiplier effect as they move through economies.

Despite this enthusiasm, there is also a much more critical view. The 'mantra' of remittances relies on a highly economic understanding

Continued

of development that repeats many of the problems of earlier development strategies and ignores the social dynamics of new power relations (Bastia 2013). Remittance senders are also involved in these complex power relations and the frequently celebratory discussion of financial remittances can overlook the often high costs on the predominantly low-paid, service workers who send a large proportion of their income to family overseas (Wills *et al.* 2008). Researchers have also begun to pay much greater attention to non-financial forms of influence, or 'social remittances', a term coined by Peggy Levitt (2001) which includes 'ideas, practices and know-how' circulating between migrants and non-migrants. Individually focused analysis of financial remittances as gendered processes on household relations highlights the ways in which new responsibilities can impact on gender (King *et al.* 2013).

There is a parallel social network literature that paints a less flattering picture of social networks. First, in a study of Algerian asylum-seekers in the UK, Collyer (2005) shows that while social networks and social capital may be important to both settled and new migrants, stricter immigration policies relating especially to asylum-seekers (such as requiring proof of longer-term financial support) may force some settled immigrants to reduce their support for new co-ethnic asylum-seekers, even if they are friends. Thus, as de Haas (2011) notes more generally, social networks can actually hinder further migration. Second, while Massey *et al.* (1987, 1993, 1998) argue that the next generation of migrants can lower the costs and risks of migration by building on existing social networks and the higher levels of human capital associated with the first generation, Reniers (1999) shows in the context of Turkish and Moroccan migration to Belgium that this may not necessarily be the case, in fact the opposite might be true. Those with the least human capital migrate first, and those with higher educational attainment, for example, migrate later.

Third, it is not simply the strong ties of family, kin and home-(town) associations which constitute the social networks of migration, but other networks that involve a whole range of actors operating both legally and illegally from large employers and their sub-contracted employment agencies, to government and private recruiting agencies, to marriage brokers, and smugglers and traffickers (e.g. Goss and Lindquist 1995; Kyle and Koslowski 2011; Krissman 2005; Lindquist *et al.* 2012). Thus, certain kinds of networks place into question 'social networks' as being intrinsically beneficial to migrants. Take smuggling and trafficking, for example: these are social networks which are worldwide phenomena but which have ambiguous consequences for migrants. For this reason, below we devote a little bit of time to exploring their implications for migration.

Smuggling and trafficking as networks

An essential place to start this discussion is with the difference between smuggling and trafficking. For Salt (2000) and Kyle and Dale (2001), smuggling ('migrant importing') occurs when someone is transported illegally (on foot, by truck, boat, etc.) across an international border. The fee is often paid up front by the migrant to the smuggler, although additional fees may be demanded along the way, either from the migrant or from the migrant's family back home. In fact, Salt and Stein (1997) see smuggling as essentially a profit-making operation, and therefore 'as a business'. Central to their account are also the multiple spaces involved in smuggling networks, including three separate stages: the stages of mobilization (that is, the spaces where the 'journey' begins), *en route* (sometimes called a 'transit country' – see Box 3.2), and insertion (the process by which individuals seek asylum and the process of obtaining necessary resources for survival) (Van Liempt and Doomernik 2006). Smuggling in this sense is common between Latin American countries (especially Mexico) and the United States (see Box 3.5 later in the chapter), but also for instance, between eastern and western Africa, the Middle East, central Asia and the European Union (see Box 3.5 and Box 3.6).

Box 3.2 WHAT IS A TRANSIT COUNTRY AND WHERE DOES IT FIT IN THEORIES OF MIGRATION?

Attaching the label 'transit' to an entire country was initially a policy tool to attempt to control undocumented migration. The landmark European Council meeting at Tampere in 1999 established

---

Continued

'partnership with countries of origin and transit' as a central objective of European Union migration policy (European Council 1999). The focus on 'transit' has remained central to EU approaches to undocumented migration. The label has spread to other regions of the world, particularly Central America, and has begun to attract more academic analysis. Papadoupoulou-Kourkoula (2009) defines transit as the stage between departure and settlement. This emphasizes the uncertain nature of transit but, as Collyer and de Haas (2012) argue, it means that the label 'transit' can only be applied to a country or a migration pattern retrospectively. This is not how it is used in policy debates. Düvell (2012) argues that in European Union policy, 'transit' is clearly a political label since it is only countries around Europe which are referred to as 'transit countries' whereas there are plenty of countries within the European Union which are also transited by migrants keen to reach other parts of Europe. This point was further emphasized in the 2015 migration 'crisis' as migrants crossed through Italy or France on their way to Germany or Sweden. Any country is, at different times and for different people, a country of origin, transit and destination. Theories are required to make sense of this complexity and highlight how and why individuals make long, frequently dangerous journeys and what the consequences are for the migrants concerned. A key characteristic of many of these journeys that helps to clarify the risks faced by migrants is their fragmented nature. Longer journeys are frequently broken up into shorter sections, interspersed with periods of informal employment during which migrants may experiment with a potential destination or subsequent stages of the journey may be planned (Collyer et al. 2012).

Smuggling is frequently attributed to war, conflict, global and structural inequalities; poverty, environmental crises, or the ruthless profit-seeking, criminal behaviour of smugglers. These are by no means unreasonable culprits to which we might assign blame, but a closer and alternative look at smuggling allows us to see more human agency, and complex networks operating through multiple spaces. For example, in their study of smuggling between Iraq, Ethiopia, Georgia, and the Netherlands, van Liempt and Doomernik (2006) agree with Salt and Stein (1997) that smuggling is a global business with both legal and illegal dimensions. But they argue that Salt and Stein neglect *changes* in state policies and that they fail to account for migrants' agency, the relationship between smugglers and the smuggled (the migrants), migrants' experiences, and their motivations, which are not strictly economic in nature. Rather than just seeing migrants as passive victims, and smugglers as 'merciless criminals' (van Liempt and Doomernik 2006: 173), these two authors want us to see migrants as purposeful agents, and smugglers as business people. In fact, they remind us that in research on the US, many of the smugglers actually own restaurants, barber shops, and the like. In their study of the Netherlands, van Liempt and Doomernik find that many of the smugglers were actually smugglers of goods but discovered they could earn more money smuggling people. They were motivated in fact by their own difficult experiences as migrants, and they knew the border regions very well.

Van Liempt and Doomernik maintain that migrants seem to have favoured destinations, and they choose particular countries because of past colonial linkages, the perception that certain countries are culturally close to their own; because they have friends or relatives there, or because a country has a positive reputation in the migrant's mind, including its ability to offer jobs, education, a career, or protection and safety. Migrants do also make economic calculations based on the cost of the trip, so sometimes cheaper destinations (based on the amount of money paid to a smuggler) are chosen. However, migrants do not always end up where they prefer. Quite the contrary; as one man reported in the authors' study, he wanted to go to Sweden to be with his relatives, but was abandoned at a service station in the city of Eindhoven, and had to claim asylum in the Netherlands. In this all too common scenario, migrants are left without valuable support networks, and have to rely on the state for survival. From this perspective, smuggling can be seen as only a troubled and dangerous business, despite any agency that migrants may have.

Nevertheless, migrants do not necessarily see smugglers as criminals and there is little stigma attached to them in the former Soviet states, according to their study of Georgians. Rather, they are seen as facilitators or 'service providers' who will bring them to their destination for a price. As one migrant put it, "I would not call him a smuggler. At the border I was allowed to walk with this man for $400 and then he was so kind to hand me over to people who brought me to the nearest city for free"

(Liempt and Doomernik 2006: 173). In fact, if the smuggler can grant the wishes of the migrant, if they can keep their word and be successful in the operation, then it makes them more desirable as smugglers. Migrants often contact smugglers through friends and relatives, and smugglers recruit migrants in order to grow their business, even offering guarantees by allowing multiple attempts for only one fee. Sometimes, they postpone demanding the fee until safe arrival in the destination, in which case the relatives in the country of origin are then contacted for the fee. In this scenario, trust is important and in the towns and villages of origin, 'good' smugglers gain a positive reputation as individuals who can provide food, shelter, rest en route, and have expert knowledge about borders and obtaining passports and visas. The smugglers themselves choose certain routes carefully based on the ease of travel, transport infrastructure, and asylum and immigration policies. But as smuggling facilitators, they are also seen as 'bad people'. For example, one man did acknowledge that migrants "were locked up in safe houses and badly treated by smugglers" (ibid.: 174), while another man acknowledges that "they have dollars in their eyes and they lie a lot to you so that you will pay them more" (ibid.). To summarize our discussion above, we suggested that smugglers are intermediaries in networks of those who enter clandestinely or for those who are seeking entry through more official channels, that is, as asylum-seekers or refugees. In many cases, smugglers will also be involved with traffickers, and it is often very difficult to distinguish between smuggling and trafficking, given that the outcome of smuggling may finish by a form of trafficking. Trafficking ('slave importing') is also a business; it usually carries with it a substantial debt burden and involves forced labour after migrating (see Box 5.2 on Trafficking in Southeast Asia in Chapter 5), often to pay back the smuggling debt. This work may last for more than several years in order to pay off individuals involved in the specific social networks of trafficking. The nature of the work performed varies enormously from case to case. It is common in most countries of the world from Africa to Asia and everywhere in between. In Myanmar and Thailand, for example Kyle and Dale (2001) discuss the disturbing prevalence of sexual trafficking in particular. In both countries, it has involved local elites (village leaders, for example), states or governments (corrupt border guards, police officers, highly placed government officials, policy-makers) trafficking agents, employers (especially brothel owners), the consumers of sex, and unsuspecting migrants (women, and sometimes girls and boys as young as perhaps 12) who are forced to perform sex and sexual work without wages.

So whether social networks involve coercion, trickery, or clandestine connections between unscrupulous agents, institutions, government officials, and customers, or whether it involves more voluntary linkages which are more beneficial to migrants' welfare, they are often called 'transnational' and we discuss 'transnationalism' in the next section.

TRANSNATIONALISM AND MIGRATION

If the concept of globalization became ever popular during the 1990s, it also seemed to deprive people of their agency and had a rather economistic, deterministic, and disempowering inflection to it. In its wake arrived the idea of *transnationalism* which soon emerged into its more migrant-focused, agent-centric and perhaps cultural successor during the same decade. For Vertovec (1999) then, transnationalism "broadly refers to multiple ties and interactions linking people or institutions across the borders of nation-states" (p. 447). It quickly became clear though that such a definition covered a vast array of issues and processes, including those that were economic in character (e.g. Bailey 2001; Vertovec 1999; Portes et al. 1999; Smith and Guarnizo 1998). In an earlier and pioneering volume and with specific reference to immigrants, Basch et al. (1994) conceptualized transnationalism "as the processes by which immigrants forge and sustain multi-stranded social relations that link together their societies of origin and settlement" (p. 7).[2] This at least had the merit of a narrower definition focused on immigrants.

At the same time, some scholars claimed that transnationalism – much like the idea of globalization – cannot be considered as new (e.g Foner 2001; Wimmer and Glick-Schiller 2002). Wimmer and Glick-Schiller make this point eloquently:

> The recent boom in research on transnational communities did not discover 'something new', but was the result of a shift of perspective away from methodological nationalism. The discovery was a consequence of an epistemic move of the observer, not of the appearance of new objects of observation. (2002: 218)

However, our concern at this point is not so much with the character of 'transnational communities' and their senses of belonging and identity (which we will return to in Chapter 7), as it is to identify the precise unit(s) of analysis involved; whether the processes associated with the idea or concept of 'transnationalism' can be used to explain migration, and whether the discourse of transnationalism is any different from 'globalization'.

To begin with, the unit of analysis in transnational studies seems to be a combination of the 'local communities' of origin and destination, and locally grounded yet globe-spanning or border-crossing post-colonial 'diasporic networks' that involve what Michael Peter Smith (2001) calls variously 'translocalism', 'transnational urbanism', or "*distanciated* yet *situated* social relations*" (p. 237). These networks are often further disaggregated or re-aggregated into other analytical units such as households, other formal or informal institutions – especially homeland associations – and economically oriented transactions such as remittances and commercial trade (Faist 2008; Smith 2005). When we refer to 'diasporic networks', it must be recognized that the concept and definition of 'diaspora' is itself contested (Blunt 2007; Brettell and Hollifield 2008a; Brubaker 2005; Cohen 1997), although diaspora might be reasonably defined as the spreading out of certain communities from an original homeland to their regrouping and the formation of new communities in a 'new' land. If we accept this working definition of diaspora, then by 'diasporic networks' we mean the social, cultural, political, and economic links that migrant communities maintain across international borders. These links may also be psychological, involving especially an emotional or imaginative attachment to a 'people(s)' or 'place(s)' of origin. The words 'people(s)' and 'place(s)' here are used purposefully for two reasons; first since diasporic networks may involve multiple places of origin, migrants may feel and practise a sense of belonging to more than one village or region and to more than one ethnic or linguistic group. Second, their attachment to, let us say, a particular village (rather than necessarily and only to a nation-state) suggests that the term transnationalism may be either a misnomer, or an incomplete understanding of belonging. That is, it is not clear whether their belonging and practices should be defined as transnational; as translocal or transurban (Barkan 2004; Faist 2008; Waldinger and Fitzgerald 2004; Smith 2005), or when it indeed involves only two countries: bi-local or bi-national. Perhaps any spatial metaphor is inappropriate, and we should instead call such networks 'pan-ethnic' (Levitt and Jaworksy 2007). Yet even the label 'pan-ethnic' may be 'essentialist' insofar as it assumes homogeneity within ethnic groups,[3] and often transnational linkages are cut across by differences in age, generation, gender, sexuality, religion, class position, and other forms of 'intersectionality'. In other words, it is difficult to see precisely how transnationalism can serve to perpetuate migration, when our understanding of the concept itself is rather insecure (Portes et al. 1999; Waldinger and Fitzgerald 2004).

Nevertheless, what might be different between an explanation of migration based on transnationalism and an explanation of migration rooted in globalization (or structuralist/neo-Marxist-inspired reading)? We can identify two related distinctions. First, Smith (2005) argues on one hand that arguments about globalization as something ominously big, economic, structural, uncontrollable, operating "behind people's backs, so to speak" (p. 236), are problematic (see also Gibson-Graham 2002). On the other hand, Smith (2005) reflects wisely that it is also a mistake to see transnational networks as evidence of a new kind of 'cultural agency', divorced from or in binary opposition to global economic (re)structuring (p. 236). For Smith then, structures are real, but they operate at both the 'local' and 'global' levels. The local drives the transnational, and the transnational shapes the local.

Brettell and Hollifield (2008a) offer a second possible distinction between an earlier more 'miserablist' migration literature, driven by the debilitating forces of globalization, and the new literature on transnational diasporic networks. This distinction rests on the 'voluntarism' of the migrants involved: they are no longer viewed as 'uprooted' (p. 120). In other words, migrants are not forced by economic and social disruption to migrate with trepidation and difficulty, but rather people who move with ease – really circulate – between different countries and cultures. That is the argument anyway, and there is plenty of evidence that at least some high-income migrants do this with relative impunity, though certainly with the obstacle of immigration visas and other checks. For example, Indian computer engineers or entrepreneurs working in the computer industry in Silicon Valley (south of San Francisco) helped to establish the Indian computer technology industry centred on Bangalore in southern India, which in turn has stimulated further migration between Bangalore and Silicon Valley, and which is supported by the Indian state (Saxenian 2005). Taking a specifically comparative approach across different nation-states, Ambronisi (2014) argues that transnationalism (measured in terms of migrants promoting the economic development of their country of origin, remittances among mothers sending money home to their children, and migrant association-sponsored development projects) differ between 'new' immigration countries such as Italy and 'old' countries of immigration such as the United States.

In contrast, Faist (2010) and Castles et al. (2014) doubt that the majority of migrants live such transnational lives, given insufficient evidence. And certainly with considerable justification, Mitchell (1997) and Michael Peter Smith (2005) rebuke what they see as this 'celebration of hybridity' or fluid transnationalism. As Smith argues:

> this serves to erase the fact that no matter how much spatial mobility or border crossing may characterize transnational actors' household, community and place-making practices, the actors are still classed, raced and gendered bodies in motion in specific historical contexts, within certain political formations and spaces. (2005: 238)

While in this book, we subscribe to Smith's more critical view, the point that we need to take from this literature is the idea of transnationalism as involving a twist on globalization arguments through its assertion of the importance of migrant networks. And it is these globe-spanning networks (whether they are cultural, economic, political, or social) that serve to stimulate, perpetuate or even impede migration.

GENDER-AWARE APPROACHES TO MIGRATION

During the 1980s and 1990s, it became something of a mantra to resent the neglect of gender (or more specifically women) in the study of international migration, and for a long time it was assumed that women only migrated as dependents (Kelson and DeLaet 1999; Kofman 1999). Consequently, feminist and other gender-aware scholars sought to compensate for the lack of attention to the active role of women in migration. Lutz (2010) refers to this as a *compensatory approach*. A second stage of research, which Lutz called *contributory*, focused on the way in which women were contributing to a range of different migrations, and there is now a rather voluminous literature that seeks to overcome the long-time male bias of migration studies.[4] Lutz identifies a third stage during the 1980s and 1990s which began to explore the power relations *between women*, in which 'intersectional' feminist research (again, how, to what extent, and in what ways, class, ethnicity, race, disability, sexual orientation, *and* gender may be significant and inter-twined), became increasingly central to migration studies. By the early 2000s, Pessar and Mahler (2003) argued that the project among critical academics to address the male bias of the literature had progressed to the point that "the pendulum . . . [had] . . . shifted so far in the opposite direction that the male migrant as study subject disappeared almost to the same degree as the female migrant had previously" (p. 814). At the same time, Pessar and Mahler (2003) claimed that the migration literature had "simply redressed the male bias by adding women; in other words, by treating gender largely as the variable sex" (p. 814). The point, as Pessar and Mahler argued, is to "treat gender less as a variable and more as a central concept for studying migration" (p. 814). In turn, Lutz (2010) claims that in fact many (but by no means all) feminist scholars of migration have now moved from a "Women's studies to a Gender studies perspective" (p. 1650). That is, they increasingly examine the difference between a person's sex and their "socially acquired and performed gender identities, way of living and role in society" (p. 1650). And instead of looking only at women or statistical differences between men and women, the literature examines how gender identities (or masculinities and femininities) are both a product of the 'social order' and produce it at the same time, especially through power relationships between men and women, the intersection with other axes of differentiation, such as ethnicity, the meanings of private and public, and the varying consequences of migration for men and women with respect to couples, (fragmented) families, and 'distant parenting' (Lutz 2010).

Related to or beyond the above, a gender studies approach involves at least five key dimensions: 1) the fundamental role of states in both encouraging differential types of migration among men and women, but also controlling men and women's unequal right to migrate; 2) the nature of intra-family and intra-community gender relations as mediated by states, and how this shapes both emigration and return migration; 3) how perceptions of more equitable gender relations abroad tend to shape migration; 4) the gendering of work and how this explains migration destinations, and in particular the migration of domestic workers and nurses in particular; and 5) the gendered character of 'marriage migration' which involves particularly the international migration of women as spouses.

Concerning the first dimension, the state in the country of immigration is seen as creating and regulating the migration of women. For example, in the UK from the late 1960s to the mid-1980s, migrant women were assumed to be dependents and prohibited from working. It was only in 1989 that migrant women could bring in their husbands and fiancés, and from that point onwards, the number of dependent migrant men increased (Kofman 1999). It is acknowledged however, that such gendered restrictions vary over space and time, and are shaped by the age, class position, ethnicity, generation, religion, and the recognizable skills of particular migrants. Yet states in the country of emigration are also actively involved in facilitating and regulating who migrates and who does not. This is accomplished, not by simply regulating migration on the basis of one's sex, but also by shaping gender relations and expectations (see for example, Silvey (2004b) on Indonesian migrants to Saudi Arabia, Tyner (2004) on the migration of Filipino women; and Yeoh and Willis (1999) on Singaporean men migrating to China with their wives remaining in Singapore).

A second dimension of this literature is how gender relations inside and outside the family in the country of emigration (as mediated by the state) combined with gender relations in the country of immigration (also mediated by the state) might explain the differential migration of men and women. In this regard, there is a rich literature on Mexican migration to the United States, for example, which documents these interrelationships. Migration from Mexico to the United States has, over much of the twentieth century, been numerically dominated by men, many of them caught in the binds of crop failure, or inadequate crop yields and low income. Gender norms and gender subjectivities (how men and women are viewed and regulated, and how they view and regulate themselves) have impelled men to migrate, and women to either join their spouses later in the US, or stay behind to tend the fields and perform other domestic tasks. This gender division of migration may itself reinforce or modify existing gender relations, and shape the socially and emotionally fraught migration process (see Box 3.3).

Box. 3.3 TRANSNATIONAL GENDER RELATIONS IN A PHONE CALL FROM EL SALVADOR?

While mobile phones, texts and social media posts might have replaced the use of traditional telephone calls over the last 15 years, Mahler (2001) recounts the fascinating subject of traditional phone calls between non-migrant wives in El Salvador and the Salvadoran men in the United States. Not only is there a problem of the limited number of phones available, but the phone company in El Salvador only allows expensive 'collect calls' to the US. If and when the wife of a migrant can reach their husband, they must plead them to accept the call. When they do, the husbands know how expensive a collect call can be and have every interest in keeping the calls very brief. The women, on the other hand, must reach out to their husbands emotionally in order to ask for more money to be sent home. This emotionally fraught situation points to the unequal relations between men who migrate and women who do not (see the discussion in Pessar and Mahler 2003: 24).

Levitt (2001) shows that men who migrate often have a higher status than those who do not, and women may consider them to be more marriageable. For the men who do not migrate, their masculinity is called into question ('be a man' and migrate), in part because of the gendered assumption that they should be able to support a family, which is difficult if they stay and remain on the farm in Mexico (Massey et al. 1987). Those who have documented status in the US are also commonly viewed as more masculine than those who have undocumented status, which may encourage lengthier migration stays and eventually settlement among Mexican men (Pessar and Mahler 2003).

In contrast, Mexican women who migrate are often accused of upsetting community norms and expectations, though such norms can certainly change over time (for example, Hondagneu-Sotelo 1994; Boehm 2008, and see King et al. 2006 on return migration to Albania). However, the result is not simply that women stay behind and men migrate; rather, as Pessar and Mahler (2003) argue, Mexican women are gaining increasing independence.[5] In that respect, some Mexican women have migrated to marry a (Mexican) man in the US, in order to become wealthier; while others have migrated because they fear the infidelity of their husbands, or insist that their husbands return home.

At the same time, many Mexican men do not choose to settle in the US, since for among other reasons, it is paradoxically also a threat to their masculinity as they 'lose control' over their wives and children in Mexico. Many prefer instead to maintain a farm back in Mexico and continue farming with their children in an environment where they can retain and even enhance their status as 'men who migrated', and therefore exercise more power over their family. Many Mexican migrant women on the other hand are more eager to stay in the United States, because although they encounter difficult conditions in the United States (including what is often their first experience with wage labour), they also feel a sense of liberation in the United States, from their husbands or other elements of gender oppression in Mexican society. In particular, they seem to have more freedom to voice their political wishes in the United States than in hometown and other political associations in Mexico. In sum, the literature on gender and migration – at least from Mexico – suggests that the propensity for return migration is shaped by the gendered experiences and differential desires to stay in the country of destination among men and women (e.g. Goldring 2001; Massey et al. 1987). The discussion of Mexican migration above is not meant to represent the nature of gender relations and their consequences for migration *everywhere* in the world. Nonetheless, this Mexican literature is supported by findings elsewhere (see e.g. King et al. (2006) on Albanian migration to Italy or Jónsson (2008), who shows that among Soniké villagers in Mali, women tease men who do not migrate to Europe as being "stuck like glue", which in turn impels young men to migrate – cited in Andersson (2014: 7)).

A third dimension is the belief that gender relations may be more equitable abroad. Here, the concern is with how the *perception* or *imagination* of gender relations in the countries of immigration shapes the propensity to migrate. This has shown to be the case for women in the Dominican Republic who see apparently happier women returning from the United States and wish to do the same (Levitt 2001). Yet their imaginations may also be disappointed if the reality of equality does not live up to their expectations, such as the case of many Asian, Latin American, and Middle Eastern women who migrate to Switzerland looking for a more equal relationship with a 'western' man (Riano and Baghdadi 2007).

A fourth dimension involves the demand for women to work in what might be called the three 'C's (caring, cleaning, and catering). This phenomenon occurs across Europe, North America, and especially in Asian countries such as Japan, Malaysia, and Singapore (for example see Yeoh and Huang 1998), but it is not by any means restricted to richer countries. The feminization of migration may be in part attributable to this increased demand for migrant women, in contrast to the mid-twentieth century, when manufacturing jobs in the rich 'western' countries were stereotyped as 'jobs for the boys' (McDowell 1991). And it is domestic workers in particular (those who care and clean, such as child-minders, cleaners, live-in and live-out maids, nannies, and so forth) that have received enormous – perhaps disproportionate – attention,[6] at the expense of women migrating for marriage who may serve the same role, but as unpaid domestic labour (Yeoh, Chee, and Baey 2013). The demand for 'care-work' and care-workers is said to have increased over the last two decades because:

1. The governments of richer and poorer countries have reduced their role in caring services – a process that is associated with neo-liberalization;
2. The average age of the population in wealthy countries has increased (the so-called 'greying' or 'ageing' of the population, especially in Asia and Europe) and elderly people require more care;
3. There has been a rise in dual earner couples, and the limits in the number of hours that men (but also women) can and/or are willing to spend on domestic responsibilities;
4. Fewer citizen workers are available and/or are willing to provide these services and/or are deemed to be too expensive relative to especially immigrant women;
5. There has been an increase in the size of people's homes at the end of the twentieth century, and an apparently growing concern about the appearance of homes (in other words, the desire to have immaculate 'trophy' or 'showcase' homes). This seemed to be especially associated with at least the UK and the US (Anderson 2001b).

6. There is a strong relationship between the demand for domestic workers and the formation of households across the world, but particularly in Asian countries, so that household formation is dependent on domestic workers and vice versa. In other words, there is a 'global care chain' (Yeates 2004), in which women from poorer countries work in richer countries as domestic workers, which in turn leaves their children in the country of origin under the care often of other domestic workers. This in turn shapes the formation or dissolution of households (Brickell and Yeoh 2014; Douglas 2006, 2012; Lam et al. 2006; Yeates 2009; Yeoh, Chee, and Vu 2013).

In other countries, and Saudi Arabia in particular, the reasons for the high demand for domestic workers are not dissimilar but also unique: the movement of Saudi women from public to private sector employment, the recognition by the Saudi Arabian government that migrant domestic labour increases labour market flexibility through the ability to hire and fire migrant domestic women according to the political and economic whims of 'Saudi society', and the decreasing contribution of Saudi women to domestic labour all contribute to a high demand for domestic labour, especially from Indonesia (Silvey 2004b). Yet the above reasons are exclusively demand-focused, and the migration of domestic workers is driven and supported by a lattice of labour export brokers and remittance agencies in the countries of origin, such as in Bangladesh, India, Indonesia, Pakistan, and the Philippines. These agencies commonly provide would-be migrants with travel arrangements, appropriate visas, and sometimes a specific position within a family, and in the case of remittance agencies, the processing of large sums of money. These agencies may be formal or informal; they may provide genuine employment and opportunities for social mobility, or they may simply offer heinous working conditions in the countries of immigration (Parreñas 2001; Silvey 2004b).

Many of these domestic workers are undocumented, and this adds another layer of explanation. Anderson (2001b) for example, suggests that it is not simply costs which explain why so many domestic workers are undocumented immigrants. Rather, it is the common practice of having to work some 12 hours a day in a hyper-productive 'any job/always on' condition. Furthermore, she adds that a 'racial ideology' persists insofar as middle class (often white) women in the richer countries feel a sense of superiority by hiring 'women of colour' from poorer countries. And this demand for undocumented female migrant domestic workers is likely to persist unless household wealth declines, unless people dramatically change their spending habits, unless there is a change in the household division of labour among men and women, unless public spending on public child care increases, unless stricter legislation on the use of undocumented migrant labour is enacted, or if labour-saving devices such as robots performing household tasks become more widespread (Samers 2005).

Yet besides domestic workers, there is a considerable demand for doctors, nurses, office cleaners, sex workers, restaurant workers and other catering positions. Untangling the reasons for this demand is not easy, since to begin with all of these positions perform vastly different functions. The buoyant demand for doctors and nurses in at least the richer countries over the last 35 years or so may relate to the purported decline in the state-financing of health care in the wealthier countries (or the rise of 'managed care' in the US) and thus cost-effectiveness and cost-containment measures that many have associated with neoliberalism. This in turn has led to deteriorating working conditions such as longer shifts, the comparatively low and declining financial remuneration which dissuades citizen nurses from training and accepting such positions, and hence the demand for immigrant doctors and nurses remains robust as they can be paid less to keep costs down (England 2015; Raghuram and Kofman 2002). But why do immigrant women 'keep costs down'? Their legal status may be more precarious (even for doctors) because of gender-biased immigration policies and/or because of sexist hiring practices in other spheres of work, or because of limited employment opportunities and the feminized inflection of poverty in poorer countries. For all these reasons, women may accept lower salaries or wages and poorer working conditions.

Yet such explanations might be limited again by their 'methodological nationalism', and England (2015) offers a novel contribution insofar as her focus is on sub-national differences in the demand for nurses in the United States. She finds that the proportion of IRN (international registered nurses) is concentrated in just a handful of US states (such as California, Illinois, New York, and Texas).

The demand for office cleaners and restaurant workers may be explained through other theories, such as the dual labour market or global city arguments, but these theories neglect gender too, and many of these jobs in services are stereotyped as 'female' by employers, or even by migrant women themselves. Together, this may restrict migrant women to certain types of jobs. In any case, many of the women recruited into such jobs are also undocumented immigrants as the pressure to keep wages low by price-competing employers and cost-conscious households encourages their hiring (Anderson 2001a).

The sixth and final dimension we will examine is that of 'marriage migration'. Like other forms of migration, marriage migration encompasses a range of movements on a spectrum from the voluntary to the forced. Such migration is defined and regulated in different ways, and as a consequence, difficult to count consistently and reliably as well. After all, someone may marry a foreigner in their own country and then migrate with their spouse later on. Thus, one migrates for marriage under the official title of a family reunification visa, but they may also have mixed motivations for migration, which range from the dangers of staying in their country of origin, escaping poverty or undesirable forms of work, genuine romantic love, 'marrying up' to support their families back home, or a disingenuous vehicle for obtaining a work or residence permit. But someone may also be called upon by a foreigner living abroad, in order to marry. This is akin to so-called 'mail order brides' as depicted in the popular media, although it now involves sophisticated communication technologies, as well as coercion, especially kidnapping, and fraud. However, once again, all of these are not mutually exclusive explanations and people's motivations are usually quite mixed (Lu and Yang 2010; Piper and Roces 2003; Chee, Yeoh, and Shuib 2012).

While there is a substantial body of work that dates to at least the 1980s in western countries, since the 2000s, this corpus of work has grown substantially, especially with respect to Asian countries. In the discussion below, we will concentrate on explaining marriage migration to Asian countries, although it is crucial to recognize that 'marriage migration' concerns virtually every country and sub-national region of the world. In fact, in singling out Asian countries here, we might give the impression that they are particularly marked by this pattern of migration, but this is not necessarily the case, nor is there much justification to group Asian countries together, other than that they might constitute a 'migration system' for this form of migration as discussed in the previous chapter. Furthermore, much of the research on marriage migration could be classified as 'methodologically nationalist', focusing as it does on specific nationalities (though often with disaggregation by gender, age, education, and other inter-sectionalities) and their incorporation within specific countries. This in itself is not necessarily problematic, since different national immigration policies shape the propensity of marriage, but certain regions may be more intensely associated with particular forms of marriage migration than others (see Box 3.4 below). Recognizing both the salience of national immigration policies, and regional specifics, we present some of the findings as they appear in national lines in the literature.

In Asian countries then, studies have highlighted the migration especially of poorer women from poorer Asian countries such as China, Indonesia, the Philippines, and Vietnam to richer countries such as Japan, Malaysia, Singapore, South Korea, and Taiwan, although South Korean women, for example, have also migrated as spouses to other Asian and western countries in substantial numbers. To provide some indication of international marriages as a percentage of all marriages and the size of 'marriage migration' (as indicated by the number of 'international marriages') in specific countries, we draw on Jones (2012a) to provide data for nine Asian countries, and for Japan and South Korea in particular (see Tables 3.1, 3.2, and 3.3 below).

Table 3.1 International marriage migration in various Asian countries (ranked by percentages of all marriages)

Country	Approximate year of data	International Marriages (in percentages)	International marriages involving a different ethnic group[d]
Singaporea	2008a	39	13b
Taiwan	2003	32	10
	2010	13	4
South Korea	2005	14	7
	2010	11	9
Japan	2005	5	5
	2010	5	5
Philippines	2009	4c	4c
Vietnamb	2005	3	3
Indonesiab	2005	1	1
Chinab	2005	0.7	0.4
Indiab	2005	0.5	0.3

Source: Adapted from Jones (2012a: 2)
a Marriages of citizens to non-citizens, including permanent residents of Singapore
b Very rough estimate
c The number for the Philippines has been increased by 30% to take into account marriage of Filipinos overseas that are not registered with the Commission on Filipinos Overseas
d Foreign spouses of the same ethnic group are Chinese in the case of Taiwan; Chinese, Malays or Indians in the case of Singapore; and Koreans in the case of Korea

Table 3.2 South Korea: international marriages as percentage of all marriages, 2000–2010

Year	No. of marriages in thousands	Total	Wife from foreign country	Husband from foreign country	Ratio (Col. 4/5)	% of foreign brides from	
						Vietnam	China
(1)	(2)	(3)	(4)	(5)	(6)	(7)	(8)
2000	332	3.5	2.3	1.2	1.5	1.3	49.1
2001	318	4.6	3.1	1.5	2.0	1.3	72.0
2002	305	5.0	3.5	1.5	2.4	4.4	65.6
2003	303	8.2	6.2	2.0	3.1	7.4	71.2
2004	309	11.2	8.1	3.1	2.6	9.8	73.6
2005	314	13.5	9.8	3.7	2.6	19.0	67.0
2006	331	11.7	9.0	2.7	3.3	34.1	49.1
2007	344	10.9	8.3	2.6	3.2	23.1	50.7
2008	328	11.0	8.6	2.4	3.5	29.4	46.9
2009	310	10.8	8.2	2.6	3.1	28.8	45.2
2010	326	10.5	8.1	2.4	3.3	n.a	n.a.

Source: Jones (2012b), based on Statistics Korea, Vital Statistics

The numbers have to be taken with some caution however. Indeed, Jones (2012a) notes that between 1990 and 2009, some 14,000 to 25,000 Filipino spouses left the Philippines each year as the spouses of non-Filipino citizens. These include those married in the Philippines to foreigners and those who left the Philippines to marry. This number represents only those officially registered, and the real numbers are likely to be much higher. In fact, unregistered marriages are quite common in Asian countries (but by no means restricted to them, and particularly in border areas (see Box 3.4 below).

While it might be assumed that poor Asian women who marry outside their country of origin would be less educated than their spouses, Kim (2008) found in Korea, for example, that many Filipino and Mongolian women marrying Korean men were in fact more formally educated than their Korean husbands, though Kim did not find

Table 3.3 Japan: trends in international marriages, 1970–2009

Year	Number of marriages ('000)	Percentage of marriages with one foreign spouse			
		Total	Wife from foreign country	Husband from foreign country	Ratio – Column 4/5
(1)	(2)	(3)	(4)	(5)	(6)
1970	1029	0.5	0.3	0.3	1.0
1975	942	0.6	0.3	0.3	1.0
1980	775	0.9	0.6	0.4	1.5
1985	736	1.7	1.1	0.6	1.8
1990	722	3.5	2.8	0.8	3.5
1995	792	3.5	2.6	0.9	2.9
2000	798	4.5	3.5	1.0	3.5
2001	800	5.0	4.0	1.0	4.0
2002	757	4.7	3.7	1.0	3.7
2003	740	4.9	3.8	1.1	3.5
2004	720	5.5	4.3	1.2	3.6
2005	714	5.8	4.6	1.2	3.8
2006	731	6.1	4.9	1.2	4.1
2007	720	5.6	4.4	1.2	3.7
2008	726	5.1	4.0	1.1	3.6
2009	708	4.9	3.8	1.1	3.5

Source: Jones (2012b), based on calculations from data in the Ministry of Health, Labour, and Welfare, Japan, 2009.

this for Chinese and Mongolian women, nor for Filipino women in Japan. The difference in Japan is explained by Filipino women entering Japan as 'entertainers', jobs that would not generally appeal to more formally educated women. This suggests that nationality and country-specific phenomena should not lead to simplistic conclusions (Jones 2012b).

Box 3.4 CROSS-BORDER MARRIAGE MIGRATION AMONG UNDOCUMENTED VIETNAMESE WOMEN AND CHINESE MEN

(From Maochun and Wen, 2014)

There is evidence of Vietnamese women clandestinely crossing the border for unregistered marriages to Chinese men as early as the 1950s, but political relations between the two countries deteriorated during the late 1970s and such migration ceased. Since the 1980s, undocumented migration *in general* has increased and so has cross-border marriage migration. But while the former has attracted the anxious attention of the Chinese media and the Chinese government, Chinese and Vietnamese authorities continued to tolerate the latter into the late 2000s. The increase in cross-border marriage migration since the 1980s is owed to the easing of geo-political tensions between the two governments, and the dearth of marriageable women in China, in part because of the one-child policy in China and widespread sex-selective abortion among Chinese couples that has produced a surplus of single men (so-called 'bare branches'). Since the 1980s, Chinese authorities sought to deport Vietnamese women back to Vietnam but found it difficult without the support of local villagers, many of whom tacitly supported or even encouraged the practice of such marriages, including 'bride kidnapping' and forced, rather than consensual marriages. Such 'support' has not lessened over time apparently. Furthermore, the sex-ratio imbalance (mentioned above) in rural border regions has been exacerbated by the migration of many younger women to China's coastal cities for work during the 1990s and 2000s. Yet, the proliferation of 'undocumented marriages' may be explained by at least three other phenomena, including cross-border marriage law and immigration policies that have remained highly restrictive and require numerous and costly official documents, which most poor Vietnamese women who enter China clandestinely cannot obtain. The physical geography of the border, with its mountains, numerous streams and trails through the dense forest hindered apprehension and drove up the costs of surveillance. The 1990s also witnessed the increasing kidnapping and smuggling of thousands of Vietnamese women across the border into China, as well as sexual trafficking in the form of forced prostitution or forced marriage with Chinese men. In Tiandeng County in China for example, it was estimated that approximately 80 per cent of 273 Vietnamese women in the 1990s were victims of trafficking, some of them more than once. There is however also some indication that this involved 'marriage fraud' on the part of Vietnamese women who would disappear after obtaining financial resources from their Chinese husbands.

While we have identified some basic dimensions of marriage migration, explaining this migration is another matter altogether. Some have attributed this migration to the racialization, sexualization, and exoticization of Asian women as 'good wives', 'sexually available', and so forth, prevailing ideas or discourses that are partly associated with what Edward Said referred to critically as 'Orientalism' (Said 1978). Yet such an explanation may better explain international marriages with western men, but explain less well, marriage between Asian men and women. Other explanations are abundant, and below we discuss at least five of them. First, some scholars attribute marriage migration to 'globalization', or more specifically, increased tourism, international study, the international transfer of businessmen and other temporary skilled migration from one Asian country to another, general business travel, and taking English language classes in other Asian countries (this seems to explain encounters and eventual marriages between less educated Korean men learning English in the Philippines and more educated Filipino women). Second, low fertility (especially but not exclusively in Taiwan), sex-selective abortion, and an imbalanced sex-ratio skewed towards men in rural areas has driven men to seek 'foreign brides'. It is not surprising then, that 'foreign brides' are over-represented in rural areas in Korea, where many younger Korean women have migrated to what are considered cities. However, most foreign women who marry settle in such cities in Asia. Third, 'ethnic' or 'cultural affinity' is yet another explanation, but this may explain more who is married to whom, than a reason for marriage migration. For example, most marriage migration to Korea is between 'ethnic Korean' women in China (that is Koreans living abroad – broadly referred to as *Josoen Jok*) and men living in Korea. Likewise, Taiwanese men seek Vietnamese women, in part because of their similar Confucian heritage. Fourth, men are argued to be disadvantaged somehow (e.g. in terms of their income and formal education) in what is commonly referred to as the 'marriage market' in their own countries and thus the search for 'foreign brides' is a deliberate strategy against a 'marriage squeeze'. However, Jones (2012b) acknowledges that this is also a stereotype, and does not reflect the diversity of men seeking foreign women. Fifth and finally, international marriage brokers and matchmaking organizations (particularly in Singapore) facilitate marriages across the border that are in turn facilitated by the role of the internet and social media (Chee, Yeoh, and Vu, 2012).

STRUCTURATIONIST AND AGENCY-CENTRIC APPROACHES

The limitations of a purely structuralist analysis of migration, a scepticism concerning the usefulness of the 'migration network' idea, and the desire to break down the distinction between determinist and humanist approaches to understanding international migration, led some scholars of migration to turn to Giddens' (1984) structuration theory (see e.g. Conway 2007; Goss and Lindquist 1995; Halfacree 1995; Mountz and Wright 1996). However, the paucity of studies that involve an explicit engagement with structuration theory is noticeable and may seem odd given that structuration seems to act as a compromise or middle way between structure and agency. Yet perhaps structuration theory's abstract theoretical constructs require a precise transfer from theory to evidence, which many observers of migration may find difficult to undertake.

Nonetheless, one of the most faithful and prominent adoptions of a structurationist approach is that of Goss and Lindquist (1995). They believe that a 'structurationist approach to migration' is applicable to rural-to-urban or circular migration within particular countries, but they use international migration to illustrate the value of the approach. Recall from the introduction to this book that Giddens views structure, not as – let us say – the 'global capitalist system', but as rules and resources, which human agents are knowledgeable about, and which they use to achieve certain aims, albeit through reflective practices ('reflexive monitoring'). In this process, the rules and resources are both reproduced and transformed. As rules and resources are repeatedly mobilized and manipulated by human actors, whether by those who have fewer resources or those who have many, social practices develop over time into 'institutions' (or in Giddens' term 'sedimented social practices'). For Goss and Lindquist then, "what has previously been identified as migrant networks . . . [should] . . . be conceived as migrant institutions" (p. 335). These migrant institutions join the individual migrant to the overseas employer and the global economy more generally. There is, as they say, an "institutionalization of migration" (p. 336). They offer this idea "as an alternative to the somewhat idealistic concept of migrant networks developed under the systems and other integrative approaches" (p. 336). Their argument that the idea of migrant networks is 'idealistic' stems presumably from their scepticism about the lack of attention to the operation of power in these networks. In other words, they see migrant networks as institutions which are rife with particular kinds of power relations, and which may not be beneficial to low-income Filipinos in particular, who are the subject of their research. Finally, in outlining their call for an analysis of the 'migrant institution', Goss and Lindquist do agree curiously that there is a 'global economy' out there, which cannot be controlled by these institutions, let alone the individual agents (migrants or otherwise) themselves. This suggests some affinity with a more standard reading of structuralism, and with the structuralist approach.

Even if few authors have rigidly mobilized Giddens' precise structurationist schema, many recognize implicitly the significance of structures, institutions, and individual agency for explaining migration.[7] This is reflected in ethnographic studies and a 'biographical approach' to migration which might be better understood as methodologies, rather than specific theories of migration.[8] These approaches or methodologies are qualitative. The more general category of ethnography entails close, daily, participant observation with research 'participants' and is concerned more with meaning than with explanation or the development of social scientific 'laws'. Ethnography has a long history in anthropology and sociology (see e.g. the reviews in Glick-Shiller (2003) for anthropology, and Fitzgerald (2006) for sociology), though during the twentieth century, this tended to explore matters of immigration (settlement) more than it did international migration (that is, the experience of moving). Beyond anthropologists and some sociologists, most social scientists (and especially geographers) ignored ethnographic approaches until the early 1990s (see the brief review in McHugh, 2000). Geographers in particular then began to explore 'biographical approaches' involving in-depth biographical or life histories and narratives of migrant 'life courses'. Often ethnography and certainly the biographical approach are 'methodologically individualist', but unlike the earlier forms of methodological individualism however, they are once again qualitative rather than quantitative and often motivated by a desire to 'de-stabilize meta-narratives' (Ní Laoire 2007: 373) such as world systems theory and other macro-explanations for migration. In other words, such approaches or methodologies seek to question the value of more determinist, all-encompassing theories.

In terms of the biographical approach, Boyle *et al*. (1998) identify three dimensions. First, migration is not to be viewed as a simple decision at a particular moment in time, such as comparing the utility of places. Rather, these decisions should be seen in relation to a "migrant's past and anticipated future" (pp. 80–81). Their 'biography' is unlikely therefore to be grasped through simple questions such as 'Why did you move', but instead through in-depth qualitative fieldwork that builds "a picture of the migration decision from a variety of angles, demonstrating how and where it fits into a person's life" (ibid.). Second, different migrations have unique yet varied causes, and it is the objective of the researcher to tease out the importance of different processes, reasons, emotions, and so on, which result in migration. The consequence of recounting these narratives leads to eclectic descriptions which may be as much about identity as behaviour, and it is therefore difficult to represent the decision-making of migrants. This difficulty of representation, however, may be a strength rather than a weakness of the methodology. Third, migration is embedded in cultural processes, and migration is therefore "a very *cultural* event" (p. 81, emphasis original). Biographical narratives therefore move beyond simply maximizing, satisficing or other formal models of decision-making.

We now move to a perspective on migration which is far from a theory and more a particular 'take' on the experience of migration. This is a certain view of the 'temporality' of migration.

The significance of temporality: another agency-centric perspective on migration

As with space, time seems so obvious, at best a backdrop to the more glaring social problems or issues that migrants face. So why bother with a perspective on time? As Cwerner (2001) argues, different conceptions of time, or what he calls 'cultures of time' are fundamental to the experience of migrants. Yet Cwerner's analysis is less an explanation for migration than a commentary on the 'cultures of time' among Brazilians already in the UK. Our concern here is how 'time' might figure in the process of migration, rather than in the experience of settlement, even if the two are related, especially when it might involve the decision to return. In any case, there are at least two ways in which we might incorporate time. The first involves an understanding of 'time-space paths' or 'life paths' (as we have seen in the 'biographical approach' to migration above). In this respect, King (2012) laments that the contributions of the well-known time geographer Thorsten Hagerstrand have largely been forgotten in migration studies (including in the first edition of this book!), although he acknowledges a renewed interest in life-path mapping as above. Life paths can be considered "a weaving dance through time-space" (King 2012, citing Pred 1977: 208), "in which people connect with each other in couples or groups ('bundles' according to Hagerstrand) at various points ('stations') and for various purposes ('projects')" (King 2012: 141). We can also extend such a time approach by incorporating a more feminist analysis in which the relationship between women, child-bearing and child care may interrupt, modify, or stimulate migration (e.g. Kofman 1999).

A second set of studies have emerged only in the last decade that emphasize the "temporal economics of illegality" (Andersson 2014), an "ethnography of waiting" (Kobelinsky 2010), or in less difficult terms, the prevalence of 'waiting'; waiting for a smuggler to guide one across the border, waiting for 'regularization', or visas or housing and work permits to be processed, and waiting for family to arrive. There is a lot of agonizing waiting *en route* and even when one arrives in the country of destination. Yet time, as Andersson puts it, has also "become a multifaceted tool and vehicle – even a weapon of sorts – in the fight against illegal migration" (p. 2). That is, governments use time to dissuade migrants from 'illegality' by 'usurping' or colonizing their time. Thus, this kind of study explores the relationship between migration policies, 'bordering' (in other words, the practice of making and maintaining borders), and the temporal experiences of migrants seeking clandestine or other modes of entry. To take one of innumerable examples, migrants from especially western Africa flock to Ceuta, a Spanish enclave on the African side of the straits of Gilbralter that is heavily fortified and a bottleneck for migrants hoping to make it to the mainland. Many languish in detention centres, protesting their imprisonment, and wait to somehow manage the crossing to the Spanish mainland, or await an unlikely visa to enter Spain. "As a consequence of such strategies, migrants are speedily diverted, deported, or left stranded, only to eventually make it back again through the rugged terrains of the pre-frontier" [the area that surrounds the border] (p. 6). In Mellila, a similar enclave to Ceuta, but closer to the Algerian border, the Spanish *Guardia Civil* arrest sub-Saharan Africans desperately trying to enter Mellila, and hand them over to Moroccan forces who then deport them to the closed Algerian border, and they return on foot to Mellila only days later. Other migrants are deported by Algerian forces back to Mali, where they wait again on the streets of Bamako, Mali's capital. The migration to Europe is in some ways never finished, only delayed.

The route to Europe almost inevitably involves waiting for a smuggler and therefore some sort of smuggling, either between African countries, or by sea across the Mediterranean. Smuggling is in fact common across most of the world's continents, and in Boxes 3.5–3.7 (see below), we look at three examples of smuggling (one between Mexico and the US) and two between Europe and the Middle East and Africa, in which we should consider migrants as active agents in the process of migration, even if they are often deceived by smugglers upon which they often so desperately depend.

Box 3.5 SMUGGLING FROM MEXICO TO THE UNITED STATES

Perhaps one of the most studied occurrences of smuggling is between Mexico and the United States. While this is often thought to involve primarily Mexicans, the number of Guatemalans, Hondurans, Salvadorans, and other migrants from Central and South America, who often travel through Mexico and cross the border 'illegally' into the US seems to have increased significantly since 2008, from 33,000 in 2000 to 145,000 in 2012, based on the number of 'apprehensions' by border officials (Massey *et al.* 2014). This is equally true of the many 'unaccompanied children' (< 17 years of age) who are smuggled across the Mexico-US border, and whose plight became well-publicized in the international media during a surge of movements (some 102,000 young people) across the border between 1 October 2013 and 31 August in 2014 (Donato and Sisk 2015; Pierce 2015). Yet whatever their age or nationality, when migrants attempt a crossing into *El Norte* (the north), this often involves the use of *coyotes* (professional smugglers) who may take an individual, a few people, a family, or even a larger group and lead them across *la linea* (the physical border between Mexico and the US). Sometimes they will accompany them to a particular town or city in the United States, but more frequently they are left to their own devices, accompanied only by a little food and water. Attempting a crossing without a *coyote* is exceedingly difficult, but it is also dangerous *with* a *coyote*. The now 'militarized' border (Andreas 2000; Dunn 1996) stretches some 1,000 miles from about a quarter mile off Imperial Beach in southern California to the desert of Arizona. The border consists of 700 miles of double fencing, infrared detection technology, specially designed vehicles, watchtowers, seismic sensors, and over 20,000 border agents (some with automatic weapons). The number of border agents has in fact increased more than four-fold, from about 4,100 in 1992 to over 20,000 in 2014 (CBP 2014). While crawling through tunnels, cutting through sections of the fences, and swimming across the Rio Grande river are all means of circumventing the carefully watched border, more likely are remote desert crossings since the events of 11 September 2001 (9/11), as border enforcement has simply pushed smuggling and other crossings further away from the cities and towns that dot the border and which are the sites of ever-present control. In the early 2010s, it was estimated that about one-third of all people trying to cross the border were caught, although it was also estimated that upwards of 92 per cent will attempt to cross again, perhaps repeatedly (Binational Migration Institute 2013). While the number

Continued

of 'apprehensions' of migrants by border control has appeared to have slowed down over the last decade (notwithstanding the increase in 'unaccompanied children' as noted above), desert crossings remain frequent. And despite efforts by humanitarian agencies to provide water stations and other assistance in especially remote locations, apparently, more than 5,000 people died between 1995 and 2012 from dehydration, heat exposure, murdered accidentally in drug trade shootouts (another part of border exchange), or even shot by the so-called 'Minutemen' (the American, anti-illegal immigration vigilantes). A sign on a Tijuana wall near the border asks *Cuantos Mas?* (How many more?) [will die] (Binational Migration Institute 2013; Nevins 2008).

Box 3.6 SMUGGLING BETWEEN MIDDLE EASTERN COUNTRIES AND THE EUROPEAN UNION: NETWORKS AND MIGRANT AGENCY ACROSS MULTIPLE SPACES

The long-brewing armed conflict in Syria that began to generate a mass movement of refugees in 2011 has created a long trail from Turkey to Greece and then on to Germany. In 2015, this trail involved new routes through northern Russia to Sweden as well (see Map 3.1). As with smuggling of other migrants or refugees from Middle Eastern countries such as Afghanistan or Iraq, the migration of Syrian refugees has also involved active decisions by migrants/refugees to choose a smuggler and a destination, even if smugglers continually disappoint them, or their destination is not reached directly or immediately. After all, desperation requires determination. However, as van Liempt and Doomernik (2006) point out, the spaces of smuggling are more complex than a route from A to B. To begin with, these networks, involve 'in-between' spaces and places such as Istanbul. In the case of migration to the Netherlands, Istanbul is a major node in migrant networks that follow the old silk road. And it is here that a 'visa mafia' exists and might provide false visas and related identity papers to migrants who have found themselves willingly or unwillingly in the city.

The route from Syria to Germany, for example, is far from straightforward. It often involves a long trek, train, or drive to the Syrian–Turkish border. From there, refugees have traversed Turkey on foot, and eventually may take one of two major routes to the EU – a sea-based route via Greece or Italy, or a land-based route to the EU through Bulgaria, that itself includes a river crossing between Turkey and eastern Greece. Here we focus on the sea route to Greece specifically, even if problems associated with smuggling along the sea-based route may increase land-based smuggling. For the sea-based route, refugees might make the long journey on bus or even foot to the city of Izmir on the Aegean Sea which, over the last decade, has become a crossroads for smuggling from the Middle East and central Asia to the EU. While sleeping in the city's streets, parks or low-cost hotels, refugees will wait agonizingly for a call from a smuggler. During this time they may quickly run out of money, as food, sometimes accommodation, and other provisions for the crossing may be bought, such as life jackets, and plastic bags to protect vital documents and mobile phones from sea water. Crossing for one person in an inflatable dinghy has been said to cost around $1,200 while crossing in a wooden boat for an entire family may be $4,000, depending on the time of year and the roughness of the sea (*Wall Street Journal* 2015c; *The Guardian* 2015b). Most refugees have taken boats further south or north from Izmir and make the six-mile crossing on the Aegean Sea to reach various Greek Islands, but especially the Greek island of Lesbos, which is only about six miles from the Turkish coast. Some 93,000 people landed on the island in 2015. Crossing the Aegean in such boats laden down with refugees can be dangerous, and at least 72 people drowned in one week alone in mid-September 2015, including many children (*Wall Street Journal* 2015b, 2015c). From Lesbos, refugees might languish a week in a refugee camp, stadium, or similar structure with food and water sometimes provided by the small NGOs present on the island. After perhaps a week, and with their fingerprints recorded and stored in

Continued

EU databases, they are likely to travel again to mainland Greece by boat, and then on to Macedonia, Serbia or Croatia, Hungary, Austria, and eventually to Germany on foot, by bus or sometimes by train. Such a journey is said to take on average about ten days from Turkey to Germany (*Wall Street Journal* 2015d; *Washington Post* 2015a). For more on the Syrian refugee crisis, see Chapter 4.

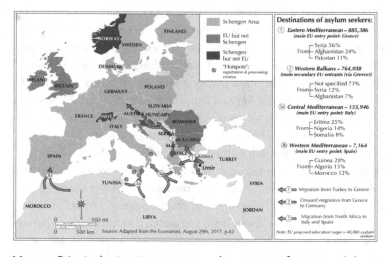

Map 3.1 Principal migration routes, and countries of origin and destination of asylum-seekers/refugees in the European Union, 2015–16. (Adapted from *The Economist*, 16 February 2016)

Box 3.7 SMUGGLING BETWEEN THE AFRICAN CONTINENT AND THE EUROPEAN UNION: CONNECTING AGADEZ, NIGER AND SABHA, LIBYA TO CALAIS, FRANCE

While the smuggling of migrants from the Middle East into Greece may have occupied the media across the world in 2014–2016, another long-standing migration has connected African countries over the last 15 years or so to the French city of Calais (Map 3.2). The story of this migration begins in eastern, central and sub-Saharan Africa. In fact, the journey can be traced to a range of countries, including for example, Burkina Faso, the Central African Republic, the Ivory Coast, Mali, Nigeria, or Senegal in central and western Africa, or Ethiopia, Eritrea, Somalia, and Sudan in eastern Africa. Like other migrations across the world, the migration of Africans may not involve a linear one, but involve multiple trajectories and destinations (e.g. Andersson 2014; Schapendonk 2011) in a form of 'fragmented' migration (Collyer 2010). Nonetheless, many are determined to reach Europe as soon as possible through the most direct route. Smugglers satisfy this desire in western Africa, by using pickup trucks filled with migrants to cross the Sahara east or northward to Agadez, Niger, an ancient trading centre, which has now become a sort of 'boomtown' for smugglers. The pickup trucks are over-laden with migrants, food, water, and eventually upright wooden sticks that are lodged between the provisions to provide something for the migrants to hold on to during the rough trip over dunes and rocky terrain. From there, smugglers will drive migrants approximately three to four dangerous days through the Saharan Desert and enter clandestinely into the western Libyan city of Sabha. Along the route, death through dehydration, kidnapping, mass-murder, rape, theft, and torture are not uncommon. While the government of Niger has passed a law that allows judges to jail smugglers for up to 30 years, it has not been enforced, and bribes (somewhere between US$150 and US$250) are offered to police and the military to turn a blind eye. Like Agadez or Ismir, Turkey, Sabha too has become one of many crossroads for smugglers, migrants, and others. Two major smuggling routes intersect in Sabha, one from Agadez and eastern and central African countries, and the other from eastern or 'the horn' of Africa (Eritrea, Ethiopia, and the Sudan). From Sabha, smugglers continue on with their human cargo to the Mediterranean coast. The Libyan government virtually collapsed in 2011, allowing the Libyan part of the North African coast to remain largely unpoliced. In such an environment, smugglers flourish. Yet, it is also

Continued

facilitated by the proliferation of mobile phone networks across African countries that connect migrants, smugglers and traffickers. From the coast, migrants will find another smuggler to bring them to the Island of Lampedusa Sicily, or mainland Italy, only 200 miles across the sea (see also the opening story in this book). After entrance into the EU, migrants (who remain undocumented) either try to claim asylum, or move on to other destinations across the continent. For some, travel may be relatively unproblematic but difficulties occur when additional border controls are imposed, as in Calais, France for those attempting to reach the UK.

The city of Calais is the location of the entrance to the Channel Tunnel which connects France and mainland Europe to the UK by train, car, or other vehicle. While many Syrians (and other migrants from the Middle East) may see Germany or Sweden as preferred destinations, for many African migrants, the UK is considered to be a sort of 'promised land', or at least, relative to other EU countries. A strong demand for low-wage workers, English as the official language, the chance to become proficient in English, and the lack of identification cards in the UK, make it attractive compared to France. Yet reaching the UK is hardly an easy task, given the difficulty of legal migration or claiming asylum in the UK (many could be considered refugees), at least from outside the country. The solution for many migrants from Africa is therefore to try and cross through the Channel Tunnel undetected, which is no simple feat given the extent of barbed wire fencing and other security measures that surrounds the entrance to the 31-mile-long tunnel, not to mention the dangers of travelling through the tunnel, either by foot, by attaching oneself illegally to a train, or jumping on a ferry. In July 2015, nine people died trying to cross through the Channel Tunnel. Since 2003, when the UK moved border controls from UK territory to Calais, migrants from all over Africa (as well as the Middle East) have amassed in various encampments in and around the port of Calais. The location of these encampments shift depending on operations by French authorities who have periodically destroyed the camps. One well known encampment in a landfill area some three miles outside Calais has become known by many as the 'The Jungle', a facility frequented by anywhere between 1,000 and 6,000 people, and which includes electricity, showers, toilets, and a daily meal, but no official accommodation for migrants. Sanitation and disease remain a dire problem. Other encampments are even less sanitary and more rudimentary. Migrants with the least resources languish in Calais in these camps, while those with more income (ironically from Iraq or Syria) can afford to pay smugglers who will take them by truck through the tunnel. Those who do not will either continue to wait for their chance to make it through the Channel Tunnel, move further along to other coastal cities to find a British smuggler, or resign themselves to living in France permanently) (for further discussion, see for example, the accounts in the *Wall Street Journal* (2015a) or the *Washington Post* (2015b)).

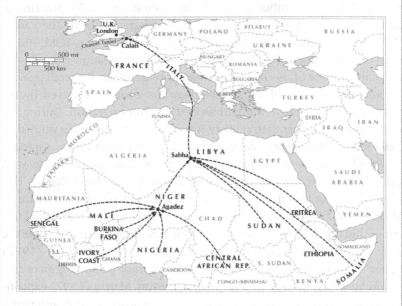

Map 3.2 Migration from eastern and western Africa to Libya and Calais, France

THE MIGRATION-DEVELOPMENT NEXUS

Exploring the relationship between development and migration is hardly a theory, but this literature has the merit of connecting the 'here' and the 'there'. And though not new by any means, there has been an increasing interest in the relationship between 'development' and migration over the last decade, what might be called the 'migration-development nexus' (Van Hear and Sørensen 2003).[9] This is witnessed in the innumerable conferences, programmes and reports by international agencies and organizations on the matter that took place in the mid-2000s (e.g. GCIM 2005; UN High Level Dialogue on Migration, 2006; Transfer of Knowledge through Expatriate Nationals or TOKTEN; Migration for Development in Africa or MIDA) (Faist 2008). Perhaps surprisingly, this interest has been sustained, as reflected in the UN High Level Dialogue on International Migration and Development 2013 and the 9th Global Forum on Migration and Development 2016. The intensity of meetings and interest, and the size of the migration and development infrastructure, has however not translated into a significant role for migration in 2015's Sustainable Development Goals (SDG). The absence of migration from the earlier Millennium Development Goals was widely discussed (e.g. IOM 2005), and yet despite the expanded number of SDGs, migration only figures explicitly in three of the 169 targets, under two of the 17 overall goals.

Our interest here is how 'development' or what is perceived by the literature to be a 'lack of it' might serve to explain migration. During the second half of the twentieth century, it was believed widely that migration could stimulate economic development in poorer countries as migrants sent remittances home or would return to the country of origin with skills, knowledge, and other financial resources (de Haas 2006). This would then have the effect of reducing emigration to the richer countries (de Haas 2007). As de Haas (2006) remarks, there is a resurrection of this sort of thinking in the twenty-first century, and migration has become the 'new mantra' (Kapur 2004) for development. In fact, Skeldon (1997) argues that migration is so closely related to economic development that it must be seen as part of the development process. Martin (1997) explains this relationship through what has become known as the 'migration hump' since plotting mobility and economic development schematically on a graph yields a bell shape, or a hump. This relationship highlights the fact that for the very poor, financial capital is often a significant constraint on mobility. When economic development provides access to more financial resources more people leave to seek opportunities elsewhere. At a certain level of development, access to financial capital is no longer such a constraint and individuals are able to fulfil more of their aspirations where they are. At that stage, international mobility begins to decline once more. This is of course a very broad generalization and there are exceptions, but it highlights the apparent paradox that economic development often encourages migration, at least at relatively low levels of development. Sassen (1998) makes a similar point, as discussed in Chapter 2. Nevertheless, as de Haas (2006) also notes, there are both 'optimists' and 'pessimists' in this debate concerning the consequences for migration and economic development in poorer countries.

The pessimists, whose outlook also dates to at least the 1960s, subscribe to a more 'dependency' argument. They focus on at least two negative consequences of migration for development. The first is the apparent 'brain drain' in terms of the loss of skilled labour, and in some cases, a loss of more manual farm labour from certain regions in the countries of emigration. Consider for example that approximately a third of engineers and researchers from poorer countries are working in the OECD countries (Faist 2008: 32). For the pessimists, when return migration does happen, it creates new divisions within the country, region, town or village of origin, which can then lead to a loss of 'community', or at least new social forms. The result is a disruption of existing and more 'environmentally sustainable' agricultural practices that depend upon long-standing social hierarchies and labour practices. Economic stagnation may result, unemployment may rise rather than decrease as a consequence, and migrants may find return migration difficult and return instead to the original country of immigration. Second, the pessimists argue that remittances may act to reduce poverty among households in the country of origin but that remittances do not necessarily lead to broad-based sustainable economic development, generally understood at the national scale. This is because financial remittances may encourage spending on imported luxury items in the country of emigration such as large homes, expensive cars, and so forth, rather than in more productive investments that would help to reinforce or build domestic and export-led industries. People then become dependent upon a remittance economy, and on occasion might even abandon productive activity in order to migrate. At the same time, investments in homes or the purchase of luxury goods leads to inflation in the national or more localized economies, which then raises prices on basic goods for those who have not migrated, and impels these individuals or groups of people to also migrate. Thus, rather than development, the result is quite the opposite (a review of this perspective can be found in Castles and Miller, 2009, Sørensen et al., 2002, and Faist, 2008).

On the optimist side, migration will lead to economic development for similar reasons that scholars claimed it would during the post-war period. Yet now, instead of there being a 'return option for migrants', there is now a 'diaspora option' (Barré et al. 2003, in Faist 2008: 33), meaning that migrants can and do foster continual ties between places of origin and destination. Circular migration is viewed as even more likely to result in 'brain gain' or 'brain circulation' (the back and forth movement of skilled people between various countries of emigration and immigration) rather than to 'brain drain'. Migrants are thus viewed as 'development agents' in a period of 'co-development' (Faist 2008). From this perspective, financial remittances are not simply used for the purchase of luxury items, but for entrepreneurial activity. Black and Castaldo (2009) find plenty of evidence for this in their study of migrants returning from Europe to Ghana or the Côte d'Ivoire (Ivory Coast), as do Maron and Connell (2008) for migrants returning to the Asia-Pacific island of Tonga from Australia, New Zealand, the UK and the US. In addition, remittances may counter problems of recession in the poorer countries and because the money is usually transferred individually or collectively through such companies as Western Union, it avoids state corruption or in many instances taxation. Perhaps even more notable now is the importance of 'social remittances'. Social remittances refer to the ideas, practices but also finances that migrants bring home or send home, and which contribute to the construction of schools, roads, religious institutions, hometown institutions, and other 'social' institutions (Levitt 1998) for the benefit of those 'left behind' (Hammar et al. 1997; Toyota et al. 2007). National governments view these social remittances as enormously valuable. For example, the Mexican government created its 'Tres-Pour-Uno' programme in 2001 – a matching programme whereby each 'migradollar' sent home through a hometown association to, let us say, build a school house in a particular village, is matched by US$1 each from the Federal, state and local governments in Mexico (Faist 2008; Orozco and Rouse 2007). And global remittances in general, as so many books, journal articles, and reports now point out, are considerable. Indeed, in 2015 remittances were more than three times Official Development Assistance (see Box 3.1).

It is increasingly common for low and middle income states with significant numbers of citizens who are resident overseas to institutionalize relations with these citizens in targeted diaspora administrations (Gamlen and Délano 2014). This may involve a special department, usually within the Foreign Ministry but in some cases comprises fully fledged 'diaspora ministries'. These government institutions, form what Gamlen (2008) has called the 'emigration state'. They have two distinct objectives, providing a range of services to emigrant citizens and encouraging citizens to feel a sense of loyalty or belonging to the state which may translate into increased remittances (Gamlen 2008). These institutions often fulfil a demand for recognition from emigrants themselves. They mark a shift in official discourse towards emigrants from one of disloyalty to the state to one of contribution, even one of self-sacrifice and heroism (Collyer 2014a). Yet there is a tension here as these institutions develop the idea of a norm of the behaviour of a 'good migrant' that draws not only on demands of the migrants themselves but transfers ideas developed elsewhere through the mechanisms of migration and development debates.

However, both Faist (2008) and de Haas (2006) find these debates overly polarized and simplistic. For instance, rather than countries suffering outright from 'brain drain', Faist prefers to speak of different stages and types of 'brain drains and gains' with differential impacts for different groups of people in the countries of emigration. He even talks of a 'global brain chain' (p. 32), when, let us say, doctors from Canada migrate to the US, and doctors from South Africa migrate to Canada. In contrast, de Haas (2006) uses insights from the 'new economics of labour migration' approach (discussed earlier in the chapter) to explore the ways in which migrant families rather than just individuals rely on remittances to improve their livelihoods and diversify risk. Leaning more towards the optimists, de Haas argues in the context of the Todgha valley in southern Morocco that instead of simply spending money on imported luxury items when migrants return to Morocco from France, Spain, Belgium, or the Netherlands, migrants do tend to invest their remittances in productive activities which can have important 'multiplier effects' that benefit both migrants and non-migrants alike. This might include investing in the installation of motor-driven water pumps in Southern Morocco as a solution to the limitations of traditional forms of small-scale irrigation, which then allows migrants to open up new farmland beyond the oases, hire paid labourers and overall contribute to the increasing productivity of agriculture. Yet it also includes the construction or purchase of 'modern' homes which provide more space to house large families. These new homes offer increased security, improvements in health; they act as an investment against loss, and the possibility of income generation through leasing.

Purchasing a home is therefore not strictly a vehicle for improving a migrant's status. While academics may not view this as 'development', de Haas believes that this "reflects a narrow view of development" (p. 575). Migrants have also invested in grocery stores, coffee houses, restaurants, taxis, delivery vans, among other ventures, and non-migrants have benefited from the businesses that these ventures create. This has led to a demand for labour among both non-migrants and internal migrants, and internal migration from other regions in Morocco has increased into the largest town (Tingha) in the Todgha Valley. Yet de Haas warns that the beneficial benefits realized through return migration and remittances are not likely to immediately contribute to the cessation of emigration. On the contrary, it is likely to increase it in the short to medium-term[10] as aspirations among would-be migrants rise.

Furthermore, de Haas (2006) admits that return migration and remittances have also contributed to social breakdown, for example between the traditional landed elite and the now emancipated share croppers, among others. Emerging disputes between these groups has led to the dissolution of village institutions which served to enforce common law – a legal system that ensured the collective regulation of land and water management. Traditional underground irrigation systems have therefore dried up from poor maintenance, which forces non-migrants to install water pumps. This then lowers water tables and contributes to the desiccation of traditional irrigation systems. The consequence has been unsustainable farming practices, the abandonment of new farms, and hence wasted investment. Those unable to invest in pumps are forced out of agriculture.

In sum, de Haas writes, "It is therefore important not to jump to the conclusion that the migration optimists were right because the migration pessimists turned out to be wrong [. . .] By postulating that migration is a household strategy to overcome local constraints on economic production and development, we should not infer that migration 'therefore' contributes to development in sending areas. This would be like falling back from one determinism to another" (p. 579). What is extremely valuable about de Haas' analysis then is his tendency to situate his analysis at multiple 'scales', from villages in the valley, to the biggest town in the valley, to the region, to Morocco as a whole, and finally to Europe.

In any case, beyond Morocco, the veracity of either view is difficult to assess, not least because it is difficult to isolate the impacts of return migration or 'brain circulation' on economic development, and there is simply inadequate research to support fully either of these arguments.

A further complication is that while return or circular migration may contribute to economic development, national data on economic development only tells us a very general story, or rather it highlights a problem of the scale of analysis. Migrants returning to one region of a country may stimulate economic development in that region, but other regions (and groups and individuals) may be left out and even hindered by the process of migration. As de Haas (2006) writes, "the fundamental question for researchers is not whether or not migration leads to certain types of development, but why migration has more positive development outcomes in some migrant-sending areas and less positive or negative outcomes in others" (p. 579).

The need to examine this highly policy driven agenda has been clearly articulated by Raghuram's (2007) questions 'Which migration? What development?' in which she re-emphasizes a deeper understanding of 'scale' (in her terminology), in order to see migration as individually developmental, as something that people do in order to enhance their own opportunities and those of their families. Other more critical participants in the debate question whether we can identify 'positive development outcomes' at all, even by alternative material measures such as owning a house. To take two examples, Lawson's (1999) study of internal migrants from rural areas in Ecuador to the capital city of Quito, and Silvey and Lawson's (1999) account of migration in Indonesia show instead that the cultural attachments to a village of origin and its way of life highlights the ambivalence that migrants feel about migration to urban centres and the 'modernization' of their lives. In this way, cultural and emotional attachments become important to the practice of migration, and seeing 'development' in one place as 'better' and the 'lack of development' in another as 'worse', represents an unhelpful binary opposition. Migrants instead feel multiple attachments across different places. Places should not be characterized as better because they are 'modern' and worse because they are 'not-modern'. This is one reason why the spatial concepts we employ are important for how we understand the causes and consequences of migration.

EXPLAINING FORCED MIGRATION

The study of 'forced migration' (sometimes called 'involuntary migration' or 'displacement') generally refers to the study of the movement of asylum-seekers, refugees and IDPs (internally displaced persons; that is persons displaced *within* a country) or individuals that are trafficked. However, below, we restrict our discussion to explaining the movement of asylum-seekers and refugees as one form of forced migration with at least four caveats. First, as noted in the Introduction, all forced migration has at least some element of 'volition' (voluntarism) in its otherwise 'coerced' character, and the distinction, as we noted in the Introduction to this book, between 'economic' (supposedly voluntary) and 'political' (forced) is problematic (e.g. Betts 2009; Black 2001; Soguk 1999; Van Hear 1998; Zetter 1988). Second, the distinction between asylum-seekers and refugees is ambiguous. After all, asylum-seekers, refugees and other migrants (if we can continue to agree to use such terms) often share the same routes, experiences, and hardships. Furthermore, so-called 'irregular secondary movements' (that is, undocumented migrants who are 'initially' refugees and who may or may not seek asylum) should be included in the movement of refugees (Black 2003; Zetter 2007; Zimmermann 2009). Such terminological ambiguity gives rise to more preferred and combinatorial concepts such as the 'asylum-migration nexus' (e.g. Castles 2003), and in any case, 'forced migration studies' is hardly looking for a single, all-encompassing theory.

Nonetheless, this should not preclude us from highlighting significant legal differences between the rights of migrants and asylum-seekers on one hand (principally regulated by national law)[11] and *prima facie* refugees (governed in principle by international law, but subject also to national law) (e.g. Hathaway 2007). A third caveat is the close relationship between internal displacement and asylum and international refugee movements, since among other reasons, IDPs may eventually cross international borders independently or certain individuals and families may be chosen by the UNHCR for refugee re-settlement in another country. Fourth, it is common in studies of forced migration, and especially in the study of asylum-seekers and refugees to discuss overlapping territories of governance, from national asylum and refugee policies to the intricacies and contradictions of international law. However, we will leave the discussion of the role of such institutions and states to Chapter 4. Fifth, although 'trafficking' is a dimension of forced migration that involves coerced work, we once again will reserve that discussion for Chapter 5 instead.

Now whatever the ambiguity, fluidity and difficulty of such terms, forced migration rose to become an especially prominent issue in the 1980s, owing partly to the immense increase in what were considered to be refugees in Iran, Pakistan, Mexico, and other Latin American, east African and Southeast Asian countries (Black 2001; Fiddian-Qasmiyeh *et al.* 2014) (see Box 3.8).

Box 3.8 A BRIEF NOTE ON MAJOR REFUGEE CAMPS

In the 2010s, there are still sizeable refugee camps spread throughout African countries especially, but also since the 1980s in the Middle East. Refugees may languish for months, and sometimes years in camps, whom Collyer (2010) calls 'stranded migrants' and Soguk (1999) refers to as the 'visibility of immobility'. Most refugee camps actually harbour refugees from neighbouring countries. We are thinking here of Afghans in Iran and Pakistan, Burmese in Thailand, Iraqis and Palestinians in Jordan, Rohingya in Bangladesh, Somalis in Kenya, and Sudanese in Chad and Uganda (e.g. Betts 2009; Fiddian-Qasmiyeh *et al.* 2014). The physical location of individuals is an important determinant of the rights they can claim from state institutions (Collyer 2014b). For example, asylum-seekers who are not officially accorded refugee status, such as those in the Calais' 'Jungle', or indeed other informal encampments in both richer and poorer countries, may be considered as part of the global refugee population, although they are not (yet) recognized as such and are not able to claim any rights.

Unlike the study of labour migration during the 1960s and 1970s however, the study of international refugee movements and other forms of forced migration grew up at a time of theoretical hesitation (a singular or unified theory of migration no longer seemed possible), ethical concerns (can or should western scholars in particular, represent the lives of refugees in the global south, and if so how should they be represented?) as well as feminist critique (how and what sort of feminism might figure in the study of refugees) (e.g. Bermudez 2013; Indra 1999; Jacobsen and Landau 2003; Hyndman 2000). It quickly became enmeshed in the study of 'development', but not in the same way that labour migration came to be understood, for example, through world systems theory' or through the 'migration-development nexus' as elaborated upon above. Rather, it began to focus instead on 'development projects' such as the construction of dams or large-scale resource extraction (for example, diamond mining in Sierra Leone) which – together with the political economy of development projects – finish by displacing thousands of people from their homes, some of whom will find their way to refugee camps across international boundaries. Yet, this sort of 'development' has intersected with other prominent explanations for refugee movements. Mazur (1989) for example, identifies both 'macro' and 'micro' explanations for refugee movements, whereas we choose to add 'meso' explanations, as well as the significance of 'territory' and 'place', without confusing 'macro' necessarily with 'structural', and 'micro' necessarily with 'agency' or 'place'. In any case, such 'macro' explanations commonly refer to environmental stress or environmental disasters (including chemical or nuclear disasters), famine, poverty, violent conflict, and war. 'Meso' explanations might include asylum, migrant and refugee networks on one hand, and trafficking networks on the other, but also importantly the practices and decision-making of institutions and organizations such as the UNHCR and other NGOs, as well as 'security disruptions' which might be decidedly 'macro' and 'micro' simultaneously. 'Micro' explanations for the creation and movement of refugees might include differential levels of fear, psychological trauma, and other torture/bodily harm, including sexual violence and forced abortion or sterilization.[12] We could add to such 'micro' explanations all of the axes of differentiation and intersectionality we have discussed throughout this book, as well as an individual or a family's religiosity/faith (see for example the special issue of the *Journal of Refugee Studies*, 24, 4), daily social interactions, including the relationship between men and women, the characteristics of the economic life of those who are displaced ('displacement economies'), and the role of communication technologies (see for example the special issue of *Journal of Refugee Studies*, Volume 26, 4).[13] Furthermore, the various reasons cited above may explain why people emigrate/migrate, but not necessarily where they end up, and that brings in a whole other set of explanations that we have reviewed earlier, from social networks to even the types of transportation available (Walters 2015) to how much money an asylum-seeker or refugee has in their pocket for a journey (e.g. Van Hear 2014).

Again, the 'macro', 'meso' and 'micro' should not be associated necessarily with any particular spatial metaphor. After all, structures can operate in single towns as much as national territories or continental regions, and violent conflicts can extend across entire countries, or be limited to only a few villages. In short, both territory and place matter for explaining migration, but they in themselves are not sufficient to explain the creation of refugees, nor of their movement. Indeed, as we will see in the next section on environmental change and migration, none of these 'phenomena', from disasters to war, necessarily translate into international refugee movements; for one, migrants may be confined to their own countries as IDPs, and individuals, families, or groups may do everything to avoid becoming refugees. In sum, consistent with the tenor of the rest of this chapter, we doubt that any single theory can explain all forced migration.

ENVIRONMENTAL CHANGE AND MIGRATION

As we noted in this and previous chapters, the environment, or more specifically, environmental change, environmental stress and environmental crises may also partially explain migration. 'Environmentally induced migration' is yet another frequently used term, and this suggests that such migration is more forced than voluntary, but it often involves substantial agency on the part of individuals, families, or even groups. Yet perhaps 'the environment' needs a bit more of our attention, before we just lump it into the cauldron of explanatory factors of (forced) migration. Let us start with the general views of the popular media and some academic analysts, which portray the relationship between environmental change and migration as a scary one. Climate change especially receives disproportionate blame for forcing migrants from their homes. This 'maximalist point of view' (Suhrke 1994), in which the environment is the main cause of migration, has led to predictions of mass migrations, including hundreds of thousands, perhaps millions, of 'environmental migrants', 'environmental refugees', 'climate migrants' or 'climate refugees'. Thus, migration is generally viewed as a threat, or at least a problem generated by climate change. This in turn has prompted governments and publics alike into 'what if' scenarios and likely responses. For example, in October 2015, the 'Nansen Initiative on Disasters, Climate change, and Cross-Border Displacement' held a global consultation meeting with the participation from the United Nations High Commission on Refugees (UNHCR) and the International Organization of Migration (IOM) that convened in Geneva, Switzerland to debate the state of climate change and policy responses (Nansen Initiative 2015). Despite the depiction of these often frightening futures, it is perhaps surprising that until the last decade or so, the principal theories of international migration that we have reviewed in this and the previous chapters have not incorporated an analysis of environmental change in any comprehensive fashion. How, if at all then, should we integrate environmental change into such theory? Before answering this question, we will begin with a set of cautionary points.

The first thing we should note is that environmental change means more than just (global) climate change. In fact, environmental disasters and other 'hazards' have shaped migration throughout history. Likewise, despite that 'the environment' is missing from the major migration theories, the study of the environment–migration relationship is hardly new, preoccupying a range of historians and social scientists since at least the early twentieth century. The second point is that the evidence for the kinds of dire predictions registered at the beginning of the twenty-first century concerning environmental disasters or climate change and migration is not very robust (Gemenne 2011). But it is probably useful to distinguish between environmental factors as 'slow-onset' stresses and 'rapid-onset' disasters'. In this way, time/temporality comes into focus again, but now in a different way. For while drought and rainfall variability (slow-onset stresses) may mean waiting to migrate and less permanent movement, rapid-onset disasters might create longer-term and more permanent movement without eventual return migration (Findley 1994; Hugo 1996; Hunter et al. 2015). Third, the simple equation that more 'environmental capital' or 'natural capital' (that is, more environmental resources) leads inevitably to less migration, and less environmental resources leads inevitably to more migration, is flawed (Gray 2009). Thus, it is important to avoid the intuitive appeal of simple arguments that emphasize population pressures on natural resources as a reason for migration. Instead, and this takes us to our last point, it may be better to adopt what a range of critical social scientists call 'political ecology', that is the relationship between politics, power, and the environment to explain the decision to migrate. In this vein, Carr (2005) elaborates on a 'minimalist approach' (in contrast to the maximalist approach noted above) in which environmental change is viewed as but one 'driver' among many. In villages in Ghana, he shows how local relationships of power (in this case along family and gender lines) along with environmental change shape the propensity to migrate (see also Zetter and Morrissey 2014). Carr's analysis raises at least two issues: there may be a need to distinguish between decisions to migrate *internally* or *internationally*, since the 'costs' of doing so are higher for the latter (Hugo 1996), and the question of power (structures) suggests that inequality may be fundamental to the relationship between environmental/climate change and migration (Hunter *et al.* 2015; Zetter and Morrissey 2014).

Given some of the above points and a long slew of studies that explore the environment–migration relationship, Black *et al.* (2011), in a highly cited study, argue that the environment is just one driver (as they, like Carr, refer to it) among many other reasons for migrating, and environmental change affects a range of drivers both directly and indirectly. For these authors, environmental drivers include both hazards (for example, droughts, earthquakes, erosion, flooding, hurricanes, landslides, pollution, sea-level rise, storm surges, etc.) and eco-system services (such as those elements of the environment that provide food and water, that provide protection, or that hold emotional value) (p. 57). Eco-system services are also referred to as 'environmental' and 'natural capital' (Hunter *et al.* 2015) as noted above. So with this in mind, Black *et al.* identify at least two basic conceptual approaches to understanding what has come to be called the environment-migration nexus. The first approach sees 'vulnerability' and 'adaptive capacity' as shaping migration, so that an environmental event can have different consequences for different people. These studies focus on hazards and individual decisions and thus have a certain affinity with the push-pull paradigm. The second approach focuses on rivers and sea-level rise. Sea-level rise is seen as more permanent than river flooding and it is for this reason that the former is more likely to produce significant displacement. But these two approaches, Black *et al.* maintain, ignore the already existing and long-standing in-migration and out-migration within environmentally stressed regions.

For Black *et al.* (2011) then, it is difficult to label a specific group of migrants as 'environmental migrants' who are in need of protection, although one can examine the contribution of environmental *change* to migration. Most studies begin with the environment as the source of migration, or imbed migration within a range of other factors. Black *et al.* reverse this investigative logic to some extent; to look at the drivers (economic, political, social, and so on) of migration and then see how environmental change affects these drivers. So what Black *et al.* do is build a model that incorporates 'structural' drivers and 'behavioural' drivers in which they distinguish between the environment and environmental change, and they see migration as part of adaptive response to change, not just a reaction to it.

Ultimately, they develop more of a framework (and less of a theory), which, instead of exploring what leads to migration, considers the "range of drivers that might affect the volume, direction and frequency of migratory movements as well as the different levels of analysis at which migration might be considered" (p. s.5). This is represented in a diagram (see Figure 3.1), which seems to have five features.[14]

- The left side of the diagram (surrounding the pentagon) takes into account five primary sets (or 'families') of 'macro' drivers of migration (demographic, economic, environmental, political, and social). These drivers are similar to the 'push-pull' variables as presented by Lee (1969) but the model they propose by no means adheres to the 'push-pull' paradigm.

- The left side (within the pentagon) considers spatial and temporal differences in the source and destination in terms of gradual or sudden and actual or perceived changes in the five drivers mentioned above and in the diagram below.

- The left side also considers environmental change and its influence on migration, but also considers the indirect influence of environmental change on the other four drivers.

- The right side considers the agency of migrants in the migration process in terms of a combination of 'personal characteristics' and 'intervening obstacles and facilitators'.

- The right side considers how points 1–3 combine so that drivers turn into the decision to stay or migrate. Reflecting on why people stay might be just as important as looking at why people migrate, since most studies focus only on those who have migrated, leading to what is called 'selection bias' (Koubi et al. 2016)

While Black et al.'s framework may seem comprehensive in its attention to questions of both space and time, it does neglect the question of 'household risk' (remember the New Economics of Labour Migration approach outlined in Chapter 2?). Thus, Hunter et al. (2015) see environmentally induced migration as a response to diversify risk within households. This is especially important in households in which their livelihoods are agricultural or natural resource-based. The point we wish to make here is that earlier theories can often be relied on in the exploration of new approaches. Combined then, the papers by Black et al. and Hunter et al. encapsulate

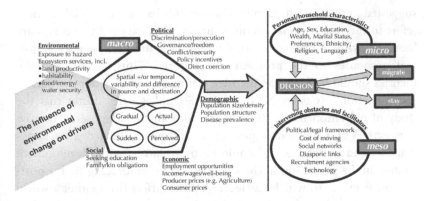

Figure 3.1 A conceptual framework for the drivers of migration

particularly well the long debate on the environment–migration nexus, and offer the broad claim that we cannot draw a simple correlation between environmental disasters or stress, and migration. We now move to our final discussion, which reviews what is called a 'social transformation perspective'.

BEYOND THEORY? A SOCIAL TRANSFORMATION PERSPECTIVE

Consistent with a creeping suspicion that began in, let us say, the 1980s or 1990s concerning the possibility of a universal theory of migration (Portes and DeWind, 2007), Castles (2010) agrees that a singular theory of migration is not only impossible, he argues that it is not even desirable. Instead he wishes to embed an understanding of migration within broader economic, political and social changes. He calls this a 'social transformation' perspective in which 'social transformation' is understood to concern "the complexity, interconnectedness, variability, contextuality and multi-level mediations of global change" (p. 1566). For Castles, there are at least two reasons that motivate his insistence on the undesirability of a singular theory of migration, and consequently, his calls for a social transformation perspective. First, migration is seen as a problem, which stems from a 'sedentarist bias', an idea that stems from anthropology and other studies of mobility (Sheller and Urry 2006; Wimmer and Glick-Schiller 2002; and earlier, Malkii 1995). As we suggested in the Introduction to *Migration*, this bias in the social sciences refers to the foundational premise that local or national territories are 'normal', while movement, mobility, and migration are seen as 'abnormal' responses to the problem of territories. Thus, social scientific theories should be based on the idea that migration is 'a normal part of social relations' (Castles 2010: 1568). Second, Castles thinks that the study of migration is too disciplinary, with all the supposedly narrow perspectives and methodologies this might entail. Relatedly, he hopes that the study of migration would be more linked with 'development studies' in the global south. The key question for embracing such a position is the extent to which Castles' argument differs from either a world systems approach or the migration-development nexus discussed earlier. The answer it seems, is that Castles wishes to move beyond the macro-economic emphasis of earlier approaches, while holding on to the power and inequality that world systems and other Marxist-inspired discussions emphasized. Faced with these challenges, Castles argues that 'middle range theory' (a term associated with the twentieth-century sociologist Robert Merton) may be one response. For instance, it might be more worthwhile to investigate particular types of migration such as labour migration between Algeria and France, asylum migration between Nigeria and the UK, or 'migration systems', such as between Southeast Asia and East Asia, or South Asia and the Gulf States. He argues that focusing in on such particular migration or migration systems "could bring together the insights provided by all the different social sciences concerned with migration" (p. 1574). In short, Castles' call is not for the end of theory but a more 'middle range theory' that tackles particular 'context-specific' migrations that can be investigated in interdisciplinary ways, and which are conscious of the complexity, inequality, and uneven interconnectedness of the world.

ASSESSMENT OF THE APPROACHES

Recall from the Conclusions to Chapter 2 that we are concerned with three elements of all these theories (or more appropriately *approaches* in this chapter): their substantive focus, unit of analysis, and spatial assumptions. In Table 3.4 below then, we illustrate the differences in the theories or approaches we have discussed based on these three elements, but we also add which social scientists might be more associated with them.

Table 3.4 Theoretical and other approaches to migration and their basic characteristics

Theory of migration	Substantive focus	Unit of analysis	Spatial assumptions	Likely scholars involved
Ravenstein's Laws and push-pull theory	Internal and international migration based on push and pull reasoning	Individuals and groups: push and pull variables	Methodologically nationalist and regionalist	Population geographers in the past. Some demographers
Neo-classical economic approach	Migration between poorer and richer countries and regions based on economic rationality	Individuals	Methodologically nationalist and regionalist	Economists. Quantitative economic geographers and quantitative sociologists of migration
Behaviouralist	Migration behaviour of individuals and internal migration based on rational cognitive 'satisficing' behaviour and place utility	Individuals	Methodologically nationalist and place-oriented	Population geographers and sociologists
New economics approach	Migration between poorer and richer countries based on the household as a collective and cohesive decision-making unit	Households	Methodologically pluralist (nation-states, regions, towns and villages)	Sociologists; some economists of migration

Continued

Theory of migration	Substantive focus	Unit of analysis	Spatial assumptions	Likely scholars involved
Dual labour market/ segmentation approach	Labour market demand in richer countries for migrant workers based on employer imperatives	'Modern industrial societies', and the 'techno-productive structures' of employers during post-Second World War capitalism	Methodologically nationalist	Anthropologists, sociologists, and economic geographers
Structuralist approaches (dependency theory, articulation of modes of production, world systems theory)	Changes in the (global) capitalist system and its effects on migration	Capitalism and migration patterns/ migrant groups divided by nationality, gender and ethnicity	Methodologically globalist, nationalist, and regionalist.	Across the social sciences
Structuralist approaches (globalization)	Effects of global flows, structural constraints and opportunities (innovations in transport and communications) on migration.	The global economy/ global economic restructuring; transport and communications innovations	Interpenetrating scales (or territories) (e.g. supranational entities, nation-states, regions, localities)	Across the social sciences
Structuralist approaches (global cities)	Effects of the global economy on the development of large, economically dynamic and diverse cities	Global economy and 'global cities'	Methodologically globalist and urbanist	Across the social scientists, but especially sociologists and human geographers

Structuralist approaches (neo-liberalism)	Changes in the global capitalist system relating to 'roll back' and 'roll forward' neo-liberalism and its effects on migration	The global economy/global political and economic transformations, especially those detrimental to poorer countries	Methodologically global, but also interpenetrating scales	Across the social sciences, but especially anthropologists, sociologists, and human geographers
Social (migrant) network theory	Migration through group, region or village, household, and individual behaviour	Networks of groups and individuals	Methodologically diverse, but emphasis on transnationalism, trans-localism, trans-urbanism, etc.	Across the social sciences, but especially anthropologists, human geographers, and sociologists
Transnationalism	Diverse cultural, economic, political, and social migrant or diasporic links which are global or cross-border in character	Diverse – diasporic networks, local communities (cities, towns and other places) migrant and immigrant groups; households; economic transactions	Diverse – transnational, trans-urban, trans-local, bi-local relations	Across the social sciences
Gender-aware approach	Diverse, but focuses on women migrants and gender relations; domestic labour and household relations	Individuals, households, groups; patriarchal structures	Methodologically pluralist (nation-states, regions, towns and villages)	Across the social sciences

Continued

Theory of migration	Substantive focus	Unit of analysis	Spatial assumptions	Likely scholars involved
Structuration theory and other agency-centric approaches	Potentially all forms of migration, but sparsely applied	Individuals, groups and institutions	No particular spatial assumptions	Across the social sciences
A temporal perspective	All forms of migration, but focuses on asylum-seekers, refugees and undocumented migrants	Individuals and groups	No particular spatial assumptions, but focuses on multiple trajectories and circuitous routes with multiple origins and destinations	Anthropologists, but also human geographers and sociologists
Migration-development nexus	Relationship between countries and regions of immigration and emigration, especially the impact of remittances, and generally restricted to labour, family, 'transnational', or 'diaspora-oriented' migration	Emphasis on nation-states within the global south, and less often, regions, towns, or villages within the global south	Leans to methodological nationalism, but diverse spatial concepts are explored	Across the social sciences but especially (development) geographers, and (development) sociologists

A theory of forced migration?	Asylum-seekers, refugees, and trafficked persons	Global structures, institutions, ethnic groups, and individuals	No particular spatial assumptions, but critical of methodological nationalism, and focus on international law	Global south-focused anthropologists, geographers, and sociologists; forced migration and refugee studies
Environmental change	Potentially all forms of migration	Individuals, networks, and institutions, especially	No particular spatial assumptions, but focuses on coastal regions, islands, and poorer countries	Across the social and some physical sciences
A social transformation perspective	Potentially all forms of migration	Global structures, institutions, migrant groups	No particular spatial assumptions, but insists on incorporating different levels; against 'methodological nationalism'	Anthropologists, geographers, historians, sociologists

Which theory is most appropriate then? By now, some readers will have made up their own minds as to the most relevant of these approaches. This is fair enough. Some might insist that no single theory can explain all forms of migration across time and space (Portes and Dewind 2007; Bakewell 2010), or that some overarching theory of migration is not desirable (Castles 2010) and these views are fair enough too. Some may choose a more integrationist standpoint. This is laudable, and one that we would encourage. However, we would also caution the reader that some of the differences between these approaches mean that they are down-right incompatible. For example, it is difficult (but probably not impossible) to combine structuralism, gender perspectives, network approaches, structuration, and biographical narratives, with a neo-classical perspective, when the former are clearly implicit or explicit critiques of a theory of migration based on individual economic rationality and push-pull factors (e.g. Boswell 2008a).

At this point, we have chosen not to critique each and every theory systematically as we would not want to subject the reader to such a laborious undertaking (a task that has already been accomplished for many of these approaches in Massey et al. (1993, 1998) and Boyle et al. (1998), and see also the suggestions for further reading at the end of the chapter). Furthermore, and as we imply above, some of the differences between these approaches are enough to suggest the limitations of each other. However, we will spend a little bit more time criticizing push-pull and neo-classical conceptions of migration, and comment very briefly on the strengths and weaknesses of subsequent approaches, and then finally call upon the necessity of a 'spatial approach' to migration.

In terms of push-pull theories, let us return to the example of African migration to Washington DC (Wilson and Habecker 2008). It is ironic that the Africans who choose to locate in that city mention the presence of international development institutions such as the World Bank as attractive, since it may be the policies of this very institution or similar institutions that are partially responsible for the migration of their more impoverished compatriots to European countries, as described for example in the opening pages of this book's Introduction. It is a shame then that Wilson and Habecker (2008) never explore the possible connections between migration to the US, and US and European-dominated economic policies in Africa.

With respect to neo-classical conceptions, they are flawed on both micro and macro grounds. From a micro perspective, the neo-classical theory works with atomized (in other words isolated), a-historical individuals who are 'rational economic actors' (*homo economicus*), and who respond to real or expected earnings differentials. Certainly, the real or expected differences in employment and wages between countries or regions might explain some forms of migration at certain times, but it seems to be the case rather that migrants are more akin to 'satisficers' than 'utility maximizers'. That is, migrants, if it is indeed economics that motivates them, are more likely to emigrate if their living conditions are so dreadful that it is difficult to stay, rather than to a comparatively small wage differential and the precise possibility of employment. The psychological and financial costs of migration may outweigh any expected gains in wages. Migrants do not always move to a region or country where wages are the highest, nor do they move where the greatest employment possibilities exist (Massey *et al.* 1998), and indeed one might find the whole approach ludicrous for asylum-seekers, refugees and those who are sexually trafficked.

From a macro perspective, the greatest number of migrants does not originate in the poorest countries and does not necessarily migrate to the richest. If that were the case, there would be large numbers of, let us say, Sudanese in Luxembourg. And the argument that migration leads to wage equalization between regions, or more absurdly between countries, is patently untrue. Wage equalization, if it happens, cannot be gauged simply in national terms, and much else, besides migration, might contribute to wage equalization (Massey *et al.* 1998).

Behaviouralist approaches have the advantage of focusing on 'satisficing' rather than 'maximizing' behaviour and the assumption that many migrants only leave when it becomes no longer possible to stay under 'humane' conditions (the notion of 'stress'), a point that Douglas Massey and his colleagues (1998) recognized as a key element of migration at the end of the twentieth century. However, the initial literature tended to neglect the relationship between migration and the structures of the global capitalist system, or at least global political and economic forces. It also neglects the significance of gender relations, migration networks, immigration policies, and a more sophisticated understanding of 'place utility' or even migrant desires and aspirations (Bal 2014). Indeed, what exactly constitutes 'place' in this context? After all, the attraction of African migrants to Washington DC is a matter of inter-locking territories: the territoriality which makes Washington attractive, and the territoriality which makes the US attractive more broadly.

If we turn to more *critical or at least non-neo-classical approaches*, each – with the exception of many structuralist and gender-aware studies – is also guilty of neglecting the role of states when these theories are used in their 'pure' form. And each mobilizes often ill-defined and problematic spatial concepts. Beginning with the more determinist theories, so-called 'new economics approaches' were a welcome corrective to the methodological individualism and rationality of neo-classical approaches, and yes, households, household budgets, and household decision-making matter, but households only rarely function as cohesively and rationally as the theory might suggest. To borrow a common psychological expression, the household acts like a dysfunctional family, or as Goss and Lindquist (1995) put it, households are not 'unified strategic actors' (p.327). There is also a need to attend to how a more thorough and thoughtful conceptualization of territories, places, networks and so forth can help us explore how families adopt different strategies, and how the outcomes of these strategies relate to particular spaces. De Haas (2006) accomplishes this to one extent or another in his study of the Todgha valley in Morocco (discussed earlier) but there is a need for more careful and explicit spatial thinking here.

The *dual labour market hypothesis and related segmentation approaches* are glaringly deficient because they focus completely on labour demand in the richer countries, and essentially neglect the countries of origin. In any case, there are not just two sectors within the world's various labour markets, and workers fit within a much more complex spectrum of jobs. States figure little in its basic theoretical proposition, and the study of 'space' (beyond richer and poorer countries) is altogether absent.

In contrast, a broader structuralist approach might seem politically attractive because it points to fundamental spatial and class-based inequalities in (what is now the disputed idea of) a global capitalist system. It highlights a new phase of globalization or a new global political economy: structural adjustment, neoliberalism and the veritable creation of migrations from poorer countries, but also richer ones. This globally induced migration also creates sub-national spaces in the form of 'mega' or 'global cities' which emerge as places with dramatic inequalities and labour market polarizations that create a demand for low- and high-income migration. Yet such Marxist-inspired analyses often lack an inadequate understanding of the role of states and their different territorial jurisdictions (national, regional, local/municipal, and so forth), and especially the role of immigration policy. They provide little room for human agency and the in-depth study of institutions, networks, families, and social axes of differentiation in different 'places' that would allow for a multi-territorial or multi-jurisdictional and more nuanced appreciation of the different forms of migration within variegated spaces of origin, destination, and between.

Arguments that employ specifically the concept of 'globalization' also suffer from a hyperbole about the decline in transportation and communication costs and the ease of movement. The cheapening of these innovations may make it cheaper and easier for some to conquer physical distance at particular times, but not for others; they certainly do not spell the end of distance (compare Graham 2002, and Doreen Massey 2005). Thus, using globalization to explain (the growth of) all forms of migration would be inadequate and misleading (Samers 2001). In fact, in the context of heightened security concerns since 9/11, it may now be generally speaking, more difficult than ever for low-income individuals to migrate to the richer countries especially, without encountering enormous obstacles.

In terms of the more integrative theories, social (or migrant) network analysis has a number of strengths. First, it provides a critique of the determination by either structural forces or individual agency, and thus points to the relevance of structuration. It questions the importance of wage disparities between countries, and argues that the social networks formed between different migrant communities in the country of origin and destination matter more than wage inequality, for example. Second, concentrating on migrant networks allows us to see how such networks transcend territories, thus highlighting the porosity of apparently fixed territories. Third, it also offers the unique insight that because of social networks, pioneer migrants lower the cost and risks of migration for newer ones. Yet there is a danger in a strict focus on networks. It might give the impression that such networks reach across space without the impediment of borders, the vicissitudes of visa policies, the trickery of smuggling and trafficking agents, the cost of airline tickets, intra-family disputes, the prevalence of racism and intra-group suspicions fuelled by the problems and sometimes scarcity of low-paid work (e.g. Goss and Lindquist 1995; Grzymala-Kazlowska 2005; Gurak and Caces 1992; Krissman 2005).

Many transnational studies provide a richer understanding of space, identity and their implications for migration, as they move beyond methodological nationalism and an exclusive focus on the region, city, town, and village in the country of emigration or immigration (methodological localism if you will). However, despite the countless books and papers which seek to complicate or re-theorize notions of space, from 'transnational spaces' to 'translocalism' and 'transnational fields', these concepts are still not satisfactory. Are certain migrant practices really reflecting a sense of trans-*nationality* or indeed *translocalism*, or something even more spatially plural? At the same time, are these practices reflecting social categories around class, gender, ethnicity and multiple ideas of nationhood even within the same 'nation-state', rather than just national or local belonging? A second problem is that transnational research recognizes the significance of states and immigration policies, yet these are generally not treated comprehensively. Third, the emphasis on social networks lends itself to a more celebratory understanding of migration, rather than on how immigration policies and the more deleterious forms of networks (trafficking, for example, and elite networks) shape migrants' lives in harmful ways. And fourth, as Castles and Miller (2009) point out, there is a question whether most migrants really live the transnational lives as described in some accounts.

From a different angle, the now abundant feminist and gender-sensitive literature on international migration has enormous value. It emphasizes how national gendered discourses, and gender relations and expectations in sub-national spaces and places shape the propensity to emigrate and return. It underscores the importance, for example, of domestic labour in the wealthiest regions and cities of both richer and poorer countries as driving a demand for female labour *because* women are stereotyped in certain ways across different places and spaces. It does however risk neglecting the structural political and economic processes operating to configure migrant networks and gender relations itself. It may also ironically privilege the study of women over men and particularly economic sectors in which women are dominant numerically.

By this point, *structuration* theories might seem an attractive theoretical friend. After all, we may find comfort in choosing a 'middle way' proposition. That structures, institutions, and actors matter is no doubt appealing, and it is an approach which we support wholeheartedly. However, Goss and Lindquist (1995) seem to overemphasize how social networks solidify into institutions. Social networks do not *always* develop into institutions, formal or otherwise. This relatively minor criticism aside, that leaves us with ethnographic and biographical narratives. Here, critical researchers on migration return to the methodological individualism of earlier approaches, although the assumptions and the methods employed are vastly different. Problems arise, however when biographical narratives and ethnographic data are to be interpreted. Here, theory should play its part; otherwise how do we make sense of migrants' stories? How do we know which data should be collected; what should be left out and what should be included when we re-count these stories? And precisely *how* and *with whom* should we conduct research, and how does that affect the knowledge we produce?

This leaves what are perhaps far more integrative 'approaches' or bodies of work that are hardly meant to yield singular or determinist theories, including the 'migration-development nexus', 'forced-migration studies', the literature on 'environmentally induced' migration, and the 'social transformation perspective'. The *migration-development* nexus has the merit, like 'transnationalism' of incorporating the 'here' and the 'there' and to link them in ways that both are mutually transforming. Yet, its focus as a body of work tends to neglect other insights from the migration studies literature, including how 'inter-sectionality' shapes migration, along with the significance of social networks.

The literature on 'forced migration' is enormously wide-ranging as the massive *Oxford Handbook of Refugee and Forced Migration Studies* (Fiddian-Qasmiyeh *et al.* 2014) demonstrates with its 53 chapters on seemingly every aspect of the refugee and forced migrant experience. The advantage of this literature is that it seems to have shifted the focus of migration studies away from explaining labour and family reunification, to the study of asylum-seekers and refugees. This has value, given that asylum-seekers and refugees in the world far outnumber those migrating under the rubric of labour and family reunification visas. However, the difficulty arises when forced migration is *not* distinguished from other forms of more voluntary migration, and the theories used to describe both become unnecessarily combined or confused.

As another element of forced migration, some of the more sophisticated overviews and reconstructions of the relationship between environmental change and migration (e.g. Black *et al.* 2011; Hunter *et al.* 2015) have remained critical of any linear relationship between climate (or other environmental) change and migration. This is an advance upon some of the overly alarmist, catastrophic, or deterministic accounts characteristic of the literature in the 1990s and 2000s. It is difficult to argue with these contributions, in part because they are so 'syncretic' or 'synthetic' – that is, they combine an enormous number of insights from years and years of migration studies. However, because they are to one extent or another comprehensive, they lack the analytical 'bite' of some narrower contributions, especially when thinking about 'space' in the ways this book has suggested.

Perhaps frustratingly for the reader, our argument is neither that we can combine these approaches or theories so easily, nor that we should not combine any of them. Should each be used in the context of a specific type of migration, or at particular moments of time? We argue that a combined set of theories – with the exception of neo-classical and push-pull conceptualizations – is necessary. Structures, institutions (especially states), social networks and social axes of differentiation along the lines of age, class, (dis-)ability, ethnicity, gender, and sexual orientation matter to our understanding of migration. However, the idea of 'structures' raises a particularly thorny theoretical problem, and we want to devote a little bit of time to the defence of the relevance of structures. It is certainly less 'fashionable' now (and perhaps it is not simply a question of fashion) to talk of 'structures', primarily because they are associated with Marxist or world-system-type analyses and 'something' 'beyond us', 'out there', often global in character which serve to disempower people. Instead, at the beginning of the twenty-first century, many critical migration researchers now prefer to talk of 'elite networks', migrant networks, migrant-based, and grassroots or community-oriented non-governmental organizations (NGOs). They also prefer to emphasize the importance of cultural, political, and social discourses, often through the concept of 'governmentality' (see Chapter 4) that shape migration. Such network and post-structuralist accounts are insightful and politically useful since it allows us to free ourselves from the potentially debilitating shackles of structures. Quite simply, if we do not conceive of structures as existing, then we can erase them from ever constraining migrant practices. This is fine, but there may be an 'intellectual masquerade' at work. Let us elaborate. Wealthy or other elites that exercise power over others through discourses or material practices used to be called up until, let us say, the 1970s and 1980s, part of the 'class structure' of capitalist societies. Class and 'class structure' largely disappeared from view in migration studies, and were eclipsed by the idea of 'elite networks'. 'Elite networks' however, sound awfully like a 'transnational capitalist class', to use Sklair's (2001) term. We can certainly think of a 'transnational capital class' as a network (that is a network of individuals associated with particular institutions and states who exercise considerable power) but as a class *as well* with certain interests. The transnational capitalist classes exercise power over other networks (or classes), especially networks composed of migrants more disadvantaged than they are. As a transnational capitalist class, their existence is predicated on the existence of the less fortunate. To put it differently, certain networks may in fact be evidence of certain structures. Furthermore, as Van Hear (2014) reminds migration scholars, 'class' (in his case, 'collectivities' which have or do not have both the economic resources and the ability to manage mobility and immobility) should be central to any account of migration. In sum then, structures, institutions, social networks, and axes of differentiation (or inter-sectionalities) such as class should be *part of* a robust approach to migration. We say *part of*, because as has now hopefully become clear, any approach to migration requires a discussion of 'space', and that is where we are headed in the next section.

TOWARDS A SPATIAL APPROACH TO MIGRATION

All research on migration is in a sense geographical, since it takes place *somewhere*, and all work in the social sciences on migration involves spatial categories (e.g. small towns, agricultural towns, cities, urban regions, rural regions, poor regions, rich regions, coastal regions and upland regions, 'developed' and underdeveloped countries, richer countries and poorer countries, global north and global south, territory, place, locale, social field, transnationalism, transnational social field, trans-local field, and so forth) (e.g. Silvey and Lawson 1999). Much of this unfathomably large body of work on migration (and not necessarily on immigration) fails to adequately define spatial concepts. When there is a genuine attempt to do so, *it is remarkable how loosely spatial terms are used, sometimes with very little scrutiny or precision*. Even 'transnationalism' which no doubt has enormous value, is, as we suggested earlier, either inadequate or flawed.

In the obsessive twenty-first-century haze of harnessing diasporas for development or looking for transnationalism, mobility and flows, the literature tends to neglect the effects of *relative* territorial fixity; in other words, territories are fixed for 'moments', not eternity, but in those 'moments', they have an effect on individuals, institutions, structures and social networks, migrant or otherwise, and this relationship is reciprocal and ever-changing. In short, territories and places of all kinds shape migrant behaviour because they have material effects; migrant behaviour shapes territories, because places are struggled over and migrants have material effects on these same territories. Perhaps the most obvious manifestations of this are international borders and immigration policies. Ironically, they are designed to have a deterrent effect on migration, but they have the perverse effect of stimulating undocumented migration for example, because 'how else are you going to get in?' The large movement of undocumented migrants and asylum-seekers or refugees, or those who overstay their tourist or student visas, may also drive worried governments to reinforce borders further. It is not just national territorialities that matter however. The particular zoning requirements, specific tax regimes, and local ordinances against the presence or activities of undocumented immigrants and others who are deemed to be 'out of place' in municipalities also matter. Here we are thinking, for example, of the many state and local ordinances in the US that seemed to proliferate in the early 2000s against either undocumented immigrants or the many informal locations where both legal and undocumented 'day labourers' gather to be hired for the day. This has the effect of dissuading further migrants from arriving in these towns and cities, or to discourage those already there from settling.

Territories are not necessarily uniquely local, regional, national, and global but they can be, and they can also entail other hybrid or combinatorial territories which cannot be captured under the usual categories of local, national, and so forth (we have in mind here refugee camps or workplaces which have their own rules and regulations, but are also subject to national and international laws). Different processes might entail different territorialities, but such territories are porous and dynamic, not fixed, and are re-shaped by the actions of migrants themselves through a multitude of structures, social networks and institutions. An attention to multiple territorialities, and places then, suggests a less spatially presumptuous approach which does not necessarily prioritize methodological transnationalism, localism, or whatever 'ism' is chosen over another. Territories matter but none are privileged before-hand. In fact, the choice of spatial metaphor is never innocent; such a choice reflects a certain politics; put differently, there is a politics of space. 'Space', then, like a picture or photograph is a way of representing the world (Dikeç 2012). If we represent national territories as fixed, frozen, and hermetically sealed, and nation-states as natural forms of territory, any movement of immigrants or refugees will probably seem an incursion or even an invasion. Conversely, if we think of borders as porous, as nation-states as heterogeneous rather than homogeneous, as cultures as fluid, and power and the law as less concentrated in, or exercised by, central governments, then we can begin to think of new spatial metaphors of migration and immigration. We hope the reader will take such a spatial imagination along into the subsequent chapters.

CONCLUSIONS

We have completed a long journey from the last chapter to the end of this this chapter. We have reviewed and assessed a wide range of approaches that attempt to explain migration. As Massey et al. (1998) and Portes (2000) noted more than 15 years ago, these theories cannot be adequately evaluated for their veracity, given that in the very least, the worldwide evidence is simply not sufficient. This might make a long review of migration theory or approaches seem fruitless, tedious, or detached from the richer ethnographic and narrative accounts of migration. This would be an unfortunate assessment. Such a review is absolutely essential, even for ethnographic research. That is, the approaches that we have explored in this chapter have illuminated different 'ways of seeing' migration. They require the employment of distinct methodologies when conducting research with individuals or groups, and raise questions about precisely whose voices should be sought out, and 'what to see', that is, how to interpret (stories of) migration through the lens of certain spatial and social metaphors. Yet rather than leave the choice completely open to the reader, we have chosen instead to point readers to certain approaches over others.

In that sense, we have shown how tempting it is to explain migration through a set of 'pushes' and 'pulls'. Yet migration is more than the summation of a balance sheet between a set of push and pull variables delineated by origin and destination countries. Instead, migration researchers have emphasized the need to connect the multiple 'heres' and 'theres' in more complex and interwoven ways. For a while, the term 'relationality' became popular in at least human geography (that is, understanding in this case the relationship between the global north and south) but other social scientists might prefer Castles' 'social transformation perspective'. In any case, we argue in this book that a more appropriate lens should involve the intersection of territorialities, structures, institutions, social or migrant networks, and a critical engagement with social axes of differentiation such as age, class, ethnicity, gender and sexuality. This has the advantage of moving beyond the stubborn grip of methodological nationalism, but also the *apparently* more critical and progressive conception of transnationalism or trans-localism. This critical engagement captures a key 'missing link' in all of these theories, namely the role of states. This will be the subject of the next chapter on 'Geo-political economies of migration control'. Let us go there now.

FURTHER READING

In addition to the literature at the end of the Introduction to this book, the following texts provide very useful overviews of migration theories, which also appear throughout this and the previous chapter:

Boyle, P., Halfacree, K., and Robinson, V (eds) (1998) *Exploring Contemporary Migration*; Massey, D.S. *et al.* (1998) *Worlds in Motion: Understanding International Migration at the End of the Millennium*; Castles, S. and Miller, M. (2003, 3rd edition, 2009, 4th edition, and 2014, 5th editions) *The Age of Migration*; Brettell, C. and Hollifield, J. (eds, 2008, 2nd, and 2014, 3rd editions) *Migration Theory: Talking across Disciplines*.

Beyond these large texts, Portes and De Wind (2007) also provide a lengthy overview of some theoretical and conceptual developments in the migration literature for the twenty-first century. Bakewell (2010) discusses the limits of structuration theory, based on a 'critical realist' perspective. Nagar *et al.* (2002) provide a critical discussion of the concept of globalization, and offer a feminist reading of it. Kofman (1999), Pessar and Mahler (2003), and Silvey (2004a, 2006) provide a summary of the literature on gender and immigration. Levitt and Jaworsky (2007) offer an overview of transnational migration studies. Goss and Lindquist (1995) and Krissman (2005) discuss critically structuration theory and networks. Lawson (2000) discusses some of the potential of biographical narratives within a geographical approach to migration. Silvey and Lawson (1999) and Silvey (2004a, 2006) explore different spatial metaphors used in migration studies. The IOM World Migration Report from 2008 devotes an entire chapter on international student mobility, including an enormous range of statistics and an outline of the causes of such mobility. These statistics can be updated with the following interactive website: http://www.uis.unesco.org/Education/Pages/international-student-flow-viz.aspx. For 'forced migration', *The Oxford Handbook of Refugee and Forced Migration Studies* provides a comprehensive volume on this subject, while those in need of shorter rendition can rely on Reed *et al.* (2015) for an accessible overview. As indicated earlier, both Black *et al.* (2011) and Hunter *et al.* (2015) provide very helpful and critical overviews of 'environmentally induced' migration.

SUMMARY QUESTIONS

1. Explain some of the limitations, but also the insights that we might gain from Ravenstein's (1885) classic text.
2. What is wrong with push-pull and neo-classical economic theories of migration?
3. Discuss at least two variants of a structural approach to migration.
4. Explain the some of the ways in which a gender-sensitive approach contributes to our understanding of migration.
5. What is meant by a structurationist approach to migration?
6. How is 'time' significant for the study of migration?
7. Why should we exercise caution instead of panicking about the relationship between 'global climate change' and migration?
8. What are the implications of a 'social transformation perspective' for constructing theories of migration?
9. Discuss some of the ways in which 'space' should figure at the heart of any approach to migration.

NOTES

1   For MacDonald and Macdonald (1964) "Chain migration can be defined as that movement in which prospective migrants learn of opportunities, are provided with transportation, and have initial accommodation and employment arranged by means of primary social relationships with previous migrants" (82).
2   Levitt and Jaworsky (2007) provide a comprehensive discussion of the different ways in which transnationalism is defined.
3   In the context of migration, 'essentialism' involves making assumptions about the cultural, political, or qualities of some supposed ethnic or national group. Such a group is supposed to have some 'essential' qualities or properties to them. There is thus a thin line between 'essentializing' and 'stereotyping'.

4   For reviews of the literature on migration and gender, see e.g. Kelson and DeLaet 1999; Hondagneu-Sotelo 1994; Lutz 2010; Morokvasic 1984; Kofman 1999; Pessar and Mahler 2003; Piper 2006, and Silvey 2004a. Gaetano and Yeoh, Introduction to the Special Issue on Women and Migration in Globalizing Asia: Gendered Experiences, Agency and Activism. *International Migration*, 48, 6 (2010), 1–12.

5   Outside of Mexican-US migration, women as 'independent' migrants should not be seen as either 'new' or limited to Mexico (Ryan 2008).

6   Other than those cited above, there was a wealth of studies from various countries that appeared in the late 1990s and 2000s. Readers who are interested might refer to Anderson (2001b), Ehrenreich and Hochschild (2004), Elias (2008), Lutz (2002), Parreñas, (2001), Pratt (1999), Reyneri, (2001), Veiga (1999), Yeates (2004, 2009), Yeoh and Huang (1998). More recent work can be found in Brickell and Yeoh, 2014; Yeoh and Huang (2010), Yeoh and Soco (2014), and Yeoh, Chee and Vu (2013). Notably, Raghuram (2008) is one of the few studies that look at migrant women in traditionally male-dominated sectors.

7   Yet another approach and criticism of structuration is to be found in Bakewell's (2010) novel discussion of a 'critical realist' approach to migration, but this will remain outside our discussion here.

8   See e.g. Boyle, Halfacree, and Robinson (1998); Lawson (2000); Miles and Crush (1993); Ní Laoire (2000, 2007); Vansemb (1995); Wilson and Habecker (2008), and on forced migration in particular, Eastmond (2007).

9   It is curious that there is very little critical discussion of what 'development' means in the migration studies literature (but see de Haas, 2006, 2007, Bakewell, 2008, Gidwani and Sivaramakrishnan, 2003; Lawson, 1999, and Silvey and Lawson, 1999). It seems to be generally understood as an increase in the standard economic and financial metrics of a country of emigration, such as Gross Domestic or Gross National Product, and the resulting benefits for poverty reduction. Very little seems to be said about more alternative, creative, cultural, or 'sustainable' conceptions of 'development'.

10 The terms 'short-term' and 'medium-term' are often used in the migration literature, but what they mean precisely in terms of years, for example, is never specified.

11 We note importantly that the EU's refugee regime shapes migration and asylum movements as well, which we discuss in Chapter 4.

12 In the US, for example, the 1996 Illegal Immigration Reform and Immigrant Responsibility Act expanded the definition of a refugee to "include persons who have been forced to abort a pregnancy, undergo a forced sterilization, or have been prosecuted for their resistance to coercive population controls" (MPI 2015).

13 The list of works that cover such explanations is innumerable. One might see e.g. Berg and Millbank (2009); Chatty (2010); Martin *et al.* (2014); McAdam (2010, 2014); Nassari (2009); Palmgren (2014); Schon (2015), and Witteborn (2015).

14 Given some ambiguity in the authors' own explanation of their diagram, our explanation differs somewhat from their explanation, and tries to provide greater clarity. We admit, however, that we may not accurately capture the authors' intention.

4

GEO-POLITICAL ECONOMIES OF MIGRATION CONTROL

INTRODUCTION

Even the casual observer at any international airport, bus terminal, or border crossing cannot help but notice the fortress-like quality of nation-states. National borders are porous, ever-changing and complex spaces (for example Anzaldúa 1987; Nevins 2002; Johnson et al. 2011) but they also symbolize the material and exclusionary power of national states. In this respect, it is a common assumption that states and their respective legal systems are essentially anti-immigration, uniformly racist, xenophobic, and exclusionary. Publics too might be expected to be univocally racist, culturalist, and xenophobic and essentially against immigration, wary of being 'swamped' (the term used by the former British Prime Minister Margaret Thatcher during the 1980s) and paranoid of 'hordes' and 'invasions' of certain kinds of migrants. There is no doubt much truth in these assumptions about states and publics, except for the adverbs 'uniformly' and 'univocally'. Actual migration, asylum, refugee, and immigration politics are far messier both in a social and spatial sense. If wealthy states are fortresses or 'gated communities' (van Houtum and Pijpers 2007), then their drawbridges are lowered for some in different times and in different sub-national spaces for quite specific purposes.

For example, Hollifield (2004) speaks of 'the emerging migration state' in contrast to the nineteenth-century 'garrison (or fortress) state'. For Hollifield, 'the migration state' is caught in a 'liberal paradox' whereby it must reconcile openness to trade, investment, and people on one hand, with security on the other. The migration state must ensure economic well-being, but also provide for the security of its citizens.

Yet states do not even meet 'their' own objectives. In fact, Cornelius and Tsuda (1994) develops what he calls the 'gap hypothesis', which suggests that the governments of wealthier countries are intent generally on restricting migration, but in reality, migration increases over time. Put differently, there are limits to governmental control and states often fail in achieving 'their' stated objectives (for example Castles 2004). Hollifield's 'migration state' and Cornelius' 'gap hypothesis' are simple conceptions of states and their (failed) objectives. For as political sociologists have long pointed out, 'the state' is not monolithic; that is, the state is not a 'thing' with one voice; it is not simply a one-room chamber of expert sages that churn out policy. Rather, states are complex apparatuses which contain many levels and different 'branches' or 'wings', often in conflict with each other. Nonetheless, Hollifield and Cornelius' arguments are stimulating enough to begin a discussion. In that regard, much has been written by political scientists in particular, to try to understand how migration politics and policies unfold. As with other dimensions of the study of migration, the multiple territories involved in these politics and policies have not been neglected, but often the critical analysis of the spatial metaphors used, or the complex intermeshing of territories remain relatively under-evaluated and hardly explicit in most studies. Indeed, Favell (2008) argues that since the 1970s, the devolution of power from national states to local levels of governments in Europe and North America have meant that local states are increasingly responsible for the regulation of citizenship and immigrant settlement. To take but one example (which we also elaborate on later), the implementation of what is called Section 287(g) of the 1996 Illegal Immigration Reform and Immigrant Responsibility Act in the United States permits individual US states and local law enforcement officials to carry out responsibilities that were previously the domain of the US federal government. However, the actual practice of these responsibilities varies greatly across localities in the US (Coleman and Stuesse 2016; Varsanyi et al. 2012), and such practices may even violate federal, state, and local laws (Heyman 1999). Thus, as Favell (2008) notes speculatively, it is in relation to these sub-national policies that an alternative and not strictly national politics of migration may erupt.

In order to explain the development of migration policies and some of their characteristics, we introduce a set of theoretical and conceptual arguments by scholars working on the (geo)-politics or (geo)-political economy of migration. We say 'arguments' rather than 'theories' since many of the supposed theories of migration policy are not explicit about their philosophical foundations, nor are they as well-developed or as clearly demarcated as theories of migration.[1] Nonetheless, enough differences can be discerned between them, and as in Chapter 3, we introduce a set of different perspectives in a systematic fashion. We begin with 'radical' ideas, including Marxist, neo-Marxist, and related approaches, such as the critical analysis of neoliberalism and its relation to the concept of 'migration management'. We follow with a brief discussion of a 'national identity approach', and then move on to Freeman's influential 'client politics' thesis. This thesis acts as a jumping-off point to explore further perspectives and elaborate on the relationship between states and migration. Although we have titled this chapter 'geo-political *economies* of migration control', many of these perspectives could *not* be classified as having an 'economistic' dimension at all, but some do. That may be a limitation of some of the arguments, but equally problematic is once again the neglect of 'space'. The purpose of this chapter then, is to expose the reader to these different approaches, as well as to emphasize the significance especially of territory and place in migration policies without discussing these policies strictly on a region-by-region or country-by-country basis, as is typical of most conventional studies. As in the previous chapter, however, we do divide our discussion of migration policies into those more relevant to richer countries (the 'global north') and those more relative to poorer countries (the 'global south'). Once again, this may seem like a crude distinction since governments in poorer countries have similar concerns, but we do maintain that there *are* significant differences in their objectives from richer countries, and it is important to underline these.

THEORIES OF MIGRATION CONTROL IN THE 'GLOBAL NORTH'

Marxist and neo-Marxist explanations of migration policy

Let us explore these theories systematically, beginning with Marxist and neo-Marxist accounts of migration policy. To begin with, a crucial point is that generally speaking, theories seem to partly emerge as a product of particular historical 'moments', and Marxist theories of migration control were by far more popular among scholars of migration in European countries and South Africa from the 1950s to the 1970s. The Marxist and Marxist-inspired literature focuses essentially on *labour* migration policies, and migrants are viewed as an exploited class, an 'industrial reserve army' – a 'relative surplus population' that can be easily called upon during times of economic expansion, and easily fired during times of contraction or devaluation (recession). Migrants are seen further as a racialized fraction of the working class, a quantitatively and qualitatively flexible labour force that can weaken 'working-class' power and force down the 'value of labour power' (the very rough equivalent of the 'cost of labour' in conventional/neo-classical economics), which would in turn benefit capitalists and the governments of particular nation-states. At the same time, migrants are perceived to lower the costs of social reproduction (refer back to Box 2.4) for capitalists and the state. This is so because many migrants will arrive in the country of immigration as working adults, and they do not need to be educated or housed by the state in the country of immigration during their childhood. Furthermore, from a Marxist perspective, migrants are seen as reducing inflationary pressures in the country of immigration since employers pay extremely low wages to migrant workers. The combined low wages and lower costs of social reproduction accelerate capital accumulation.[2] It is not surprising then that Marxist scholars spoke of the 'structural necessity' of migrant labour for European industry (for example Castells 1975; Castles and Kosack 1973; Miles 1982).

Finally, in Marxist thought, the state is considered the 'executive committee of the bourgeoisie';[3] in other words, the state is said to represent the interests of the capitalist class in either the country of emigration or immigration, but more likely the country of immigration. Since the 1970s, few people have adopted a Marxist approach to explain migration policies. This might be explained by the withering of Marxist analysis in general and the decline in (industrial) labour migration to countries in northwestern Europe during the 1970s and 1980s. Another reason is that Marxist approaches have never involved a very sophisticated treatment of the state with respect especially to different kinds of migration and they therefore fail to capture comprehensively the relationship between states and migration. Some scholars (for example Cohen 1987, 2006; Samers 1999, 2003a; Sivanandan, 2001) do try to construct a nuanced neo-Marxist appreciation of migration policy by incorporating a more complex analysis of the state in terms of migration, but most people have abandoned such an approach. Nonetheless, the Marxist perspective seems to show up implicitly in the critique of neoliberalism.

From a neo-Marxist approach to the critique of neoliberalism

In the 2000s, a renewed academic focus emerged on the relationship between government policies and especially *labour* migration, and this has been couched typically in terms of a critique of neoliberalism. We have already elaborated on the idea of neoliberalism in earlier chapters so we will not reiterate its contours here. We will point out simply that governments have come under critical scrutiny by academics in particular for privileging the liberalization of the movements of *some* people at the expense of others, hence the implicit (and sometimes explicit) critique of the class dimension of migration policies within the critique of neoliberalism. In fact, even when 'low-skilled' migrants are accepted, their social entitlements are severely limited and their economic rights curtailed. As Smith and Winders (2008) insist, neoliberal policies, programmes, practices, and discourses expect migrants to be "hyper-mobile, reliable, disposable, productive, affordable, implicitly youthful and male" (pp. 63–4).[4] Similarly, scholars have relied critically on different versions of the idea of 'neoliberal governance' or 'neoliberal governmentality' (more on this later) to explain migration policies in the wealthier countries in particular. For example, Bauder (2008) explores how a neoliberal-oriented media reinforced the restructuring of welfare in Germany and the privileging of highly skilled migrants in migration policy. In contrast, Coleman (2005) and Sparke (2006) explore the contradictions of neoliberal governance on one hand with security concerns on the other in the US. They speak of the contradictions of the North American Free Trade Agreement (NAFTA) after the events of 11 September 2001 (9/11) when the neoliberal geo-economics of maintaining the flows of goods and services across the Canadian–American border in the context of NAFTA bumped up against geo-political demands to securitize the border in the context of the 'war on terror'. Ultimately, Sparke argues, the geo-economic imperative trumped the 'geopolitical imaginations of homeland fortification' (2006: 12), and in many ways, the 'securitization' of the border actually *facilitated* economic flows, rather than impeding them.

If we can agree that neoliberalism or 'neoliberalization' has remained a strong feature of governance since the 1980s, then the essential argument is that governments have selectively liberalized migration (or neglected to some extent stopping those who are entering clandestinely) because:

- Low-wage migrants perform jobs that *allegedly* citizens do not wish to perform under the prevailing conditions and/or low wages that are offered, and by providing cheap labour, they benefit capital accumulation (economic growth). At the same time, they are only desirable in the eyes of national governments insofar as they do not create unusual burdens on now skeletal welfare systems. Sciortino and Finotelli (2015) call this the 'immigration as welfare burden' perspective (see also Bommes and Geddes 2000) but there are other logics at work, that may contradict such a view, as noted immediately below.
- Governments (especially those of the European countries and Japan) perceive there to be
  a 'demographic deficit' and are worried about the effect that fewer domestic workers will have on the sustainability of social programmes in the future, especially pension schemes. Sciortino and Finotelli (2015) refer to this view as 'immigration as a welfare resource'.
- Governments believe that highly skilled immigrants (from business managers to computer scientists and doctors and nurses) either stimulate capital accumulation through innovation and investment, or save cash-strapped health and other social services by providing necessary skills.
- Governments agree that international students provide financial resources directly for universities, and often indirectly for governments and businesses. They see international students as ensuring that science and engineering departments receive adequate students and that national competitiveness is secured through scientific and engineering innovation.

Neoliberalism or 'migration management'?

As mentioned briefly in the Introduction, governments and some less critical scholars began to speak of the importance of 'migration management' in the 1990s instead of, or alongside 'neoliberalism'. 'Migration management' or 'managed migration' refer to a set of policies across the wealthier countries that are designed to regulate carefully the type and number of migrants, whether it concerns asylum-seekers, family members, or workers. However, migration management is aimed especially at ensuring that the 'right' kind (and often the 'right' number) of migrants meet specific labour market demands (for critical discussions, see e.g. Ashutosh and Mountz 2011; Morris 2002; Geiger and Pecoud 2010, Kofman 2010; Scott 2016). This is accomplished through Canadian- and Australian-inspired 'points systems' or similar tiered migration management systems that either involve different 'tiers' or skill categories or attach a certain number of points to a migrant's education, skills, finances, and so forth. They allow entrance into a country based on the total number of points, usually out of 100. These labour market demands are closely monitored and various national agencies and programmes have proliferated since the early 2000s to streamline labour migration and recruit more highly skilled migrants, while carefully restricting others (such as asylum-seekers, less-skilled workers, and refugees). Here we are thinking of institutions such as the Migration Advisory Committee in the UK or Canada's Foreign Credential Referral Office, but the list of such institutions across richer countries is substantial. While these kinds of recruitment or vetting organizations are hardly new, what is new is the complex and often sophisticated ways in which different kinds of migration are assessed for their 'value', especially to labour market needs and national economic competitiveness. For example, the UK has developed an Australian-type tiered migration management system, Germany fashioned a special 'Green card' for information technology specialists in the early 2000s, as did Ireland for highly skilled migrants in general, and Denmark implemented a 'foreign experts' visa (Pellerin 2008; OECD/SOPEMI, 2008). Compared to the 1980s and early 1990s, migration management also involves an increasing reliance on temporary labour migration to satisfy labour market demands, again with close monitoring of the 'temporariness' of migrants (Pellerin 2008; Samers 2016).

In this respect, 'migration management' as a set of discourses and policies is certainly compatible with neoliberalism, in that it is designed to promote economic competitiveness on the one hand and restrict access to those who might be a burden on stripped-down welfare systems on the other. At the same time, paradoxically, it requires a small army of state workers to manage such migration. And perhaps more importantly, migration management reflects government's concern with other issues including security, the dictates of foreign policy, trade liberalization, promoting the international standing of a country, creating or maintaining ethnic or racial homogeneity, nation-building (see for example Walsh (2007) on Australia or Spoonley and Bedford (2012) on New Zealand), and advancing 'an imagined future' (Walton-Roberts (2004) on Canada). Equally, governments seek to control the growth of informal labour markets, stop smuggling and trafficking, sustain pension systems, meet or address humanitarian obligations such as family reunification, and regulate acceptance and settlement of asylum-seekers and refugees. By now it should be clear, however, that the balance at the beginning of the twenty-first century has shifted on the one hand to the selective and sometimes temporary recruitment of highly skilled labour, a restrictive stance towards family reunification (for example, Kofman, 2004), tightening the criteria by which asylum-seekers and refugees are accorded a haven, and reducing low-wage migration through bureaucratic and highly scrutinized temporary labour migration schemes ('guest-workers programmes') (e.g. Samers, 2015b) for certain low-wage economic activities that will be discussed in Chapter 5.

Those critical of the discourse of 'migration management' from both the 'left' and the 'right' in the countries of immigration stress its effect on creating undocumented migration by at least two means. The first is through the use of regularization programmes (also called amnesties or legalization programmes) such as in France, Italy, Spain, Switzerland, and the US over the last 20 years. Right-leaning political parties tend to view such programmes suspiciously on the grounds that it encourages others to migrate and 'wait it out' in an irregular status until the next 'amnesty' (regularization). This has been noted for the surge in the number of unaccompanied minors migrating from Central American countries to the US between 2013 and 2014 (e.g. MPI 2014b), but some research from Italy, Spain, and the US questions the idea that amnesties/regularizations lead to further migration (e.g. Orrenius and Zavodny 2003; Finotelli and Arango 2011). In any case, right-leaning political parties also see regularizations as a signal that a government has lost control of migration. In contrast, those more left-leaning citizens may favour certain kinds of regularizations in certain countries at particular moments. However, they also see the tightening of asylum policy as creating undocumented migration because by making it more difficult to enter through the legal channel of asylum-seeking, migrants are left with little choice but to enter clandestinely. Regularization programmes, along with other polices, seem to highlight not only the many unintended consequences of managed migration and greater restriction (for example Castles 2004; Massey et al. 2002), but also the class dimensions of migration management (Datta et al. 2007; McGregor 2008). Nevertheless, most critical discussions of migration management do not couch it in a language of class. It is more common to see 'migration management' as a means of selecting more or less 'desirable' migrants through 'exclusionary' processes that involve a fortress-like mentality (for example, the idea of 'Fortress Europe' that emerged in the 1990s). Nevertheless, Marxist arguments still seem remarkably useful, but changed social and political economic realities, academic fashions, the desire to provide fresh perspectives, and a focus on asylum-seekers and refugees and those marginalized from the world of work altogether, have all but buried a Marxist analysis of migration policy.

THE NATIONAL IDENTITY APPROACH

In what Meyers (2000) calls the 'national identity' approach, migration policy can be understood through national models of citizenship, understandings of nationhood among governments and citizens, including myths of national homogeneity and identity, and the character of political and legal culture in a given country.[5] He recalls the now widely read works of Higham (1955) and Jones (1960) who link a sense of 'American identity' to US immigration policies during the nineteenth and early twentieth centuries. As Meyers (2000) explains, both Higham and Jones claim that "social cleavages, social unrest and industrial unrest within American society foster fears of losing national identity and of national breakdown" (2000: 1253). These processes in turn lead to nationalism and nativism (xenophobia). Nativism is seen as a "psychological phenomenon: a decline in American confidence in the country's unity produces nativistic outbursts; an optimistic mood limits nativism" (ibid.).

Other 'classic' texts within a national identity approach divide the wealthy countries of immigration into 'settler societies' (for example, Australia, Canada, and the US); 'ethnic states' (for example, Austria, Germany, and Switzerland); homogeneous and heterogeneous states (see Castles and Miller 1993), and countries which subscribe to a jus soli (law of soil) or a jus sanguinis (law of blood) conception of citizenship (we will discuss this distinction further in Chapter 7). Still others, such as Faist (1995), speak of only two forms of citizenship today in the wealthy countries: 'ethno-cultural political inclusion' and 'pluralist political inclusion'. Certainly, we could carry on ad infinitum with these typologies, and given the common thread of critiquing methodological nationalism in this book, it would be easy to poke fun at such a way of thinking. However, Meyers (2000) suggests that this approach has three strengths: first, it examines national cultural idioms and traditions which frame what is politically possible. As he argues: "State policies are not constructed in a vacuum, but rather are influenced, to some degree, by the history and traditional ways of thinking of a society" (p. 1255). This might be viewed in a positive light by some, but more critical scholars have stressed how governments or feelings among citizens that a country is being 'swamped', 'invaded', 'culturally diluted', or that 'national cultures' are 'in decline', can and have driven migration policies (e.g. Guiraudon 2000). A second strength for Meyers is that it explains why some states prefer temporary migration (e.g. Germany) and others permanent (e.g. France). For example, Hollifield (2000) argues that France's migration and immigration politics are still shaped by a peculiarly French notion of republicanism, in other words, a universalizing vision in which all 'foreigners' are 'good' insofar as they assimilate or integrate into French political culture (that is, a 'culture of democratic politics'). Only permanent immigrants, successive French administrations claim, will embrace this political culture (see Brubaker 1992). Third, conflict along ethnic, racial, and religious lines, such as between Anglophones and Francophones in Quebec, and Jews and Palestinians in Israel, can drive what kind of migration or immigration is desirable. A fourth possible strength of this approach, and related to the second and third points immediately above, but not explicitly discussed by Meyers, is how certain migrants based on their ethnic background may be preferred over others. Thus, 'national myths', especially the belief in ethnic homogeneity by either governments and/or citizens has an effect on who is preferred or accepted into a particular nation-state. For example, during the 1980s and 1990s, the German government liberalized migration and made special provisions for Aussiedler (ethnic Germans), while the Spanish government began in the 1980s to privilege labour migration from Spanish-speaking Latin America (especially Ecuador and the Dominican Republic). The South Korean government also began to favour the migration of ethnic Koreans (*Josean Jok*) to Korea in the 1990s, who were commonly living in China or Japan.

Thus, such preferences, and much of the 'nativism' (anti-immigrant politics and restrictionist policies and practices) that we associate with a post-9/11 era were in fact enacted and played out well before the security-obsessed restrictionism after 9/11. We only need to think of the seemingly now unfathomable 'Chinese Exclusion Act of 1882' in the United States, the 'White Australia' policy during much of the twentieth century, or the 'moral panic' against Asians in New Zealand in the early 1990s. These are simply three examples of the many moments in which nationalism, national identity, racism, and economic uncertainty have mixed together in the most toxic and exclusionary ways in various countries. Even the criminalization (or what Palidda, 2011, calls the 'racial criminalization') of asylum-seekers and undocumented migrants (in other words associating asylum-seekers and undocumented immigrants with criminals and treating them as such) that we might associate with the first decade of the twenty-first century was visible in the early 1990s across the wealthy countries. Again, to take a few examples, the Australian government enacted a law on mandatory detention in 1993, in which any migrant without the necessary papers could be expelled or imprisoned (Hyndman and Mountz 2008). Similarly, the militarization of the southern US border began in earnest with the Immigration Reform and Control Act of 1986 (IRCA), and intensified with the Illegal Immigration Reform and the Immigrant Responsibility Act of 1996 (IIRIRA) in response to fears of terrorism, among other anxieties (e.g. Massey *et al.* 2002). In the EU too, most European countries drastically lowered their 'acceptance rates' for asylum-seekers (i.e. the percentage of asylum applications governments approve) during the late 1990s and early 2000s. While acceptance rates have risen since then, the twenty-first century has entailed an intense suspicion and scrutiny of asylum-seekers, refugees, undocumented migrants and similar denizens in the wealthier countries, especially in both richer and not-so-richer countries from Malaysia to South Africa.

There are in any case, limits to the national identity approach.[6] Indeed, instead of taking 'national identities' as influential in determining immigration policies, others have explored the way in which national discourses intersect with age, gender, sexuality, ethnicity, 'race', and so forth, to construct multiple affinities, identities, and state policies of migration and immigration. And whatever the motivation for 'humanitarian' policies that address asylum-seekers and refugees, they do seem to bear little resemblance to the 'ethnic constitution' of at least western states. Consider for instance that despite the criminalization and subsequent incarceration of undocumented immigrants in the US or failed asylum-seekers in the UK, the acceptance rate in 2015 for countries with the largest number of asylum-seekers (namely Bhutan, the Democratic Republic of Congo, Iraq, Myanmar, and Somalia) in settlement countries, that is those countries which accept asylum-seekers as proposed by the UNHCR, increased from an average low of about 55 per cent in 2005 to an average high of about 80 per cent in 2015 (UNHCR 2016a). The point is that countries oscillate between liberalization at some times (for some) and restriction and criminalization at other times (for others), and sometimes both occur simultaneously. How do we explain these oscillations, or even the fact that the number of immigrants reached unprecedented levels in many richer countries in the 2000s? (see 'Introduction', Tables 1.1–1.3). A national identity approach seems to fall short here as a means of explanation. In the subsequent sections then, we will explore a number of different 'liberal' (that is, non-Marxist) theories that seek to elucidate the oscillations between expansionary policies (that is, those geared towards increasing migration), and restrictive migration policies.

FREEMAN'S 'CLIENT POLITICS' THESIS

In Freeman's (1995) influential notion of 'client politics', migration policies are a product of different 'clients': the "small and well organized groups intensely interested in policy ... [who] develop close working relationships with those officials responsible for it" (p. 886). Migration policy will be shaped disproportionately by the strongest clients. Freeman argues that since migrant and related ethnic minority groups are increasingly making their voices heard among governments and since certain employers will lobby for more liberal migration policies because they depend on migrant labour, migration policies are likely to be 'expansionary'. The 'clients' can be divided into two groups: those that support migration (business leaders, immigrant and ethnic minority groups, pro-migrant NGOs, etc.) and those who, generally speaking, do not support migration (nativist; in other words, nationalist or anti-immigrant groups and their related NGOs). Freeman argues that migration policies will be expansionary because the benefits of liberal immigration policies are concentrated among employers and pro-migrant groups, while the costs are diffused among everyone else who makes up the voting citizens of a nation-state. Drawing on the work of Ruggie (1982) and later Hollifield (1992), a final claim of Freeman is that the 'advanced liberal democracies' of North America and western Europe are composed of an 'embedded liberalism' which prevents the selective exclusion of migrants based on racial or ethnic grounds. He calls this an 'anti-populist' norm, and concludes therefore that a consensus will be sought across the political spectrum, instead of allowing debates on immigration to seep into party politics.

JOPPKE'S 'SELF-LIMITED SOVEREIGNTY' ARGUMENT

Freeman's analysis received considerable attention and set a certain agenda for research, while quickly attracting critics. And many of his direct critics were to be found in Europe, or at least they used European countries to explore the veracity of Freeman's claims. To begin with, Joppke (1998a) argues that Freeman's analysis, which arose out of the lobbying climate of the United States, made little sense in the context of what he called 'unwanted family reunification' in European countries after the early 1970s, when Belgium, the Netherlands, France, Germany, and the UK had halted labour and family migration. They only reluctantly agreed in the late 1970s and early 1980s to once again accept family reunification owing to the *moral and legal rights of labour migrants*. But then this was not a result of lobbying by 'client groups', but rather by the decisions of liberal courts. In fact, Joppke points out that Freeman paid little attention to the judicial dimension of states, and how the *legal process* within countries also serves to create expansionary immigration policies. Specifically, Joppke argues that judges are immune to the demands of 'clients' (e.g. anti-immigrant groups), and are therefore more preoccupied with constitutional laws and statues. Thus, rather than Freeman's 'clients' shaping migration policies, Joppke insists that the legal process (moral obligations and legal constraints) figure prominently in expansionary immigration policies. He does acknowledge, however, that these obligations and constraints change over time and *national differences* in the moral responses to 'unwanted immigration' matter, especially between northern and southern European countries. Italy, in particular, only began experiencing immigration in the mid-1970s. Yet Freeman concludes his analysis that courts' liberal jurisprudence prevents states from barring entry to 'unwanted family migrants' and/or to deporting migrants that governments see as undesirable or 'illegal'. Thus, national states limit their own sovereignty – their own ability to control increasing migration. He calls this 'self-limited sovereignty'.

Joppke (1998a) agrees further that Freeman's anti-immigration clients still matter to one degree or another to the politics and policies of migration (what he calls *political process* rather than *legal process*, as above), and thus migration policies will not always be 'expansionary'. In fact, Guiraudon (2000) argues that the 'costs' of an expansionary migration policy are not as diffuse as Freeman claims, and even Joppke's (1998a) argument that the *legal* process is expansionary could be seriously questioned by various states' reactions to the 'unwanted'. As we saw in the 'national identity' approach, governments may be as concerned then with satisfying nativist (i.e. anti-immigrant) political parties and their constituents as they are with meeting employers' demands for a certain number of low-paid or highly skilled workers. This translates into a problem for the governments who seek to legitimate more liberal migration policies to citizens in the country of immigration, and as a consequence migration policy becomes more restrictive (e.g. Boswell 2008a; Samers 1999). We will certainly revisit these issues later in the chapter, but let us first turn to the extent to which pro-migrant clients actually can and do exercise their power by drawing on a very different and unusual critique of Freeman's work from a political sociology perspective.

A POLITICAL SOCIOLOGY OF MIGRATION POLICY

Writing from the context of Italy and Europe more broadly, Sciortino (2000) raises a number of objections to Freeman's analysis, but let us specify just two of these reservations. First, for Sciortino, migration policy is only rarely shaped by resident foreigners. That is, they are not very powerful 'clients'. This is even more the case with undocumented migrants and asylum-seekers who, for Sciortino, are hardly lobbyists because they are outside the formal political system. Second, in his rejection of the political economy approach to migration policy (whether it be Marxist, client politics-based, or otherwise), Sciortino argues that political decision-makers and legislators do not simply process 'information' from 'clients' (let us say, a demand for amnesty or regularization for undocumented migrants on the part of migrant NGOs, or a demand by NGOs to liberalize family migration). Rather, he advocates a general political sociology approach, and specifically, a 'sociology of decision-making' whereby we should study how policy-makers actually do relate to lobbyists, and whether they care what lobbyists have to say (see also Guiraudon 2003; Boswell 2008a). However, whether or not policy-makers are in fact influenced by lobbyists, policy-makers are themselves shaped by class, racial, and other gender identities which in turn form what is and what is not acceptable migration policy (Samers 2003a). Furthermore, against the claims of Freeman and Sciortino, Tichenor (2015) argues that migrants themselves do in fact shape migration policies, a point he insists has been over-looked by political scientists. We will return to this question later too, but let us consider the diversity of employers who are out there.

Political sociology and Freeman's 'client politics thesis': Are all employers similar 'clients'? Government responses to the employers of undocumented migrants

In Freeman's thesis, the category of 'employer' is too simplistic. After all, some employers claim that other employers are using undocumented migrant workers unfairly ('social dumping'). 'Unfairly', meaning that for those employers or entire sectors which do not rely on undocumented migrant workers, their competitiveness is threatened by those who do. The question is then, do employers of undocumented migrants lobby governments for expansionary migration policies in order to maintain an 'unfair' advantage? Perhaps, although there is little research on this issue, in part because politicians and policy-makers may be unwilling to acknowledge that they are favouring those economic interests which rely on undocumented workers. To put it differently, are employers that use undocumented workers appeased by governments? Freeman completely ignores that most governments in the rich world are concerned increasingly with the relationship between undocumented migration and informalized employment over the last two decades, either because of the implications of 'national' labour markets becoming 'disorganized' (Wilpert 1998), because publics associate undocumented immigrants with security risks, or because governments feel the pressure among worried, anti-immigrant citizens that the government should once again be 'doing something' to stamp out informalized/illegal work and undocumented migration. Work raids and deportations act as effective symbols of this capacity to regulate both. However, the politics around informalized/illegal employment and undocumented migration are also contradictory. The lack of personnel to scrutinize errant employers, the difficulty of imposing fines and the obstacles to deporting workers, the periodic shortages of migrant workers during moments of crackdowns, and the apparent benefits to national competitiveness that stem from paying workers extremely low wages and little or no benefits, might go a long way in explaining the contradictions of regulating informalized employment and undocumented migration.

FOUCAULDIAN AND ACCOMPANYING CRITICAL APPROACHES TO MIGRATION AND IMMIGRATION CONTROL

For more than 15 years, a range of critical scholars have oriented themselves to critiquing the outcomes of national or local state policies for migrants and immigrants by tightly or loosely employing a 'Foucauldian perspective', based on the writings of the historian and philosopher Michel Foucault (Gill 2010; Kuusisto-Arponen and Gilmartin 2015; Samers 2015a). Perhaps one of the reasons for this reliance on Foucault is the widespread 'criminalization' of asylum-seekers, refugees, and undocumented migrants since especially 9/11. In this section then, we explore some difficult Foucauldian ideas, provide a brief mention of the variety of studies that have used his concepts over the last decade or so, and offer an even terser overview of other studies that have extended Foucault's concepts.

Particularly popular for thinking about the political geography of migration and immigration since the early 2000s have been Foucault's conceptions of governmentality, neoliberal governmentality, biopolitics, biopower, and discipline or disciplinary power. Governmentality is a somewhat murky concept. Foucault suggests that it is not simply the actions of government, but it involves a range of practices that include analyses, calculations, procedures, reflections, and tactics that allow for the exercise of complex forms of power and have 'population' as their object (Foucault 2007). That is, it emphasizes how governing is not simply associated with the state, but with other actors beyond the state such as airlines and airport officials, ferries and other carriers, international human rights organizations, labour intermediaries, privately run data centres and surveillance software, privately run prisons and detention centres, private security guards, and various forms of (social) media. Foucault was particularly concerned with what he called 'neoliberal governmentality' (Foucault 2004 [2008]); that is, how markets (such as the way in which private detention centres function) come to govern certain 'populations'. Yet these 'populations' such as 'asylum-seekers' or 'economic migrants' are actually *invented* populations. In other words, 'biopower' refers to the operation of power that labels, measures, and ultimately *creates* populations ('populations' that are considered to be healthy, worthy, and permitted to enter a country, or conversely, 'populations' that are considered unhealthy, unworthy, or considered a threat to, let us say, national security). Such biopower is said to be exercised through sophisticated techniques or policies (what Foucault calls 'technologies of government' or 'technologies of power'). These technologies include anything from the use of visas to deportations, in order to mark, code, measure, and regulate migrant populations. Governmentality and biopower operate through biopolitics; the rationalities of, let us say, anything from governmental departments that collect statistics on migration to the discourses and representations of the media (e.g. Geiger and Pecoud 2013). Alongside governmentality, biopower and biopolitics are 'discipline' or 'disciplinary' forms of power. For Foucault, discipline had at least two meanings: the use of particular forms of knowledge such as economics or public administration in order to govern 'bodies' (and not necessarily entire populations), but also discipline as in forcing bodies (often with punitive measures) to behave in certain ways out of vulnerability and fear. To put it differently, migration control is 'embodied' (Mountz 2004; Hyndman and Mountz 2008), and much feminist geopolitics often uses Foucault's emphasis on the governance of bodies, and focuses on how practices both within and beyond the state actually affect people they are designed to target (e.g. Hiemstra 2012).

Foucault's concepts, and especially 'biopolitics', have been mobilized in the context of the question of vulnerability and sexual trafficking (FitzGerald 2016); unauthorized migrant workers in Israel (Willen 2010); asylum-seekers in Ireland (Conlon 2010); borders (Johnson et al. 2011; Walters 2006), asylum accommodation and detention in the UK (Darling 2011); family detention in the United States (Martin 2012b); islands and migration control in the Australasian and Mediterranean regions (Mountz 2011); recession, Europe, and migration (Bailey 2013); human smuggling in Canada and the Netherlands (van Liempt and

Sersli 2013); marriage migration and the 'emotional technologies' of love in Denmark and the UK (D'Aoust 2013); surveillance and death in the Greek-Turkish 'borderzone' (Topak 2014); and in terms of the Darwinian character of EU immigration policies (Rajas 2015). Yet, despite Foucault's concern for the way in which 'bodies' were disciplined, measured, and coded, he has been called on his inattention to the category of gender within migration studies specifically (FitzGerald 2016), and we will return later in Chapter 5 to see how this shapes the discourses and practices of trafficking.

Such critiques of migration and immigration policies have been accompanied by other Foucauldian-inspired critical readings of migration and immigration policy that involve the study of so-called 'thanatopolitics' or 'necropolitics', that is a politics of who gets to live and who gets to die (Mbembe 2003). Necropolitics (as a form of biopolitics) reduces asylum-seekers and undocumented immigrants to just 'bare life'; to getting by on just enough reluctant hand-outs by the state and/or NGOs, often in camps, detention centres, squatter settlements, or the street, as Darling (2009) shows in his analysis of asylum, destitution, and 'compassionate repression' in the UK. Similarly, Vaughan-Williams (2015) has both mobilized Foucault's concepts and moved beyond them to show how 'irregular' migrants are animalized — what he calls a 'zoopolitics', which we will return to briefly later in the chapter. In sum, the strength of Foucauldian-based or Foucauldian-inspired studies are their attention to not simply how certain migrant groups are governed, but how they are created in the first place through a range of processes that cannot be limited to strictly government policies.

THE SECURITIZATION OF MIGRATION?

Those interested in a Foucauldian approach have often mobilized his work to both explain, and importantly critique the consequences of what are perceived to be the increased 'securitization' of migration and immigration since 9/11 (e.g Bigo 2002). Yet, the literature on the relationship between security and migration is diverse (Bourbeau 2013 and Messina 2014 provide reviews). Certainly, many studies are not Foucauldian at all (see e.g. Givens et al. 2009) or despite an emphasis on discourse and actors beyond the state, do not refer to Foucault's work explicitly (e.g. Boswell 2007b, 2009: Guild 2009; Hampshire 2009). But most studies in this vein implicitly call into question Freeman's (1995) neglect of the question of security for states, whether this is a matter for local governments, national states, international fora, or private bodies. Likewise, 'embedded liberalism' might seem like a mirage for those who are subject to dawn raids, placed in detention and forcibly returned to their country of origin. But what is meant exactly by the 'securitization' of migration? While it might seem obvious, it is difficult actually to pin-point an abstract definition of security or securitization (Guild 2009). Following the so-called Copenhagen School of critical security studies, Boswell (2007b) defines 'securitization' as the links made between migration and terrorism in political discourse, rather than just highly restrictive migration policies or securitarian practices.[7] In deploying Boswell's definition, 9/11 has heightened the securitization of migration policy in the United States, but not necessarily in European countries (see Box 4.1), and the securitization of migration (in Boswell's terms) in wealthier countries certainly did not begin in 2001 (see e.g. Bigo 1998; Huysmans 2000; Tirman 2004; Waever et al. 1993; Weiner 1995). Yet Hampshire (2009) claims that 'securitization' is not just "simply about making discursive linkages between a given policy issue and security. It is additionally about justifying extraordinary measures to combat the threat thus identified" (p. 118). In contrast to Boswell (2007b) then, Hampshire (2009) argues in the context of the UK that "a government-led securitization of migration has indeed occurred in Britain since 9/11" (118), but he cautions that this is not the defining logic of migration policy in Britain. Indeed, states may not want to 'frame' migration as a 'security problem' since it would create "unfeasible expectations about the state's capacity to control migration" (119).

Though most migrants have felt the force of securitization practices (if not discourses) in one way or another, this has had consequences in particular for Muslims throughout the West, who, as individuals or as a group, have been stereotyped as potential terrorists. True, the aftermath of 9/11 may have had little effect on the restriction of asylum-seekers and refugees from Muslim-majority countries (Salehyan 2009), but both within the West and all across Africa, Asia, and even the Middle East, immigration and other authorities have 'cracked down' on Muslim-dominated organizations, towns, or neighbourhoods. Innocent Muslims have been murdered or swept up in raids, deportations, closures of Mosques and Islamic organizations of every stripe and color, confiscations of property, cultural and religious humiliation, and the loss of jobs and income (e.g. Howell and Shryock 2003; Iyer 2015). Others, such as Latinos in the United States, have felt the 'collateral damage' of securitization after 9/11, even if it has been aimed principally at Muslims (Givens *et al.* 2009; Messina 2014; Waslin 2009). What is evident then, is that securitization viewed from the eyes of the state implies *insecurity* for innumerable migrants themselves.

Box 4.1 THE SECURITIZATION OF MIGRATION AND IMMIGRATION IN THE UNITED STATES AND THE EUROPEAN UNION?

The US seems to be the paradigmatic example of how political discourse and practices have been securitized and central to migration and immigration policy-making. This has entailed new laws, institutions, and programmes, including the integration of the Immigration and Naturalization Service (now US Citizenship and Immigration Services) and the US Immigration and Customs Enforcement (ICE), into the new Department of Homeland Security (2002); the creation of the US Patriot Act, the Anti-Terrorism and Effective Death Penalty Act (AEDPA) which abolished more or less "the judicial review for all categories of immigrants eligible for deportation" (Hagan *et al.* 2008: 65); the Enhanced Border Security Act, the Visa Entry Reform Act, and the implementation of the US-VISIT biometric system for non-clandestine entry (fingerprinting, body and retina scans, 'dataveillance' in the form of national digital intelligence databases, and so forth). Together, these are designed to create so-called 'smart borders' which integrate sophisticated military technology into the policing of entry at borders and transportation hubs, fast-tracked entry for the 'right' people (through the NEXUS, FAST, and CANPASS programmes) in-country surveillance, and racial or ethnic profiling based on country of origin and other criteria. At the same time, all of this, at least until the mid-2000s, was accompanied by a record number of deportations, though this is often not related to terrorism charges, but rather to the associated criminalization of migration (e.g. Amoore 2006; Coleman 2005; Cornelius 2004; Nevins 2008; Sparke 2006; Tirman 2004).

Like the US, a whole range of European-level institutions, policies, programmes and practices similar to those of the US and designed to perform the same basic functions of 'security' have proliferated. And yet, this seemingly obvious securitization of migration in the EU, which is re-told in the 'critical security studies literature', has been questioned by Boswell (2007b) as noted above in the main text. Thus, in the context of the EU after 11 September 2001, Boswell argues counter-intuitively that neither in France, Germany, the UK, nor the EU as a whole is there much evidence that migration has been linked to terrorism in political discourse. Once again, this is an important critique, but as she herself recognizes, there are national differences in how political discourses integrated questions of security *at particular moments*, and migration and security practices do become fused together to one degree or another. Likewise, as mentioned in the main text above, Hampshire (2009) argues that there *has* been a securitization of migration in the UK, especially since 9/11, both in terms of political discourse and in terms of practices, and he cites several episodes, policies, and practices that demonstrate this. For Hampshire, although the 'tabloid media' have frequently posed migration as a security threat, as has the immigration-skeptical NGO 'MigrationWatch', it is a government-led securitization that has used 'security moves' to justify for example, "increased powers to detain and deport foreign nationals, restrictions on asylum claims, a lower threshold for deprivation of citizenship, and new border control measures, including biometric technologies" (2009: 119). Such policies and practices were inscribed after 9/11 in the 2001 Anti-Terrorism, Crime and Security Act (ATCSA) of 2001. Part 4 of ATCSA authorized the Home Secretary to designate a 'foreign national' as a 'suspected international terrorist' which would then enable indefinite detention without a charge or a trial. After a series of protests, appeals, and counter-appeals, Part 4 of ATCSA was overturned by the House of Lords on the grounds that it was discriminatory towards 'foreign nationals' and thus violated the European Court of Human Rights legislation. As it became apparent in the early 2000s that 'acts of terrorism' were as much likely to be perpetrated by British citizens as

Continued

'foreign nationals', the securitization of migration waned, only to be reinvigorated ironically with the 7 July 2005 bombings in London and the 2006 Immigration, Asylum, and Nationality Act (IAN act). We say 'ironically', since those who committed the violence on July 7 were in the majority British citizens. Ultimately, the repeal of Part 4, counter-appeals by the government, and the passing of the IAN act signifies both a certain 'embedded liberalism', and a 'self-limited sovereignty' (discussed earlier in the text), but also the securitization of migration discourses and practices in the UK (see also Bourbeau 2013, on 'securitization' in France).

A second body of work, loosely in the frame of this securitization literature, interrogates the symbolism or significance of 'home' in political discourse. Sparke (2006) points out that even the word 'homeland' in the United States' Department of Homeland Security seems to signify a new form of excluding those from outside the United States. Similarly, Walters (2004) speaks of the 'domopolitics' of security in which the Latin word *domo* (meaning 'to tame' or 'break in') is closely related to *domus*, meaning home. The emphasis on 'home' is crucial, insofar as 'home' is a place of 'family', 'refuge', or 'sanctuary', and governments control migration and immigration in the name of protecting the 'home' (see also Darling 2011). As Walters remarks critically and sarcastically:

We may invite guests in our home, but they come at our invitation; they don't stay indefinitely. Others are, by definition, uninvited. Illegal migrants and bogus refugees should be returned to 'their' homes. (Walters 2004: 241)

A third body of work that also loosely connects 'security' and migration since 9/11, has focused on the upward trend in deportations in many wealthy and less wealthy countries from the 1990s to the mid-2000s (for a review, see Coutin 2015, but also e.g. Anderson *et al.* 2011; De Genova and Peutz 2010; Gibney 2008; Hagan *et al.* 2008; Hedman 2008; Hiemstra 2012; Hyndman and Mountz 2008; Mountz 2010; Schuster 2005; Schuster and Majidi 2015). In fact, deportations increased so rapidly in the 2000s that Gibney (2008) spoke of a 'deportation' turn, and De Genova and Peutz (2010) of a 'deportation regime'. While deportation may be as much associated with the 'criminalization' of immigrants as it is with 'securitization', deportation is part of the means by which the mobility of asylum-seekers and undocumented migration is controlled, and the privilege of citizenship and the 'bad illegal alien' is re-affirmed (Peutz 2006). It is important to recognize however, that despite the ubiquity of deportation as a 'technology of government', the majority of undocumented immigrants are not deported owing surely to the cost, because of logistical difficulties, and because of opposition to such practices by a range of pro-migrant groups and civil rights organizations (Chauvin 2014; Hampshire 2013). Yet why did deportations increase so precipitously and why did they become so widespread? Gibney and Hansen (2003, in Schuster 2005) argue that states see deportation as both ineffective yet absolutely necessary. 'Absolutely necessary' because it sends a number of signals to both citizens and to asylum-seekers, namely that: 1) it placates a public who may be quite vocal about the failure of government to control the number of asylum-seekers; in other words, the state has to be seen to be 'doing something' (what we referred to earlier as the problem of legitimation); 2) deportation acts as a disincentive to other would-be asylum-seekers; and 3) states can apply pressure on failed asylum-seekers and others to leave voluntarily by the threat of forced removal. Whether it actually has these effects is quite another story, which we address very briefly further below. In any case, deporting migrants is difficult and expensive. Migrants may not have the proper documents, or their actual country of origin is unknown (perhaps because they purposely destroyed their own documents), or they actively resist this deportation. And in what has come to be known as 'extraordinary rendition' in the US, migrants may be removed immediately from their places of home or work with only the possessions that they have on their bodies, placed on special unscheduled flights registered under the name of 'dummy corporations' from private (military) airports, often obscured from the glaring eyes of the media and the wider public (Peutz 2006).

In the United States, forced deportations ('removals') increased from about 40,000 annually from 1990 to 1995, to 438,000 in 2013[8] (Hagan et al. 2008; Department of Homeland Security 2013). In the UK, deportations ('forced removals' in official language) actually declined steadily between 2004 and 2014 (Blinder 2015). However, when they reached an apex in 2003, and the UK deported more than 17,000 asylum-seekers from a total of just over 61,000 applicants, the Labour government under then Prime Minister Tony Blair saw these figures as an achievement, rather than a cause for alarm. The responses have been similar in other wealthier countries. Joint bi- or multi-national deportations have become part of the internationalization of the landscape of control (Schuster 2005).

It is all too easy to see those who are deported as 'obvious criminals' and dangerous to the security of liberal western republics. Indeed, for Hyndman and Mountz (2008), public discussion of migration lumps together all migrants into a feared group consisting of 'terrorists', 'refugees', and 'economic migrants', and as a consequence migrants are stripped of their individuality. An attention to the individuality of the deported is tackled in Peutz's (2006) evocative, eye-opening and path-breaking field-work among Somalis deported from Canada and the US to Somaliland. Peutz (2006) calls for an 'anthropology of removal' to try to understand the lives of deportees in what might be called an 'industry'. Her 'anthro-pology of removal' explores the lives of the deported in both the country from which they are deported and the country in which they are forced to return. Many Somalis in the United States and Canada are convicted of very minor offences that might hardly even warrant jail time for a citizen, and we have already mentioned how Somalis may be 'kidnapped' by US immigration authorities and immediately sent on a plane back to Somalia. But what happens when they arrive in the country of origin, especially in a 'failed state' such as Somalia? Among the Somalis that she interviews who were returning to Somaliland (the northern part of Somalia), they are largely unwelcome in their country of origin, and they face the shame of returning without evidence of 'success', strangers in a strange land whose language(s) and culture(s) they have forgotten. Indeed, one of the individuals that she interviews tells of the need to find support networks upon their return that are different from those in the United States. "No more American mentality, no more Western mentality now. This is Somali [sic] now, now everybody's got to go with his own clan" (cited in Peutz 2006: 223). As Peutz puts it, "many deportees are 'returned' to a certain place and time in such a way that it can never be a homecoming for them, only another arrival" (225). In fact, Schuster and Majidi (2015) show that Afghani migrants deported from the UK and returned to Afghanistan will attempt to return to the UK because of poverty, debt, family obliga-tions and the stigma of having failed to settle successfully the first time. Even for those who are not, indeed can never be returned, the potential threat of deportation, or 'deportability' provides a strong disciplinary force (De Genova 2002).

TERRITORIALITY AND MIGRATION AND IMMIGRATION POLICIES: BEYOND 'METHODOLOGICAL NATIONALISM'

Having reviewed Marxist accounts, the 'liberal' debate on the politics or political economy of migration control, as well as 'Foucauldian' and 'securitization' approaches, we will need to switch focus to say considerably more about the political *geography* of migration control. In this case, we are concerned with territorialization (or changes in, or the effects of) the 'scales' of migration control. Recall from the Introduction to this book that we use 'scale' partly as a synonym for 'territory', and 'territorialization' as a synonym for changes in the way in which multiple territories (or scales) interact. Such 'territorialities' (from what constitutes Australian waters to what is policed as Canadian or Libyan waters) are always unstable and even regions that we might take for granted, such as 'East Asia', 'Europe', 'the European neighbourhood', 'the Mediterranean region', and 'North America', are all social constructions created by migration management strategies (Casas-Cortes et al. 2013; Mountz and Loyd 2014). These strategies are reinforced by uncritically accepted maps or unacknowledged 'cartographies of illegal immigration' (Walters 2010) which in turn serve to make some regions and borders seem natural to us, pregiven, and thus legitimate. Nonetheless, territory and physical borders matter, whether they are unstable and fluid, and they have often disastrous consequences for migrants.

Let us start with what Guiraudon (2000) and Guiraudon and Lahav (2000) call 'up-scaling', 'down-scaling', and 'out-scaling'. They formulated these terms in the context of changes in migration and immigration policy in the EU, but we can in turn employ their ideas to examine changes across the world. We will divide the remainder of the chapter into two sections: 'up-scaling' and 'down-scaling', while weaving a discussion of 'out-scaling' throughout these two sections. The term 'up-scaling' is consistent with arguments concerning the globalization or internationalization of politics (or 'global governance') of which so much has been written about in the last three decades. If we are thinking about the EU, then up-scaling involves the shifting of decision-making on migration and immigration policy from national states to supra-national institutions such as the European Commission or the Council of the European Union. This is often called 'supra-nationalism', or if it occurs outside the EU, 'global governance'. 'Down-scaling' entails shifting power or decision-making to local governments (let us say, state and local governments in the US, provincial and local governments in Canada, or regional governments in Germany and Malaysia), as well as NGOs, charities, and other bodies. 'Out-scaling' (or in a more common terminology, 'privatization' or even 'outsourcing') shifts control to private actors but also to NGOs and similar institutions. Out-scaling is not necessarily a geographically specific metaphor, but it does have geographical implications. At any rate, it should be familiar to the reader by now, since Foucauldian approaches have emphasized such non-state, quasi-state or private actors in the making and maintaining of migration and immigration policies. Let us explore and elaborate upon each of these three processes ('up-scaling', 'down-scaling' and 'out-scaling') then in the following sections.

Up-scaling

Up-scaling takes various forms, but it is often called 'global governance' or in the context of the EU specifically, 'supra-nationalization'. Below we will review five forms of up-scaling as follows:

1. The global governance of migration in general;
2. The global governance of asylum-seekers and refugees;
3. Related to 1 and 2, the 'border externalization' of asylum, refugee, and migration control;
4. The supra-nationalization of migration policies in the EU;
5. The implications that international trading blocs and institutions such as NAFTA and GATS have for the development of national migration politics and policies.

1. Is there a global governance of migration?

We are often asked by students whether there is some sort of global migration law that regulates immigration. The answer is no, if by that, one means a formal framework of multilateral (i.e. multi-state) global governance, but that does not mean global governance is completely absent either. In fact, the global governance of migration has risen to the top of national government agendas over the last three decades (Pecoud 2015), and in its wake, a sort of 'thin multilateralism' (Betts 2012) has developed, in which some forms of migration are subject to more multilateral governance (e.g. refugee movements or trafficking) than others (e.g. labour migration and family reunification). Yet even for labour migration, which might be assumed to be the singular purview of national states, there are international conventions, especially those of the International Labour Organisation (ILO) and the United Nations (UN). Major international legal instruments include the ILO's Conventions on the Rights of Migrant Workers (ILO 97 in 1949, and ILO 143 in 1975), and the UN Convention on the Rights of All Migrant Workers and Members of their Family (1990, but effective only in 2003), as well as two protocols on trafficking (the 'Palermo Protocol', signed in 2000 and implemented in 2003) and smuggling ('The Protocol against the Smuggling of Migrants by Land, Sea and Air', signed and entered in force in 2004) (Balch 2015; Pellerin 2008). In the case of the trafficking protocol, it is overseen by the United Nations Office on Drugs and Crime, but it is not involved in enforcement. Indeed, there are limitations to this global, or more accurately, *international* human rights regime. First, there is no means of enforcing adherence to these legal instruments, only a general agreement that their principles should be followed. Second, in the case of trafficking, prosecution of the perpetrators does not necessarily always translate into protection for victims, and may even lead to the deportation of these victims (Balch 2015). Third, ILO and related conventions only cover workers whose jobs "are defined as continuous and indefinite", which clearly does not cover the growing number of temporary labour migrants and their respective families. Considering the above, it seems that such conventions have only limited effects on regulating migration and immigration (Pellerin 2008: 32).

2. The global governance of asylum and refugee protection?

The apparent global governance of asylum-seekers and refugees (the 'international refugee system' or 'international refugee regime') is often considered a form of 'formal multilateralism', and the only form of formal multilateralism in the global governance of migration more broadly (Betts 2012; Goodwin-Gill 2014; Hampshire 2013; Loescher 2014). The international refugee system began with the Office of the United Nations High Commission for Refugees (the UNHCR) in 1950 as a temporary organization to address persons displaced after the Second World War and the desire to offer shelter to those fleeing persecution by nominally

Communist regimes. It therefore arose as a response by especially the US government to particular circumstances, and the US and other 'western' governments wished to limit the scope and power of the UNHCR and protection to pre-1951 *European* refugees (e.g. Loescher and Scanlan 1986). In this global refugee system, the key legislation as we noted in the Introduction to the book is the Geneva Convention (1951) and the 1967 (New York) Protocol, while other regional agreements govern poorer countries (we will discuss this later in the chapter). In the twenty-first century, the international refugee system is overseen by the UNHCR, which in 2015 had offices in 125 countries, a staff of 9,300 people and a budget of some $7 billion (UNHCR 2016b).

The principal element of the Geneva Convention is *non-refoulement* – that is the right of refugees not to be returned to a country where they risk persecution. This is inscribed in international refugee law, but the international law of refugee protection is complex, composed of innumerable (regional) treaties, the standards and practices of international organizations – not least the UNHCR, and national laws. As such, the Geneva Convention is subject to a range of exceptions (see Goodwin-Gill 2014). In any case, western governments increasingly questioned the sanctity of the Convention in the early 1990s as refugee populations began to grow across the world, owing not simply to refugee crises within Africa and the Middle East, but also the rapid rise in asylum-seekers from within Europe itself, especially Bosnians fleeing war in the former Yugoslavia. Since the early 1990s, richer states have remained reluctant to provide funds and resources to the system, have tightened asylum rules and regulations, and remain extremely cautious about accepting refugees as requested by the UNHCR. In fact, while Europe hosts some 3 million refugees (about 20 per cent of the world's refugees), many European states do not accept any refugees, and it is in the poorer world (especially in many African, Asian and Middle Eastern countries) where 80 per cent of the world's refugees (12 million) reside (UNHCR 2016a).

The relative weakness of the international refugee system and in particular the Geneva Convention has been, not surprisingly, the subject of considerable criticism. To begin with, if the 1967 protocol dissolved the racist and exclusionary character of the Convention for persons claiming asylum, it tended to privilege civil and political rights to refugees in *Europe*, rather than to economic and social assistance in poorer countries (Hyndman 2000). Furthermore, in the early twenty-first century, states have interpreted the Geneva Convention very strictly, and social protection once in Europe and North America for example, is rather limited (Hyndman and Mountz 2008). And while aid and humanitarian-oriented assistance is pursued in many signatory countries (countries that have signed the various conventions since 1951), it does not meet the needs of the people it intends to serve, or it has counter-productive effects, or it is often caught up with foreign policy concerns (Greenhill 2010). Finally, despite the Geneva Convention as international law, it is routinely violated because there is little enforcement and asylum-seekers have become subject to security concerns (Hyndman and Mountz 2008).

3. Border externalization

Zolberg (2002) uses the term 'remote control' to refer to the general process of preventing migrants from ever reaching the borders or shores of the wealthier countries *before* they can claim asylum. While the term 'remote control' may no longer be helpful since actual control involves a whole range of practices at the cartographic borders of states, including the actions of border guards, this 'externalization' of asylum control out to 'countries of transit' (see the explanation in Chapter 3) or the countries of origin has been the subject of intense discussion over the last 15 years. Such processes have been described in many ways, including the 'external dimension' (of the EU) or 'externalization' (Boswell 2003; Collyer 2007; Lavenex 2006a; Mezzadra and Nielsen 2013); as 're-scaling' tied to development assistance (Samers 2004a); as 'interdiction', 'excision', 'excisement', and relatedly 'neo-refoulement' (Hyndman and Mountz 2008); as an 'archipelago of enforcement controls' or the 'transnationalization of the state' (Mountz 2010); as 'border externalization' (Bialesiewicz 2011, 2012; Casas-Cortes, Cobarrubias, and Pickles 2013; Hiemstra 2012; Lavenex and Schimmelfennig 2009; Menjívar 2014; Mountz 2010; van Houtum 2010); as 'off-shoring' and 'outsourcing' (Bialesiewicz 2012); or as 'extra-territorial projections' (Vaughan-Williams 2015). While this idea has been developed extensively in the context of the EU, it may be employed in relations between the US government and Latin American states and migrants, the Australian government and Southeast Asian migrants, incursion into Canadian waters, and many other countries of immigration and emigration. For example, the Canadian government passed a law in 1987 (Bill C-84) that prohibited the assisting of "any

Canadian-bound migrant who was not in possession of valid travel documents" (Dawson 2014: 6). This in turn permitted the government to prevent ships from entering a zone of up to 12 miles beyond Canadian territorial waters, provided that they reasonably suspected a ship contained passengers without the right to stay in Canada. More broadly, by the 1990s, the government began increasing the number of immigration offers stationed overseas in the name of 'interdiction' in order to prevent people from entering Canada without the appropriate documents.

For asylum-seekers in particular, Hyndman and Mountz (2008) usefully refer to it as 'neo-refoulement' (as mentioned above) which "refers to a geographically based strategy of preventing the possibility of asylum through a new form of forced return different from non-refoulement, the strictly legal term that prohibits a signatory state from forcibly repatriating a refugee against its commitment codified in Article 33 of the 1951 Refugee Convention" (p. 250). Neo-refoulement involves therefore the return of asylum-seekers and other migrants to transit countries or regions of origin before they reach the sovereign territory in which they could make a claim (p. 250). As Hyndman and Mountz recognize, this is not new, but they add that it deserves more attention given how common a practice neo-refoulement has become. This neo-refoulement is manifested in a number of different countries, but it is particularly noticeable in Australia's 'Pacific Solution' (see Box 4.2), and it is also visible in the European Union's asylum and refugee regime (see Box 4.4) and over the last few years, in the management of the Syrian refugee 'crisis' (see Box 4.5).

Box 4.2 THE *TAMPA* INCIDENT AND 'THE PACIFIC SOLUTION' OF THE AUSTRALIAN GOVERNMENT

(From Hyndman and Mountz 2008[9])

Hyndman and Mountz (2008) recount the history of an Indonesian ship (the *Palapa*) with 433 (largely Afghan) asylum-seekers on board, which began to sink in the waters between Australia and Indonesia in late August 2001. The ship was heading for Australia's Christmas Island so that migrants could claim asylum there. In trouble, the asylum-seekers were rescued by a Norwegian Vessel (the *Tampa*). Upon the request of the asylum-seekers, the captain of the Norwegian ship brought them towards Christmas Island, but the Australian government denied the ship access to the territorial waters surrounding the island, warning that the captain would be charged with 'people smuggling' if he tried to land the boat. The health conditions of the asylum-seekers (including 47 children) worsened on the exposed decks of the *Tampa*. Diarrhoea, dehydration, limb injuries, skin diseases, insufficient toilets, and hypothermia presented the captain with a dilemma. Requests for medical assistance from Australian authorities were ignored initially and the government insisted that medical help would only be accorded if the *Tampa* stayed outside a 12-mile exclusion zone around the island. After three days, the captain attempted to enter territorial waters and dock the boat, but was denied access. In this stand-off within the 'get tough on immigration' re-election campaign of then Conservative Prime Minister John Howard, the Australian Navy eventually seized the boat and on September 3, the asylum-seekers were moved to the island of Nauru. As Hyndman and Mountz write, this

> signaled new realms of cruelty in the detention regime in Australia with the introduction of what was called the Pacific solution. Australia refused to land migrants arriving by sea. Instead, detention and processing was subcontracted out to small, poor islands north of Australia, including Manus, Papua New Guinea and Nauru. (Hyndman and Mountz 2008: 259)

The Pacific solution entailed the 'power of excision', whereby the Australian parliament declared that certain outlying islands of Australia were no longer part of national territory in terms of migration law. This included Christmas Island and robbed the migrants of their ability to claim asylum. And so began the start of a two-tiered strategy which involved *interdiction*, that is, intercepting migrants at sea so that they could not reach the Australian mainland. This might also involve towing boats to Indonesia (Australia has signed agreements with Indonesia and other counties to halt

Continued

smuggling operations) which is not a signatory to the Geneva Convention, and thus migrants would be unlikely to find much safe haven. The second 'tier' of the Pacific Solution involved holding asylum-seekers and other migrants in *detention centres* on one of the islands. This meant no access to lawyers or Australian legal procedures. If asylum-seekers were ill, they might be flown to the mainland, but they would remain unable to claim asylum. Shockingly, the International Organisation for Migration (the IOM) ran the detention centre on the impoverished island of Nauru, in which asylum-seekers had only sporadic access to semi-potable water; disease moved from toilets to food, and they had no means of contacting their families. As one asylum-seeker put it, "The detention camp is a small jail and the island is a big jail. All of the island, same jail. I want to get freedom" (Gordon 2005, cited in Hyndman and Mountz 2008: 261). The Australian government is accused of violating international law, and numerous elements of the Convention on the Rights of the Child. Furthermore, the United States High Commission on Human Rights has raised its objections to Australia's actions and policies in public fora. Eventually, the fate of the asylum-seekers was resolved in part by the UNHCR which stepped in to negotiate the re-settlement of these asylum-seekers with other countries. In the end, 131 of the asylum-seekers on board the Tampa eventually found refuge in New Zealand and the remaining were dispersed among various countries, including Canada.

Although we can gather a number of insights from the event discussed in Box 4.2, the point of this discussion is to highlight two processes: first, the suspension of 'normal' legal procedures, and second, a certain 're-territorialization' of the problem of asylum control to islands that do not fall under the jurisdiction of a country's migration law.

4. Is there a supra-nationalization of migration policies to and within the EU?

We return to the question of governance beyond the state, but in this case at the level of the EU, since it is the EU which comprises the only supranational asylum and migration regime in the world. It has been evolving into such a regime since at least the Treaty of Amsterdam in 1997, but the Lisbon Treaty of 2009 (and the Stockholm Programme, 2009–2014) solidified what had previously been only a creeping supra-nationalism into a more truly 'supra-national regime' where individual states cannot veto the decisions of a sufficient number of member states. In the wake of the Lisbon Treaty, the EU authorities established the CAMP (Common Asylum and Migration Policy) in 2011, which created a regime in which member states must comply with the CAMP, however limited this CAMP might be in terms of 'competencies' (specific elements of migration policies that EU institutions have the right to rule over) (Hampshire 2013; Maas 2016). Let us examine the CAMP more closely, which contains the following basic attributes

a.   National governments still determine the *number* and *type* of labour and other migrants to their respective countries from outside the EU including family members (but see Box 4.3) and asylum-seekers and refugees. However, in principle, for all these 'categories' of migrants, and especially for asylum-seekers and refugees, EU countries are bound to the CAMP. The case of asylum-seekers and refugees is discussed in more detail in Box 4.4.

b.   Initially based on the 1985 Schengen Agreement, *land* border controls within European countries were eliminated in 2007, except for the UK and Ireland, where their respective governments have instead developed a 'Common Travel Area'. Bulgaria, Cyprus, and Romania are also not included within this intra-European space of Schengen countries, but they are 'candidates' as of 2016. And despite open land borders, airports continue to separate passengers coming from Schengen countries from passengers originating in non-Schengen countries. Furthermore, the liberalization of internal border controls are being threatened by ongoing refugee movements and terrorism in Brussels, Paris, and so forth in 2015 and 2016.

c.   The creation in 2009 of a European-wide policy to impose sanctions on employers who hire 'third country nationals' (migrants from outside the EU) without the specific right to work in the EU. Not all countries have complied with this Directive.[10]

d.   The beginnings of a European-wide 'integration policy', in which the European Commission has proposed an EU-wide policy that would create a universal European permit in terms of working and residence rights.

e. The creation of a European 'blue card' (a special residence and work permit) for highly skilled migrants and offering privileged status to them, by "creating a harmonised fast-track procedure and common criteria (a work contract, professional qualifications and a minimum salary level)".[11] Thus, admission to an EU country depends on a work contract provided by that country's government. It facilitates the process of labour migration and "entitles holders to socio-economic rights and favourable conditions for family reunification and movement around the EU". It also "promotes ethical recruitment standards" to limit – if not stop entirely – active recruitment by EU States in developing countries already suffering from serious 'brain drain'.[12] Its period of validity is between one and four years, with possibility of renewal.

f. This is designed to increase the 'competitiveness' of the EU vis-à-vis other regions in the world (e.g. Asia and North America). Yet the 'blue card' has been deemed a failure by many, in large measure because national states wish to dictate the regulation of their own labour markets and are therefore sensitive to EU-level interference (e.g. Cerna 2010).

Box 4.3 FAMILY MIGRATION POLICIES IN EUROPEAN COUNTRIES: FROM NATIONAL DIFFERENCES TO THE EU DIRECTIVE ON THE RIGHT TO FAMILY REUNION

Before the 'enlargement' of the EU in 2004, and well before the Lisbon Treaty of 2009 and the development of the Common Asylum and Migration Policy, Kofman (2004) offered a unique and detailed account of family migration policies in a wide range of EU countries. She argued that the lack of existing studies that addressed such migration policies in the EU was on one hand surprising, since from the 1980s to the mid-2000s, family migration represented the majority of all legal migration from outside the EU (this had fallen to about one-third in 2011) (OECD/SOPEMI 2013). On the other hand, she argued, this lack of attention was not surprising, since there was an emphasis on labour migration in which men were associated with economic life and the public sphere, while women were associated with social life and the private sphere (Kofman 2004: 256). Whether or not such a view still prevails in lingering state policies and practices, deserves far more attention in light of the EU Council Directive on the Right to Family migration (2003), which gradually came into effect over the latter half of the 2000s, and which was not covered in Kofman's assessment. In this brief discussion, we do not address whether or not such gendered national policies have persisted nor whether the restrictive definition of 'family' has remained unchanged, precisely because we want to focus on the above Directive, which *in theory* supplants to one degree or another, national policies. The Directive created EU-wide rules for family reunification in 25 EU countries, and *in principle* determines 'the conditions' that must be met for family reunification to occur and the rights of such families. In the Directive, non-EU, but legally present residents in an EU country can bring their spouse, children (usually below the age of 18), and the children of their spouse to the EU state in which they reside (other rules exist for EU citizens wishing to bring in non-EU nationals). EU member governments can, *at their discretion*, also permit reunification with 'an unmarried partner', 'adult dependent children', or 'dependent older relatives' (EU Commission 2016).[13] If family members are brought to the EU, they are entitled to receive a residence permit and obtain access to education, employment and to vocational training on the same grounds as other non-EU migrants. After five years of residence, family members may apply for independent status, provided those 'family links' still exist in the EU. There are however, further stipulations, such as the 'respect for public order' and 'public security'. Individual states can require non-EU migrants to have 'adequate accommodation', 'sufficient resources', 'health insurance', and a 'qualifying period of two years'. Family reunion can also be denied to spouses under 21 years of age. Only one spouse can apply for reunion (thus polygamy is not recognized), 'integration measures' may be imposed (such as language requirements), and penalties for marriage fraud or 'marriages of convenience' have been imposed (European Commission).[14]

Continued

Despite the Directive, much discretion exists however, and the Directive only determines the *conditions* under which migrants have the right to family reunion, but that does not mean they will be granted family reunion. In fact, by 2008, 19 infringements had been noted by the European Commission, but rather than pursue 'non-compliance proceedings', the Commission provided further guidance on compliance through a 'Green paper' and a 'Communication' to the Council of the EU and the EU parliament in 2011 (Caviedes 2016), and a similar Communication in 2014 (European Commission 2016).[15] However, the European Court of Justice (ECJ), has intervened in striking down national provisions which seemed too restrictive and which violate the European Convention on Human Rights which is designed to protect 'fundamental rights' concerning family life and the 'best interests of the child' (European Commission 2016).[16] Ultimately, some countries were impelled to change their policies to reflect the Directive, while considerable autonomy persists in actual decision-making on individual family cases. Kofman's detailed paper on national policies therefore, still seems to have considerable relevance.

Box 4.4 CONTROLLING ASYLUM-SEEKERS IN THE EUROPEAN UNION: THE DUBLIN CONVENTIONS, CAMP, AND 'EXTERNALIZATION'

Beginning in the 1980s, increasing national restrictions on asylum-seeking were simultaneously accompanied by the implementation of the 1990 Dublin Convention to restrict and 'ease and share the burden' of asylum-seeking at the European scale (Thielemann 2004). The 1990 Dublin Convention consisted of two principal elements: 'confinement' and 'refoulement'. Confinement sought to distinguish between 'economic migrants' and 'true political refugees' by imposing visa restrictions on countries which would potentially be the source of asylum-seekers, and preventing them from ever reaching the shores of the EU – a form of 'border externalization' discussed earlier. Development aid and 'technical assistance', especially with borders and surveillance, would be increased, and border guards would be posted in far-flung regions such as the Ferghana Valley of eastern Uzbekistan where smuggling and trafficking operations in the early 2000s were developing (Samers 2004a). This was to be enhanced by 'outsourcing' and more specifically the use of 'carrier sanctions' – that is, fines which were to be levied on ferry companies, airlines, and other 'carriers' (transport companies) for bringing in an undocumented migrant (Samers 2003a). However, this use of private actors to undertake migration control has been gradually complemented or replaced by state agents placed at various points within transport infrastructures (e.g. Collyer 2007).

The second feature of the Dublin convention consists of 'refoulement', which itself has three dimensions. First, an asylum-seeker's claim must be processed in the first country where the asylum-seeker claims asylum, which is designed to reduce 'asylum-shopping'. Second, asylum-seekers are to be returned to the first 'safe country' or 'safe country of transit' (which is often not safe at all). In other words, if the Ukraine, for example, is deemed a 'safe country' by the EU, then an asylum-seeker from the Ukraine cannot claim asylum in the EU. Similarly, if a migrant from Uzbekistan travels through the Ukraine and later reaches the EU, the migrant may also be returned to the Ukraine. Such refoulement practices intensified throughout Europe during the 1990s (Samers 2003a, 2004a) and have now been extended to Syrians and others in terms of being returned to Turkey.

Alongside the Dublin Convention, the EU authorities established the European Refugee fund in 2000 to which EU institutions pledged a sum of €630 million for the period 2008–2013 in order to share the financial costs of settling, 'integrating', and voluntarily repatriating refugees.[17] As asylum-seeking climbed the agenda of EU governments in the early 2000s, Dublin II replaced Dublin I in February 2003 by building on Dublin I's principles. Dublin II was designed to clarify and instruct European countries on which country should be responsible for a decision on an asylum request. The responsible state would also have to accept the return (within a specific and limited time period) of an asylum-seeker residing 'illegally' in another

Continued

member state. For this purpose, Dublin II added the EURODAC (European Automated Fingerprint Recognition System), as well as only two weeks after 9/11, the approval of the VIS (Visa Information System, which includes such measures as iris scanning, finger printing and face recognition) to dissuade 'visa shopping' among other objectives. This was to be joined by at least two other measures and institutions or measures: SIS II (a second generation of the Schengen Information System) which involves countries in the EU that have eliminated their internal border controls (Boswell 2007b; Samers 2004a; Walters 2008), and the creation in 2004 of FRONTEX (European Agency for the Management of Operational Cooperation at the External Borders of the Member States of the European Union). Following the Treaty of Lisbon and the CAMP (Common Asylum and Migration Policy), these measures and institutions were extended with at least three policy instruments: the creation of Dublin III in June 2013 (which is applicable in 32 countries including Iceland, Liechtenstein, Norway, and Switzerland); second, the implementation in December 2013, of EUROSOR, a €250 million surveillance system: "designed to improve the management of Europe's external borders, it aims to support Member States by increasing their situational awareness and reaction capability in combating cross-border crime, tackling irregular migration and preventing loss of migrant lives at sea" (FRONTEX 2015a); and third, the replacement of the European Refugee fund by AMIF (Asylum, Migration and Integration Fund in 2014).[18] To add to this panoply of EU institutions, the Italian navy created *Mare Nostrum* ('our sea') in October 2013 which sought to control smuggling operations on the Mediterranean Sea while at the same time preventing the very tragedies recounted at the opening of this book. *Mare Nostrum* would soon be replaced by FRONTEX's 'Operation Triton' in October 2014 with a monthly budget of € 2.9 million, the agency that controlled the operation of a small number of airplanes, helicopters, and patrol boats (FRONTEX 2015b). Ultimately, such policies are contradictory, since on one hand, there is an apparent humanitarian mission of saving lives at sea, or reducing misfortune on land, but on the other, a desire to tackle 'irregular migration'. Yet tackling irregular migration may actually increase 'asylum-seeking' or even the evasion of authorities as a means of entry, which in turn leads to countless deaths in the Mediterranean. Such policies, which seem to work at cross-purposes, illustrate well the contradictions, inconsistences, and even the hypocrisy of governing Europe's malleable external borders.

How does such a system confront something as grave as the Syrian refugee crisis? Box 4.5 explores the development of this crisis in more detail.

Box 4.5 THE MANAGEMENT OF THE SYRIAN REFUGEE CRISIS IN EUROPE

We have to be clear about the term 'Syrian refugee crisis'. Whatever the arguments of national politicians and the mainstream media in Europe, the 'crisis' is less a crisis for *Europe* and *Europeans*, than it is a crisis for *Syrian refugees* in Europe, as well as in *Syria and neighbouring countries*.[19] It is especially in Syria where, since the beginnings of the armed conflict in 2011, citizens have been haunted on a daily basis by illness, malnutrition and even starvation, where violent death is always around the corner, and for those more 'fortunate', where education has been put on hold, where skills have been eroded over time by economic decline, and life has been interrupted. Even if we concede that countries such as Germany or Sweden have faced challenges in settling so many refugees, some 4–5 million Syrian refugees are also found in Egypt, Iraq, Jordan, Lebanon, and Turkey – countries with arguably less resources for doing so, and where in many cases their lives have only marginally improved, if at all. In fact, the number of refugees in Turkey by 2014 had reached about 1.8 million, compared to approximately 200,000 asylum applicants in Germany in the same year (Dahi 2014; *The Guardian* 2015a), or even compared to the additional 577,000 registered refugees in Germany by September 2015 (*Der Spiegel online* 2015a). Crucial to the regulation of the movement of Syrian refugees have been EU officials and authorities, the Turkish,

Continued

Greek, Hungarian, and German governments (not to mention other European states such as Croatia, Macedonia, and Sweden) as well as local governments and citizens (especially in German towns which have settled comparatively large numbers of refugees), the UNHCR, other NGOs, and private actors. Let us begin with the European Union. We have already discussed the role of 'externalization', which broadly speaking prevents refugees and other migrants from claiming asylum outside the cartographic borders of the EU. This means that Syrian refugees and other migrants with few resources have at least three likely choices: becoming a refugee, in Jordan and Lebanon especially, often in camps; residing in a refugee camp in Turkey or settling semi-permanently and 'illegally' outside camps (here we are thinking of the 'Little Syrias' that have sprouted in Ankara, Gazientep, Istanbul, or Izmir); being re-settled from Turkey by the UNHCR in another, probably European country, or clandestinely entering the EU where they can apply for (but not necessarily receive) asylum directly once they reach a European country. We focus on this European dimension, beginning with Greece. After crossing from Turkey to Greece, migrants have encountered the state and NGOs, especially in 'hotspots', such as the island of Lesbos (see also Box 2.3). Here, the Greek government has interviewed and screened migrants – checking IDs for fraud, and finger-printing them as part of standard EURODAC procedures. In contrast, NGOs have provided some assistance to refugees and other migrants on Lesbos and other Greek islands, but their resources have been stretched thin. Now, if migrants have applied for asylum in Greece, they must remain in Greece (or Italy if they have travelled further by boat) because of the Dublin Conventions (discussed in Box 4.4 above). However, the likelihood of receiving a positive decision on their asylum claim in Greece has been slim. After all, Greece had the second lowest acceptance rate of asylum-seekers in the EU (only 14.3 per cent) according to 2014 data (*The Economist* 2015c). If their claim is not accepted, they are either deported back to Turkey (which remained rare until Spring 2016), or thrown into the mass of failed asylum-seekers, who were likely to head north to Germany or other northern EU countries. As refugees and migrants moved from Greece in 2014–2015 on through Bosnia or Serbia, and then Croatia, they eventually reached Germany, largely on foot or by trains. As a way of addressing the large numbers of refugees amassing in southern European countries, the European Commission floated the idea in May 2015 of relocating approximately 40,000 migrants from Greece and Italy to other EU countries based on a complex algorithm, but Spanish and many Eastern European governments rejected the plan. Instead, member states agreed to voluntarily accept about 32,000 asylum-seekers, mainly from Eritrea, Syria, and, to a lesser extent, Iraq. At the same time, a number of EU governments worked with the UNHCR to re-settle about 22,000 people living mainly in Jordan, Lebanon, and Syria in the EU. These numbers were however extremely diminutive compared to the approximately 4–5 million Syrian refugees outside Syria (*The Economist* 2015e).

By September 2015, the German Chancellor Angela Merkel made it clear that Germany should welcome asylum-seekers and that Germany's basic right to asylum had no upper limits (*Der Spiegel online* 2015b). Acting quasi-unilaterally, she called for thousands of migrants in Hungary to be re-settled in Germany. Images of German citizens cheering the arrival of refugees in Munich's railway station were widely circulated in popular media. Asylum-seekers have been housed in a variety of accommodations from designated shelters to other make-shift premises, and social services – wherever available and however inadequate – have been mobilized to assist them with settlement. If asylum-seekers' claims were approved and they were given refugee status in Germany (a process which could take at least six months after arrival in Germany), they would in principle be entitled to work immediately. By October 2015, the German and Swedish governments had accepted the greatest number of Syrian asylum-seekers and refugees.

Yet for those migrants/refugees who had not yet reached Germany, increasing obstacles would lie in the way of their trek northward: they would soon confront hastily constructed fences by Hungarian authorities at the Hungarian–Serbian and Hungarian–Croatian borders in September and October 2015 respectively. The construction of these fences is not entirely surprising, given that

Continued

the then Hungarian Prime Minister Viktor Orban argued that 'European values' were in danger of being eroded by people with 'different civilizational roots' and the "'intellectual derangement' of liberals" (cited in *The Economist* 2015e: 42; *The Guardian* 2015c). As a consequence, thousands of migrants were unable to cross into Hungary and reach Germany but this has simply displaced their movement elsewhere, as described in Chapter 3. Yet, in this mixture of policies and fences, Afghanis, Iraqis and Syrians were allowed to move on to other European countries, but the window for Afghanis and Iraqis began to close by the winter of 2015, and tempers frequently flared between Syrians and others in Greece and elsewhere in the Balkan states, as Syrians were viewed as privileged relative to other nationalities. Such discrimination in favour of Syrians may in fact be 'illegal' as it violates non-discrimination clauses in the Geneva Convention, the European Convention on Human Rights, and other treaties.

In Germany, the reaction of the public and politicians has always been mixed, some supporting Merkel's relative embrace of refugees in the summer and autumn of 2015, while many opposed it. However, as early as November, even Merkel began to retreat from her earlier, more open position, perhaps for a number of reasons: strong Conservative opposition, opposition from within her own party (the CDU); vocal opposition from towns apparently stretched by their resources to provide for large numbers of asylum-seekers and refugees; and later, the Paris attacks on 13 November 2015. Reactions by the public soured further, especially after an incidence where many – apparently mainly Moroccan and Algerian (rather than Syrian men) – had sexually assaulted German women in front of Cologne's main railway station on New Year's Eve in December 2015 (*Der Spiegel online* 2016).

By early spring of 2015, migrants moving through Greece were encountering yet another obstacle, the border between Macedonia and Greece. Here, EU, Greek, and Macedonian officials were allegedly only allowing through Syrians (and to a lesser extent Afghanis and Iraqis) who were more likely to be accepted as asylum-seekers in northern European countries. As a consequence, some 12,000 migrants appeared to be stuck at the border by April 2016, and in early April, migrants were clashing with Macedonian border officials as tear gas and rubber bullets were fired at migrants attempting to break through a section of fencing. For those who could not move on, they remained in 'limbo' in Greece, some camping out in the port of Piraeus or sleeping in the streets of Athens (NY Times 2016).

In April 2016, the EU and Turkey signed a 'one-for-one' deal that 'rewarded' Turkey for reducing migrant flows by taking in one refugee from the Greek islands, while one refugee would be re-settled from a Turkish refugee camp in a European country. This included a pledge of €6 billion (more than US$6 billion dollars) to Turkey to assist refugees, and by ensuring visa-free travel to the EU for Turkish citizens and renewed talks on Turkish accession to the EU. While many politicians may have applauded this deal, or at least saw it as a necessary solution, critics saw it as a violation of the right to claim asylum protected under at least the European Convention on Human Rights.

5. Do international trading blocs and institutions have an effect on national politics of migration?

Do global or international 'neoliberal' institutions such as GATS (General Agreement on Trade in Services) and NAFTA (North American Free Trade Agreement) shape national migration politics? In other words, is there a supra-nationalization of migration policies? Let us begin with GATS. Pellerin (2008) focuses on so-called 'Mode 4' of the World Trade Organization's global-oriented GATS. Mode 4 pertains to the "temporary presence of physical persons of one member state in order to supply services to another member state" (p. 31). Services have become among the most propulsive sectors of global capital accumulation (economic growth), and poorer countries (especially those who are chief exporters of service labour such as India) are interested in liberalizing trade in services, but only if this includes Mode 4. While richer countries agreed to this stipulation, they insist that they should be able to regulate such limited mobility as they see fit, and should be able to intervene if such mobility threatens the 'territorial integrity' of their country. Furthermore, it should not be underestimated that this limited mobility is geared towards mainly high-income/highly skilled workers. In Canada for example, this translates into three categories: business visitors, professionals and intra-company transferees who receive preferential treatment such as waiving the requirement that an employer verify if a 'native worker' is available for their position. A similar set of selective criteria are to be found in negotiations between Japan, the European Union, and other poorer countries. In sum, Pellerin (2008) argues that what is called 'lex mercatoria' (commercial law), which is part and parcel of GATS, has a distinct effect on shaping (if not determining) both the definition of a migrant worker and temporary migration policies. She argues in this context that even highly skilled workers are often not entitled to benefits accorded to citizens, and human rights are dispensed with. The effect of GATS on shaping who is and who is not a migrant worker and temporary migration policies are accomplished through its legally binding measures that must be incorporated into national migration policies (Pellerin 2008).

We now turn to NAFTA. NAFTA is an example of a regional trading bloc, and was signed between Canada, Mexico, and the US in 1994. Unlike the EU, it comprises quite limited provisions for labour mobility under Chapter 16 of the agreement. Four categories of temporary migrants are involved: business visitors, intra-company transferees, professionals, as well as traders and investors. These categories only apply to citizens of the signatory countries (i.e. Canada, Mexico, or the US). For these categories of individuals, TN (or Trade NAFTA) visas are issued with a limit of three years, and renewal is possible. TN visas provide Canadian and Mexican professionals (this includes about 60 occupations) with a faster track to temporary employment in the US without having to apply for a conventional visa, at least for many Canadians (more on this in a moment). Those taking advantage of TN visas grew from 66,000 in 2004 to 99,018 in 2009. We would note however that the spouses of those on TN visas are not provided with any entry privileges and have to funnel through the usual migration channels. Furthermore, the granting of visas is discriminatory on the part of the US. In fact, from 1994 to 2004, the US government imposed regulations on Mexican professionals that were not demanded of Canadian professionals. This has been exacerbated by the comparable American and Canadian education systems and the common use of English, which may disadvantage Mexican workers (Gabriel 2010: 13; Samers 2015b).

Let us use the example of nurse migration within NAFTA to illustrate some of these points (Gabriel 2010: 2013). For Gabriel, the story of NAFTA is a lesson in the combined internationalization, neoliberalization, and the gendering of migration policy (see also Samers, 2015b). NAFTA fits within the neoliberalization of Canada's economic development, which has included both trade liberalization and 'continentalization' on the one hand and changes in Canada's economic policies on the other, including marketization and privatization. These processes and changes included substantial financial cuts in Canadian hospital funding and financial austerity that led to the intensification of work and a loss of control by health care providers. The overall result was deterioration in the working conditions for Canadian nurses, and many migrated to the US, where conditions were arguably much better. While data is problematic, US Immigration and Naturalization Service data suggests that in 1991, a total of 2,195 visas were granted to Canadian nurses for working in the US, but this had reached 6,809 by 1999 (of which most − 5,975 − were women). However, while Canadian nurses enjoyed fast-tracking privileges to working in the US during the 1990s, under the 2003 US policy called 'VisaScreen', Canadian nurses also had to meet similar credential requirements to those of all other foreign nurses (including English language proficiency and qualifying exams), and these requirements can even vary by US state. NAFTA itself does not contain any provision for the mutual recognition for nurses' qualifications, and despite that NAFTA functions to liberalize the movements of people to some extent, it does this selectively and runs up against the power of national (and sub-national) migration policies (Gabriel 2010).

Ultimately, NAFTA has liberalized to one extent or another the movement of *some* categories of workers (professionals rather than the less-skilled, the poor, and so forth) and discriminated in favour of Canadian nurses for example, until 2004. NAFTA is therefore an illustrative example of the limited supra-nationalization of migration policies, although within an 'asymmetric relationship' (Delano 2009) in which the US seems to dictate the supposedly supranational character of such migration policies (Samers 2015b).

Though with differences in the contours of labour mobility, other regional trading agreements also have labour mobility provisions alongside GATS, including ANZCERTA (the Australia-New Zealand Closer Economic Relations and Trans-Tasman Travel Arrangement), the Common Market for Eastern and Southern Africa (COMESA), Protocol II of

CARICOM (the Caribbean Community and Common Market), the Japan-Singapore Free Trade Agreement; various agreements between the ASEAN bloc countries (a large group of East Asian countries), APEC (Asia Pacific Economic Co-operation Forum), MERCUSOR/(MERCOSUL) (Argentina, Brazil, Uruguay, and Venezuela, and a number of associated countries), and SAARC (South Asian Association for Regional Co-operation). Lavenex (2007) argues that in their various ways, all of these regional trading agreements in conjunction with GATS not only demonstrate the liberalization of the movement of the highly skilled, but also how states' migration policies are shaped, but not determined necessarily, by international trading blocs and related agreements. As Lavenex makes clear, states control their own movements of highly skilled migrants but they must meet the dictates of GATS or regional trading agreements when relevant. A final point to be added here is that alongside this qualified supra-nationalization of mobility, private employers now play a substantial role in the movement of highly skilled workers, and illustrates the 'out-scaling' or 'outsourcing' that Guiraudon and Lahav (2000) have written about. For example, the UK government launched a pilot scheme in 2000 for corporate multinational firms to 'self-certify' work permits for intra-company transferees to the UK (Lavenex 2007). Yet these 'temporary workers' are a far cry from undocumented migrants and distant family members, and in their case the state wields its powerful presence together with other private actors too numerous to mention here.

Down-scaling migration control

Most studies of migration control tend to be methodologically nationalist or supra-nationalist, or they neglect the role of NGOs as gatekeepers in actually regulating entry (here, the refugee literature is an exception in its focus on NGOs). Fortunately, a whole range of social scientists have over the last 15 years emphasized the significance of *local* dimensions of restriction, securitization and criminalization, as well as the effect of NGOs.[20] Our task here is at least two-fold; first to explore more sub-national dimensions of migration policies and practices, especially whether the national identity approach is relevant in particular national contexts, including the practices of non-state entities, and second, to examine the local geographies (or the localization) of restriction, securitization and criminalization, including detentions and dispersals.

Wright and Ellis (2000b) have added an explicitly spatial argument to the well-worn critique of the idea of 'national identity' as the source of migration and immigration policies (see also Money 1999). In a unique paper, they contend that the politics of migration and immigration in the United States will be shaped increasingly by the city-region, state-based, or regional character of migration settlement in an extraordinarily diverse United States. 'Assimilation' (see Chapter 6 for an explanation and discussion) is not dead according to these authors, but the idea of an American national identity centred on 'whiteness' is to one degree or another no longer tenable, if it ever was. They caution, however, that such regionalized migration politics will increasingly be the product of the relationship between a new white minority living in regions dominated numerically and politically by immigrants, and the politics of immigrants in these same regions.

From a different but similarly spatially sensitive angle, national identity and a nativist (i.e anti-immigrant) response appear as more localized and focused on 'culture' rather than explicitly on 'race'. For example, Smith and Winders (2008) note that in Nashville, in the American state of Tennessee, the resentment among many citizens towards undocumented Latino migrants is couched in the 'neutral language' of 'illegality', but it is also about the defence of place, heritage, and 'culture', where migrants are perceived to be a threat to an 'American culture', represented presumably by 'the culture' of Nashville. Campaigns in the southern United States to make English the official language are common, but it is not just a struggle over culture; there is resentment on economic grounds as well. One 'black worker' who they interviewed resented migrants for taking 'all the benefits' while blacks[21] were citizens and received nothing. Similarly, Smith and Winders cite the Governor of Georgia who complained that "It is simply unacceptable for people to sneak into this country illegally on Thursday, obtain a government-issued ID on Friday, head for the welfare office on Monday, and cast a vote on Tuesday" (Office of the Governor, 2006, cited in Smith and Winders 2008: 67). In light of such a statement, Smith and Winders (2008) protest that, "Moving through illicit border crossings, identification theft, stolen public resources and voter fraud in one breath, these allegations of criminality invoke fear that not only Georgia but also the nation is threatened and encourage militaristic defences of 'America' at multiple scales" (p. 67). At other times, this localized politics can be confused and contradictory. For example, the Governor of the state of Arkansas supported the deportation of 'illegal aliens' but insisted also on free prenatal services for undocumented pregnant women because it promoted a 'pro-life'/anti-abortion agenda (Smith and Winders 2008). Looking beyond Nashville and the southern US, Box 4.6 continues an examination of the criminalization of Latino-origin undocumented migrants across the United States.

Box 4.6 FEDERAL POLICY CHANGE, WHITE/ANGLO IDENTITY AND THE LOCAL CRIMINALIZATION OF LATINO MIGRANTS IN US CITIES AND TOWNS

The Illegal Immigration Reform and Immigrant Responsibility Act (usually abbreviated as IIRIRA), in the United States became effective in 1996. Section 287(g) within this act allows the Department of Homeland Security to authorize state and local law enforcement officials to carry out duties that were previously accorded to only Federal officials, subject to mutual agreement between all the different levels of government and subject to 'appropriate training' and that they 'function under the supervision of sworn U.S. Immigration and Customs Enforcement (ICE) officers'.[22] This involves the issues of gang or other organized crime activity, human smuggling and trafficking, money laundering, the movement and sale of narcotics, sexual-related offences, other violent crimes, and material and other support for tackling problems in remote areas.

Yet after 9/11, Varsanyi (2008) shows how the city of Phoenix, Arizona declined the invitation of the Federal government to become partners in Federal law enforcement. The police in particular have rejected this invitation on the grounds of the excessive costs of implementing such a policing strategy. Instead, local voters have urged their governments to undertake immigration policing 'through the back door' by enacting local ordinances that prevent certain kinds of behaviour among undocumented day labourers, such as waiting at hiring locations in order to be picked up by employers. All of these ordinances are designed to rid certain neighbourhoods of undocumented migrants as 'dirty' and 'disorderly human beings' who are 'out of place' in the landscape.

This immigration policing 'through the back door' is witnessed throughout the United States. Among the most infamous cases is that of Hazleton, Pennsylvania (a small city of about 30,000 people in the northeast part of this US state). In July 2006, the city council approved the Illegal Immigration Relief Act Ordinance which stipulated that businesses and landlords would be fined for hiring and renting housing to undocumented migrants, and declared English the official language. The Mayor argued that illegal immigration had caused overcrowded schools, hospitals and social services and increased crime. A Federal District court judge ruled that the Hazleton ordinance was unconstitutional. This has had repercussions for other towns across the United States who are seeking to pass, or are considering passing similar ordinances. In the ruling against the city council, the Judge argued that "Federal law prohibits Hazleton from enforcing any of the provisions in its ordinances," "Thus, we will issue a permanent injunction enjoining their enforcement" (*New York Times* July 26, 2007).

A similar tension has arisen in the city of Carpentersville, Illinois (about 40 miles north-west of Chicago). There, the number of migrants from Mexico and Central America rose by about 17 per cent between 1990 and 2007, and about 40 per cent of the town's 37,000 residents are of Latino origin. An immigration ordinance is being sought by many of its 'white' citizens where two residents have been elected on to the village board with the support of two local newspapers brandishing an anti-illegal immigration agenda. It has divided the city, not just between 'white' citizens who reject the presence of the large numbers of undocumented migrants, Spanish, and the changing cultural landscape, but also between 'white' citizens, some who rely on immigrants as employees in their businesses. Some police leaders too are worried about the potential effects of this discriminatory legislation on their relationship with Latino residents. After a long battle, the English-only law passed in a 5 to 2 vote in June 2007 and, undaunted by any compromises in their legislation, the two residents of the village board vowed to plough on for their demands for crackdowns on employers and

Continued

landlords. Apparently such measures are pending in about 35 towns across the US (see 'All immigration politics is local (and complicated, nasty, and personal)', *New York Times Magazine* 2007). Nonetheless, as Varsanyi *et al.* (2012) show in a later paper, there is considerable variety in the way in which Section 287g is applied in the context of different cities, with some either being more reluctant to check the immigration statuses of individuals stopped for a misdemeanour, or lacking in clear procedures for when, where, and how to do so.

The regionalized restriction and criminalization discussed above can even be carried out by non-governmental organizations (NGOs), although this may be manifested in informal policies and practices, rather than ones necessarily officially approved of by states and their policies. Take the 'Minutemen' (MCDC or Minutemen Civil Defense Corps), a group of 'citizen volunteers' that patrol the US–Mexico border. Comprised allegedly of some 350,000 members (DeChaine 2009), its purpose is "to secure America's sovereign territory against incursion, invasion and terrorism."[23] It has been especially concerned with the so-called 'Border Fence Project' which has consisted of a steel security fence all along the US southern border. The group has sought to "continue to stand watch at the border and report illegal activity, build border fencing, urge local and federal officials to enforce the law and push for the enforcement of our laws to keep our country and your families and children safe" (Interview with the leader of the MCDC, cited in DeChaine 2009: 57).

Beyond a 'local' critique of a national identity approach, and the regionalization of restriction and criminalization, one might also imagine a different 'scale' of local control, and this is the perspective of Gill (2009, citing Back 2006) who points to a micro-territoriality of the securitization of asylum in the UK's Lunar House (the headquarters of the Immigration and Nationality Directorate in Croydon, south London). This is where asylum-seekers often report to claim asylum or address similar matters. Gill notes that asylum-seekers, after standing for hours in a long separate queue from other migrants, finally reach the decision hall, whereby they confront a protective plastic screen and bolted-in chairs, which do not allow them to approach the interviewer. "This means that they often have to raise their voices in a public room in order to recount their cases for asylum, which can include harrowing accounts of their experiences in their countries of origin" (p. 225). 'Security' is the explanation given for the protective plastic screen and the bolted chairs.. While these measures of 'security' are not designed to protect Home Office officials against acts of terrorism, which are highly unlikely anyway, they *are* designed as a measure of security and as a not-so-subtle indication of distrust (Gill 2009). He argues in his study of four types of asylum sector intermediaries (i.e. those working in detention centres, the National Asylum Support Service, asylum case-workers, and immigration judges) that they all have a certain discretionary authority, which, if it is directed against the wishes of asylum-seekers, is not because of the legal process or financial constraints, but because they are *steered* in a number of different ways. "This steering does not operate through the disciplining, sanctioning or threatening of subjects (although the findings of this article do not rule it out)" (p. 20); rather, he argues that these actors with their discretionary authority treat asylum-seekers in deleterious ways because state power depicts asylum-seekers in "damaging and defamatory ways, thereby depicting them as a population that is deserving of particular treatments" (Ibid.). In short, these actors are not *coerced* into treating asylum-seekers 'badly', but rather *seduced* through 'damaging and defamatory' depictions into treating them in this way. These 'depictions' are not just through discourse, however, but through the terrible 'spaces' such as Lunar House (p. 229).

Detentions

The last 20 years have witnessed an increase in the detention of asylum-seekers and undocumented immigrants in richer countries, and this has now been addressed in an extensive body of work (e.g. Athwal 2015; Dawson 2014; Dines et al. 2015; Gill 2009; Hagan et al. 2008; Hubbard 2005a, 2005b; Hyndman 2012; Martin 2012a, 2012b; Moran et al. 2013; Mountz 2010; Schuster 2005; Silverman and Hajela 2015). This vast array of critical studies range from assessing the value of Giorgio Agamben's notion of 'bare life' (briefly mentioned earlier in the chapter, and elaborated upon a bit more below), to even likening the governmentality of asylum-seekers and electronic detention in the US to the cartoon character 'Venom' ('Black Spiderman') (Koulish 2015). Others have more broadly referred to the seemingly ever-growing practice of detention as 'carceral geographies' or 'carceral spaces' (Moran *et al.* 2013). In Canada, Dawson (2014) has shown how the use of the word 'hotel' by the Minister of Citizenship and Immigration between 2008 and 2013 to describe detention facilities ('the Immigration Holding Centers' in Montreal, Toronto, and Vancouver, as well as provincial prisons) is contradictory, since the word hotel relies on a traditional notion of Canada as 'hospitable', yet detention facilities across Canada are clearly not hotels in the common understanding of the term. On the contrary, they operate as prisons, often with razor (barbed) wire, 24-hour surveillance, frequent handcuffing, metal detectors, confiscation of most personal items, separation of families by sex, solitary confinement, and strict rules for meal times. Detention centres in the UK and the US are similar to Canada in their restrictive and securitized physical and social architecture (e.g. Hayter 2004; Martin 2012a, 2012b; Schuster 2005).

In the UK, Gill (2009) demonstrates that judges, lawyers, and case-workers (including interviewers and translators) are overwhelmed by asylum cases. Under rules enacted in 2004, legal aid has diminished, and lawyers are forced to limit the time they can spend on any asylum case. Consequently, the proportion of unsuccessful initial claims and failed appeals rose noticeably, and the number of incarcerated asylum-seekers increased dramatically from about 250 in 1993 to about 3,400 in 2014 in the UK (Bacon, 2005, in Gill 2009; Silverman and Hajela 2015). Detention in Canada is also widespread and has increased over the last decade. Nearly 10,000 people were confined to detention in 2011–2012, an increase of some 1,000 people in 2010 (Dawson 2014), and in the United States, detentions reached a record high based on 2012 data: 32,953 people were held in 'administrative custody' before being removed from the US (Coleman and Stuesse 2016).

In light of the chronic use of detention, Vaughan-Williams (2015) has gone so far as to refer to the use of such detention as 'zoo-politics', and the 'animalization' of 'irregular' migrants. He has in mind the case of the transformation of a zoo – literally a former zoo in Tripoli, Libya, into a detention centre for predominantly Ghanaian, Nigerian, and Chadian migrants heading north to Europe. Vaughan-Williams' perspective is rooted in a broader set of studies that rely on the work of the Italian philosopher Giorgio Agamben who speaks of 'states of exception', in which states suspend prevailing laws, and asylum-seekers are reduced to 'bare life' in camps and similar installations (think of the Australian island detention centres mentioned earlier, but also on the island of Lampedusa)

(Dines et al. 2015). Others such as Ong (2006) question whether laws are completely suspended, preferring to call them 'spaces of graduated sovereignty', in which there are degrees of national sovereignty and local, regional, or privatized control and regulation. In fact, Guild (2009), among others, protest that asylum-seekers and refugees are not victims, but rather are 'struggling for their rights' (p. 25) in these and similar circumstances. In any case, some of these 'detention centres' are used immediately to house asylum-seekers before their claim is assessed. Some are used as 'holding centres' while a migrant's claim is undergoing appeal or has been rejected, and an asylum-seeker or other category of 'non-citizen' is awaiting deportation (e.g. Hyndman and Mountz 2008). In fact, some who have actually been granted refugee status after a history of torture, rape, or other forms of violence are also frequently detained before they are 'released' into 'the world outside'. Many countries have time limits for how long an asylum-seeker can be detained, varying from anywhere between 32 days in France to usually about six months in Germany. This is not true for Denmark, Greece, Ireland, or the UK however, where asylum-seekers can languish indefinitely. Conditions vary, and among the worst are the zones d'attente (waiting zones) at Charles de Gaulle Airport in Paris. Access to necessary information, legal support, even adequate food and sanitation is far from guaranteed. Conditions may not always be as terrible as this in other centres, but they are nevertheless prison-like structures. In many cases, governments find it difficult to even construct these detention centres, given citizens' aversion to the very presence of asylum-seekers (see Box 4.7).

Box 4.7 LOCATING AN ASYLUM 'ACCOMMODATION CENTRE' IN ENGLAND'S COUNTRYSIDE

(from Hubbard 2005a, 2005b)

Detention centres are placed somewhere by governments, and their emplacement can encounter local resistance. In 2002, the British government chose three locations as possible sites for asylum 'accommodation centres': Bicester in the British county of Oxfordshire, Throckmorton in Worcestershire, and Newton in Nottinghamshire.

Continued

Hubbard (2005a, 2005b) explores the politics surrounding the proposed accommodation centre at Newton. The site was the location of a former Ministry of Defence installation which included an airfield and military housing in a fairly isolated rural setting. The town of Bingham was about 1km away. Like the other two sites, Newton was to be an 'open' facility where asylum-seekers could leave and move about as they pleased, unlike the Oakington reception centre mentioned in the Introduction to this book, and some of the more prison-like installations discussed above. And like the other proposed sites, the Newton site would be designed to contain the full range of services in order to avoid a 'burden' on local social services, while housing about 750 asylum-seekers. This would include religious spaces for Christian and Muslim individuals, a health centre, nursery, and education centre. The British government recognized that its location was likely to stir up opposition and so it conducted a closed meeting with local government officials. The plans for the proposed site were publicized however, and a local group (the Newton Action Group) mobilized a sizeable grass-roots opposition to the plans. The local council argued that an asylum accommodation facility at the site was 'inappropriate and incongruous' in Nottinghamshire's countryside because of the visual impact of the new centre; because it would prevent the construction of homes that were needed by first-time buyers; that the centre would lead to increased traffic; and that asylum-seekers would be more appropriately housed elsewhere where they could access a range of services. The political arguments of local residents were varied, and in most cases seemed to hide racist attitudes. Some involved a dubious regard for the welfare of asylum-seekers, claiming that asylum-seekers would not have access to sufficient services and the support of friends, relatives or other community members. It is worth noting however that the Refugee council also shared these concerns. Other arguments focused on the opposition to such a centre on 'green-belt land' and the countryside more generally, as the Council argued. Hubbard notes however that there was remarkably little opposition to the earlier proposal to construct residential homes on the Newton site, but this changed with the plans to build an accommodation centre.

In fact, Hubbard explores critically the justifications of the local council and residents against the site, and he insists instead that racism and notions of 'whiteness' figured in the protest among 'white' citizens against the proposal for the Newton site. Hubbard argues that the countryside was seen as the location of a pure, white English identity in contrast to the potential arrival of 'not quite white' bodies that were deemed to be dirty, dangerous, sexually predatory, and ultimately threatening to local residents, and in particular 'white' women. The fear of trespass, rape, and other forms of violence permeated the arguments of Bingham's residents as to why the asylum-centre should not be located at the Newton site. Ultimately, the British government chose not to construct the accommodation centre, arguing that it was not suitable given problems with its 'site sustainability', including its accessibility by transport, and its location on 'green-belt' land.

The case of the Newton site underlines the significance of the meaning of places discussed in the Introduction to this book and how a local politics of migration may be just as significant as a supposed 'national' politics of migration, although in Hubbard's study, notions of 'national identity' are also invoked to exclude asylum-seekers from rural areas.

Dispersals

Let us now turn our attention to the dispersal of asylum-seekers and refugees. This is a particularly spatial tactic or strategy (what again Foucault would call a 'technology of government'). As a tactic or strategy, it is used by governments to 'lessen the burden' on particular areas of countries and/or to avoid the 'concentration' of particular nationalities in certain regions, cities or towns. It can also be used for the questionable purpose of trying to avoid racial or ethnic conflict between indigenous citizens and new arrivals, as if asylum-seekers were the cause of racism! This is simply another example of 'blaming the victim'. Often, dispersal has unintended consequences. For example, during the 1970s, Vietnamese refugees (the so-called 'boat people' fleeing the end of the Vietnam War) were taken in by the French government under the auspices of the

UNHCR and were dispersed from the Paris region to western France. After about a year, most of the refugees chose to re-settle in the Paris region in order to be closer to other Vietnamese and Asian immigrants (White et al. 1987). In 1999, the UK sought to manage the so-called 'asylum crisis' by replacing welfare payments with vouchers for clothing and food, and dispersing asylum-seekers away from their predominant settlement in London and the greater south-east to northern British cities. This was justified on two grounds. First, cheaper accommodation could be found in northern areas with a large amount of empty or very low-rent housing. Second, it would reduce the likelihood of reducing potential conflict between asylum-seekers and some of the nativist, xenophobic, and racist citizens living in a few poor southern British coastal towns, no longer attractive to domestic or international tourists, and where migrants were increasingly housed in bed and breakfast-type hotels (Audit Commission 2000; Schuster 2005). Ironically, the decision to disperse asylum-seekers and refugees to cities and towns outside the south-east led to conflicts in these new areas of settlement. For example, approximately 3,500 Kurds were settled in the Sighthill section of Glasgow, and in July 2001 a Turkish asylum-seeker was killed in clashes with local citizens (Hubbard 2005a; Phillimore and Goodson 2006).

What this section on restriction and criminalization tells us is that Freeman's use of the idea of 'embedded liberalism' appears as myopic, and is not limited to strictly national governments. Sure, there may be moments of 'humanitarian' concern, or a legal, policy, and voluntaristic commitment to a humane settlement and treatment of undocumented migrants, asylum-seekers, refugees, and others. And sure, there may be stories of refugees pleased generally with their new-found surroundings in wealthy countries. Yet these instances may seem more of an exception, rather than the rule (think of all the migrants/asylum-seekers and refugees who are now confined to Greece and awaiting deportation back to Turkey). So having explored the contours and consequences of restrictive migration policies and the criminalization of migrants, let us turn to the related issue of the securitization of migration policies.

The localization of resistance to the criminalization and securitization of migration

While certain publics and their sub-national representatives have accepted this criminalization and securitization of migration, it does not proceed without contestation. Indeed, academic attention to only top-down policy measures constructed by local, national, or international states and other actors, such as NGOs, would miss a grassroots geography of opposition (variously described as 'contentious politics', 'social movements', 'resistance', or most recently by Tichenor 2015, as the agency of immigrants) to criminalization, securitization, and more restrictive migration and immigration policies in general (e.g. Gill and Conlon 2015). Migrants are not simply the victims or passive receptors of state actions and policies. Rather, migration shapes the development of states and state policies themselves. Below we provide evidence from South Korea, France, the UK and the US.

In 1994–5 in South Korea, migrant workers along with labour organizations and religious groups (Catholic, Protestant, and Buddhist) protested for basic rights for migrant workers in the context of the then labour government. In 1995, 11 Nepalese workers continued this two-year long outcry, by protesting in front of the MyongDong Cathedral in Seoul in 1995, the traditional location for workers' struggles during military rule. Relying on both the local symbolism of the Cathedral, and perhaps more importantly the successful history of opposition to military rule by Korean workers in the 1980s, the labour government in the 1990s would eventually agree to reforms of the migrant worker 'trainee system', which sympathetic human rights organizations decried as 'exploitative'. The Nepalese in particular lamented being treated like 'animals' (recall the discussion of animalization earlier in the chapter) and demonstrated how an urban and even highly localized movement could shape the national politics of migration (Chung 2014).

In July 1996, the French government experienced some resistance from a group of undocumented Malian migrants who began protesting their impossible status as non-residents, non-refugees, but also as non-deportable persons as well. They lived in limbo. Staging a live-in hunger strike in the St. Ambroise and St. Bernard churches in Paris, they were joined by an entourage of academics, celebrities, clergy, NGOs, and a 10,000-strong march of supporters. The police eventually evicted the migrants from the two churches in order to send a message to other would-be migrants to France that they "will have no luck in France" (President Chirac, cited in Chemillier-Gendreau 1998, n.p.). However, it did prove an embarrassing headache for President Chirac's government, and the migrants sent a reciprocal message in return: this will happen again (Samers 2003a). Several years later, the French government did indeed encounter similar protests, this time around the Sangatte refugee centre (opened in 1999) near the Channel Tunnel port of Calais, which eventually contributed to the closure of the Centre in 2002. Certainly there were other reasons for its closure, such as the objections of the British government which saw it as facilitating illegal migration to the UK through the Channel Tunnel. Yet migrants did play their part in lamenting publicly the inadequate conditions of the Centre and treatment by local police. By the end of the 2000s, further protests would again erupt at the 'Jungle' in Calais, not far from the original site of the Sangatte refugee camp, and particularly from 2009 onwards. This time, it would be led by the NGO 'No Borders UK' and the No Border network, with participation by migrants themselves, against destruction of the camp by French authorities (see e.g. Millner (2011) for an academic analysis).

In the UK, the British government has faced its own troubles in acts of resistance at the then Oakington Reception Centre (closed in 2010) and the Campsfield Removal Centres near Cambridge and Oxford in the UK (Hayter 2004; Schuster, 2003). In fact, Campsfield has been the site of hunger strikers, rooftop protests, public appeals, self-harm, and suicide since the 1990s (Gill 2009).

In the United States over the last decade, migrants have become especially vociferous in protesting against US immigration policies. This has been demonstrated time and time again, but perhaps most colorfully in the 'Immigrant Workers Freedom Ride' of 2003. In September of that year, some 1,000 immigrant workers boarded 18 buses from 10 cities and set out to Washington DC with numerous stops along the way. The idea for the Freedom Ride began in Los Angeles as a burgeoning capital for trade unionism and other allied movements along the lines of race, gender, ethnicity, and so forth. In particular, the Hotel Employees and Restaurant Employees International Union, in which thousands of migrants work, became the engine room of the Freedom Ride. Washington, as the seat of national political power, became an obvious destination for the buses, given its strategic and symbolic importance (Leitner et al. 2008). As Leitner et al. (2008) point out, the immigrants involved hoped to voice certain demands in Washington DC, which included:

a. The legalization of undocumented migrants, particularly those who were working and paying taxes;
b. Easing restrictions on obtaining citizenship;

c. Demanding rights and the reform of US immigration policy;
d. Bringing back workers' rights that had been decimated under 'neoliberal' policies;
e. The respect for civil liberties and rights for all.

In the "'safe' space-time of the buses", far from the grip of political and police power, migrants shared stories of crossing the border, of the fear of deportation and the experience of discrimination (Leitner et al. 2008: 167). They sang songs, practised civil disobedience tactics, and formed a collective political identity. The various stops along the way allowed the migrants to connect with a variety of organizations, such as the local branches of unions, faith-based organizations, student organizations, job organizations, and local community groups. In this mobile movement, all of these organizations lent support, provided camaraderie, formulated politics and built solidarities in this mobile movement. They used every manner of protest available in the cities and towns through which the buses travelled, from religious services to marches. When they faced polit-ical leaders in Washington DC, however, their collective identities began to fracture between more radical and more reformist lines, and their oppositional discourses changed from one of human rights to a more mainstream motto of "hardworking, tax-paying, 'play-by-the rules' immi-grants" (Leitner et al. 2008: 168). While the Freedom Rides meant that the Immigration and Customs Enforcement office cracked down on migrant organizations such as the 'New American Opportunity Campaign' (Ibid.), it has also no doubt shaped the debates around regularizing the estimated 12 million undocumented migrants in the US in the 2010s.

A more permanent feature of this resistance in the context of the US is San Francisco's 'City of Refuge Ordinance' (Ridgley 2008). Ridgley explores and outlines what she calls the 'insurgent genealogies of citizenship' in US sanctuary cities, by which she means the history and emergence of an alter-native vision and path of citizenship. Sanctuary cities emerged in the 1980s to protect the rights of Central American refugees in the US against police enforcement of US immigration laws, and have to one degree or another successfully challenged the criminalization and securitization of migration. Similarly, in the context of the UK, Darling (2010) and Darling and Squire (2012) elaborate on the 'spatial politics' and practices of the UK City of Sanctuary movement that began in the city of Sheffield in 2007, and which – as both a movement and a network – has sought to shield asylum-seekers and refugees from at least immediate deportation. Drawing on Castree (2004), who distinguishes between a 'place-based' and a 'place-bound' politics, they show the movement served to build a place-based movement in Sheffield, and a network across the UK.

Finally, immigrants resist in perhaps less dramatic ways by choosing or not choosing certain forms of mobility. In a novel analysis, Stuesse and Coleman (2014) show how the necessity of 'automobility' in southern US states may lead to eventual arrest and deportation from the 'Secure communities' programme (2008–2014) which is an extension of section 287g of IRRIRA (discussed in Box 4.6). Through this programme, an individual who was arrested and in violation of an immigration regulation (and even some US citizens) could be provisionally detained and then eventually sent to federal prison, with the possibility of deportation. Thus, in order to avoid immobility through the fear of stops by police while driving (whether these were random or not), undocumented immigrants developed what they call an 'altermobility' that uses surrogate drivers (carpool networks and formal and informal taxis), social media and a texting 'app' called PaseLaVoz, to warn about traffic stops and share information about police checkpoint activity. As the immigrants who used these services have proclaimed, the only 'secure community is an organized one'.

While these examples may not convince the reader that asylum-seekers, refugees and undocumented migrants can exercise much claims-making power, historical evidence suggest that migrants do shape government practices – sometimes these practices have damaging consequences for all migrants, sometimes they may raise consciences and lead to humanitarian gestures in the most imperceptible and unpredictable ways.

By now then, we should have a broad understanding of migration policies in wealthier countries, and some of their contradictions and consequences. We now devote our attention to poorer countries whose migration issues and migration policies are similar in many ways to those of wealthier countries, but different enough to warrant a separate discussion.

MIGRATION CONTROL IN POORER COUNTRIES

It would be a mistake to assume that the governments and publics of the poorer countries within Africa, Asia, and Latin America are not concerned with controlling migration, but there are remarkable differences between richer and poorer countries, which we will consider in a moment. In the meantime, we should point out that poorer countries are not countries of emigration only. In fact, among the 20 countries with the greatest number of immigrants in 2015,[24] the Ukraine ranks twelfth with a migrant population of 4.8 million accounting for 10.8 per cent of the total poplation, Thailand thirteenth (3.9 million, and 5.8 per cent), Pakistan, fourtheenth (3.6 million, and 1.9 per cent) and Kazakhstan, fifteenth (3.5 million and 20 per cent). The Syrian refugee crisis (see Box 4.5) has also made a demographic impact as Turkey has become a significant country of immigration for the first time, with 2.9 million migrants (most of them, presumably, refugees) making up 3.8 per cent of the population (See Introduction, Table 1.5).

Following our earlier presentation of the 'national identity approach' towards the beginning of the chapter, one of the questions that we first ask is whether the supposed 'ethnic homogeneity' of poorer countries compared to the 'diversity' of western states, matters? Let us be clear though, we are not about to debate the relative 'ethnic homogeneity' of certain poorer countries; rather, what is important for us instead is to think about how the assumption of homogeneity might feed into migration and immigration policies. Box 4.8 considers this question.

Box 4.8 CONTROLLING THE MIGRATION OF BANGLADESHI WOMEN TO MALAYSIA: A QUESTION OF NATIONAL IDENTITY?

(Adapted from Dannecker 2005)

Can the politics of emigration from Bangladesh and the politics of migration in Malaysia be understood from a national identity perspective? In the early 1980s, certain Asian countries became major destination countries for migrant women from other Asian countries, including Bangladesh. Both Bangladesh and Malaysia are predominantly Muslim countries and Malaysia, like the Gulf states, has sought Muslim migrants as low-paid workers. However, a male-dominated migrant workers' organization in Kuwait and an Islamic organization in Bangladesh pushed the Bangladeshi government to

Continued

halt the emigration of Bangladeshi *women* in 1981, and the government began to allow only professional Bangladeshi women to emigrate. They justified this decision on the argument that "women's honour could only be protected if women were not allowed to leave their families, their communities and their 'home'" (Dannecker 2005: 657). In 1988, the Bangladeshi government reversed the 1981 decree, and migration increased significantly. However, in 1997, a new and even more stringent ban was implemented, and professional women were also prevented from migrating. Women were not allowed to leave Bangladesh without a man, and this time, the Bangladeshi government justified its policies ironically on the research of human rights organizations which warned of the dangers that migrant women face while overseas. The Human Rights organizations denounced in turn the use of their research in this way. To understand the 1981 and 1997 emigration controls, we need to look more closely at Bangladeshi society. In fact, the restriction on the emigration of women could be viewed from the 'outside' as a matter of a national (Islamic) identity, but in fact, it is as much shaped by gender relations between men and women.

Many Bangladeshi men view Bangladeshi women abroad as 'loose', sexually promiscuous, and unable to control their wishes for consumer goods and other desires when in Malaysia. This is part of a larger perspective among Bangladeshi men who see the freedom of migrating women as a violation of *purdah* (a common practice in Islamic countries whereby women are physically separated from men other than their husbands, and their bodies and faces are covered and veiled). *Purdah* remains a powerful idea and practice in Bangladesh. All of the men that Dannecker interviews refer to *purdah* as the ideal gender order, and as one of the male migrants in Malaysia claims:

> It is not good for Bangladeshi women to come to Malaysia and work here. They are without guardians. Therefore they often behave wrongly. They have contact with men, they do not dress properly and they spend their money on consumer items instead of sending it home, therefore we keep away from them. (Dannecker 2005: 660)

Bangladeshi women may have different views. Consider the thoughts of one Bangladeshi woman in Malaysia:

> See, Malaysia is a Muslim country. Nevertheless, women can work here. They can earn their own money and people do not gossip about them. On the contrary, their husbands support them and even help them with their housework. In Bangladesh, men do not work in the house, they just leave their *lungis* [a man's garment in the form of a skirt] wherever they are. Men are idle in our country. At home, we are not supposed to work and even if we do, people will say bad things. Our husbands will never support a working woman, even if he has no job. In Bangladesh, they say that a good Muslim woman does not work outside the house, but the Malays are also good Muslims, aren't they? (Ibid.: 667)

Male migrants to Malaysia bring back this image of the 'loose' Bangladeshi woman, and that shapes the opinions of even those who have never migrated. As Dannecker notes, "The bad reputation of migrant women in Bangladesh is a result of the successful transnational networking of male migrants, Islamic organizations and intellectuals" (p. 662). These organizations have gained in strength while pushing for an Islamic identity. They create a transnational space by regular visits of representatives of their organizations of migrant workers in Malaysia and the Middle East. So whereas men display with pride the material goods in their homes which they have earned from migration, Bangladeshi women tend to avoid this. As one Bangladeshi man claims:

> We are very embarrassed that our daughter went to Malaysia even though we need the money she earns. We all know how the women behave there and that the environment is not good for them. Probably she will not find a husband after coming back. (Ibid.: 660).

As a consequence, Bangladeshi men tend to avoid migrant women either in Bangladesh or Malaysia. Thus men mobilize the notion

Continued

of *purdah* to exclude women from the spaces of migration and in doing so construct a particular kind of Bangladeshi nationality and community. And in Malaysia, Bangladeshi men present themselves as representatives of Bangladeshi culture, a sort of national identity, if you will. This supposed 'national identity' is really a reflection of Bangladeshi men's wishes to change the 'gender order', as well as the practices of the Bangladeshi and Malaysian government. It is therefore a strategy to maintain male privilege. That the Bangladeshi government-implemented bans on female migration made the Malaysian government increasingly reluctant to allow Bangladeshi women to enter Malaysia. However, regardless of attempts by Bangladeshi men to impede migration, protests from civil society organizations in Bangladesh and recruitment agencies forced the government to change its mind on the ban. Still, the procedures for the migration of women are much more complicated than for men.

The national identity approach, as we have hopefully shown in the context of richer countries, suffers from 'methodological nationalism', and many studies on migration control in poorer countries continue to be methodologically nationalist, including through the collection of data that we ourselves just presented above! To be fair, it is not that national governments or national statistics do not matter in the regulation of migration and immigration; quite the contrary, and we in turn will highlight the significance of such control. In fact, El Qadim (2014) has stressed that most studies of migration from richer and poorer countries focus disproportionately on the state in the country of origin, rather than on the degree that national governments in the countries of origin can shape the migration and return of their citizens, and regulate the flow of remittances. So let us explore the national regulation of emigration and immigration in poorer countries. Like richer country governments, poorer country governments seek to balance a number of competing objectives in the context of crippling debt, poverty, unemployment, and the crisis of social reproduction across the poor world (Silvey 2009). These contradictory objectives might include:

1. Encouraging emigration for the purposes of obtaining foreign exchange through remittances. Remittances are encouraged as a means of development in lieu of overseas aid or other state-led or international programmes of assistance. Indonesia and the Philippines have special government-regulated labour export agencies to facilitate economic development (such as the Philippine Overseas Employment Administration). Some governments, such as that of Algeria and Morocco, require returning migrants to purchase local currency at customs, in order to obtain foreign currency reserves, and in Chapter 3, we mentioned the Mexican government's Tres-por-Uno programme to encourage remittances from the United States especially. Furthermore, as we showed in Chapter 3, remittances (besides being used for the construction of homes and the purchase of luxury goods) can also be used to develop a range of more socially oriented projects ('social remittances') such as for schools, religious institutions, roads, community centres, and the like.

2. Related to point 1 above, encouraging emigration as a questionable strategy to reduce unemployment (a so-called 'safety valve'), and in order that a country's emigrants will return with new skills for accelerated economic development.

3. Cooperating with the countries of origin through bi-lateral (between two states) and multilateral agreements in order to realize the benefits of the two objectives above. This may involve unwelcome and dissatisfactory compromises in foreign policies, whether economic or otherwise, in the name of economic development.

4. Discouraging the emigration of certain categories of workers in order to avoid what is perceived to be 'brain drain' or at least the depletion of particular skill-sets for specific economic sectors. A prominent example here is the South African government's concern about the loss of doctors and nurses to Europe and North America especially.

5. Protesting against the illiberal migration policies of wealthier countries, which might impede migration, and therefore hinder the objectives of points 1 and 2. The Mexican government has been particularly vocal in this regard in relation to the restrictionism of US policies towards Mexican migrants, even in the context of NAFTA (Delano 2009). More broadly, poor country governments need to balance political and economic negotiations with the countries of

immigration and the demands of its own citizens overseas in these countries. After all, a range of 'diasporic communities'[25] from Chinese to Dominicans to Filipinos shape to one degree or another the policies of the countries of emigration by a 'transnational politics' of voting and other actions (more recently, see e.g. Lafleur 2013; Pearlman 2014).

6. Curbing the continual migration of asylum-seekers, refugees and undocumented migrants from neighbouring and often poorer states as a means of quelling unrest generated by increased unemployment, or as a perceived means of preventing 'terrorism'. In terms of asylum-seekers and refugees, we have mentioned the Geneva Convention and the 1967 (New York) Protocol. However, as of 2015, very few Asian countries, in contrast with most Latin American countries, had signed both the Geneva Convention and the 1967 (New York) Protocol. Only five Asian countries had in fact ratified either the Convention or the Protocol: Cambodia, China, East Timor (Timor Leste), Japan, and the Philippines (Hedman 2008; UNHCR 2015a). There are however also regional conventions that in theory govern their movement, even if they rarely offer sufficient protection in practice. These include the 1969 Organization of African Unity Convention, and the Cartagena 1984 Declaration in Latin America. Even if states tacitly accept asylum-seekers and refugees, as in South Africa where government practices are relatively protective, there is widespread xenophobia among South African citizens (Gordon 2015).

The relative weaknesses of actual international practices in terms of protection, combined with limited national protection (despite the South African government's more welcoming approach) is extended to the lack of protection to undocumented immigrants as well. Somaliland (northern Somalia), which is not simply a country of emigration, is actually home to thousands of migrants from neighbouring Djibouti, Ethiopia, and southern Somalia. In October, 2003, three European 'humanitarian workers' were murdered in different cities in Somaliland. Although it was unclear who the perpetrators were, the government then cracked down immediately on 'illegal' migrants and deported 77,000 migrants within the span of 45 days. These 'illegal' immigrants were seen as the source of 'black magic', HIV/AIDS, drugs, and 'spoiled (i.e. inappropriate) culture' (Peutz 2006). Undocumented Zimbabwean workers have also been routinely deported from Botswana (Galvin 2015), and Southeast Asian countries such as Malaysia and Thailand have also deported large numbers of migrants, especially since the 'Asian financial crisis' of 1997–8. For example, in May 2003, the Thai government allegedly deported some 10,000 Burmese migrants back to Burma through a single border checkpoint at Mae Sot, and Malaysia also 'encouraged' or deported some 400,000 Indonesian 'illegal migrants' in 2004 (e.g. Ford 2006; Hedman 2008). Deportations and related violence is also 'out-scaled' as we saw in a previous section of the chapter. Box 4.9 below provides a discussion of this in the context of Malaysia.

Box 4.9 THE 'OUT-SCALING' OF MIGRATION CONTROL IN MALAYSIA

(From Hedman 2008)

Like the United States, Malaysia too has a certain version of the 'Minutemen' – the 'Rela' (*Ikatan Relawan Rakyat*) or 'People's Volunteer Corps'. It began as an 'auxiliary enforcement unit' – a para-public organization – set up by a 1964 government act. Its initial mission was to "help maintain security" and "the well-being" of the people of Malaysia (cited in Hedman 2008: 375). Consisting of somewhere between 340,000 to 475,000 people (in about 2006–7), this number far exceeds the number of Malaysian police and other armed officers. Over the last decade, Rela has practised what are effectively illegal 'Rela raiding parties', rounding up 'illegal migrants' in cities such as Kuala Lumpur. For example, a 'raiding force' of 300 people, joined by the Police and City Hall officials and the Immigration Department, swept through shops and restaurants in the Jalan Masjid India area in January 2001. Rela's powers were expanded in February 2005, and it has experienced closer government involvement and tacit support for its orientation towards policing 'illegal' immigration. After the 2005 expansion of powers, Rela had the right to enter people's homes without warrant, use

---

Continued

firearms, request people's immigration papers at will, and manage detention centres, which have been noted and publicly decried for abuses and their frequent use of torture. Rela members are largely insulated from government prosecution for its actions. Worse, members would be paid RM80 (about US$22) for each apprehended undocumented migrant, and not surprisingly, the number of detained immigrants has doubled from about 17,000 in 2005 to at least 34,000 in 2007. Evictions of migrants and destruction of their homes have occurred in villages outside Kuala Lumpur. Raids have continued and on one night in February 2006, a similar 'round-up' of 'illegal migrants' occurred in Selayang Baru in Greater Kuala Lumpur, where beatings by the Rela were reported. Five migrants were found dead. And often the people who are 'rounded up' are refugees recognized by the United Nations High Commission for Refugees. It is not surprising then that Rela have been accused of excessive force, arbitrary arrest, and stealing illegal immigrants' possessions during these raids. The government's Human Rights Commission have been aware of this problem and recommended ways of reforming Rela, acknowledging that some members of Rela were "taking the law into their own hands" (cited in Hedman 2008: 374), and that "We cannot condone abuse of power by Rela members and we will take steps to prevent a repeat of such incidents" (Ibid.). It appears, however, that this has been little more than government rhetoric.

7. Placating already impoverished citizens against further immigration and periodic bouts of nativism and xenophobia. For example, Klotz (2000) identified a 'new non-racial xenophobia' in South Africa in the shadow of apartheid. Indeed, in 2007, Zimbabweans became the targets of violence, simply for being migrants, and Indonesian and other migrants have become the target of anti-illegal immigrant violence in Malaysia as discussed in Box 4.9. Thus, poorer country governments also face issues of political legitimation (that is, they have to be seen to be 'doing something' about immigration to satisfy their respective citizens).

A more notorious dimension of migration policies is its gendered character, and thus the final discussion of this chapter will explore the intersection of gender and migration policies in Asian countries.

Box 4.10 MIGRATION POLICIES AND GENDER IN ASIAN COUNTRIES

As with other macro-regions or countries, laws and policies in terms of both emigration and immigration in poorer (or at least *less wealthy*) countries are either clearly discriminatory or they are 'gender-neutral' but have gendered consequences. There are commonalities in terms of emigration and immigration policies across Asian countries, again recognizing that we are examining these issues through a national lens.

Migration policies *to* Asian countries can be characterized overall as limiting quantitatively the number of labour migrants, limiting the duration of migrants' residence, and limiting the ease with which they can settle through the absence of 'integration' policies, especially those relating to family reunification (Piper 2004, 2006; Seol and Skrentny 2009). From their research in Japan and South Korea, for example (clearly some of the wealthiest countries in Asia), Seol and Skrentny (2009) point out that there is in fact little settlement in Asian countries, because of the significance placed on economic growth rather than rights, the lack of family reunification policies as noted above, and the perception that there is considerably more immigration control in Asian countries. Likewise, if one includes Singapore here (again, recognizing its comparative wealth in relation to other Asian countries) then policies sort migrants' entry and settlement rights based on *skill*, certainly echoing many 'western' countries as well (Yeoh and Lin 2013). Most legal migration for women *to* Asian countries is limited to often two-year domestic labour contracts tied to a single employer, reflecting the stereotyping of women as domestic workers or carers. Since their contracts are not permanent, and women work in the

---

Continued

private sphere, national labour laws do not apply. This, as we will see in Chapter 6, can have the most detrimental consequences for migrant workers (Piper 2004, 2006).[26] As Silvey (2004b) notes, Asian governments have remained relatively silent on the abuses that occur in the private sphere, and it has been left to NGOs to address their often muted suffering. More generally, the policies and societal practices in both the countries of origin and destination in Asian countries have meant that the choices for migrant women are very constrained (Piper 2004, 2006). Think, for example, of Bangladeshi women in Malaysia as discussed in Box 4.8.

We can divide a discussion of migration policies into state and non-state actors (including labour brokers) and their interests; national and international human rights instruments, and the role of NGOs (Piper 2004; Lindquist et al. 2012). Concerning the first, that is, state actors, most migration policy seems to be delivered 'top-down' by governmental elites. Piper argues that the majority of government officials cannot relate to the lives of migrant women and especially poorer migrant women, and officials are viewed as 'aloof', 'arrogant', or 'disinterested' (p. 221). States are heavily involved (as noted in point 1 of the 7 general characteristics of state objectives in migration policy above) in organizing the emigration of female migrant workers to countries such as Saudi Arabia, but illegal 'labour export agencies' also exist, and most Asian governments, especially in Indonesia, either do not have the resources to clamp down on these informal agencies, or simply tolerate them since they seem to serve the aims of national economic development, and/or conform to what is *generally* accepted treatment of poor women in both the country of emigration and immigration. Often police and other officials are involved in illegal cooperation with these informal labour export operations (Silvey 2004b; Rodriguez 2010; Rudnyckyj 2004). Alongside and within these 'illegal export agencies' are 'labour brokers' who consist mainly of men, but also increasingly women with a history of labour migration as domestic workers themselves (Lindquist et al. 2012). Such 'actors' in the governance of migration should remind us of the Foucauldian literature which focuses on non-state actors. In any case, the result of these quasi-state and private actors is anaemic protection for emigrant women overseas, and again the conditions of Indonesian women in Saudi Arabia comes to mind (Silvey 2004b; Rudnyckyj 2004). Although governments are generally concerned about the welfare of their citizens overseas, they are usually in a weak bargaining position vis-à-vis the country of destination. Bi-lateral agreements are rare, and if there is mistreatment of a government's citizens abroad, there is little they can do, especially if the country of immigration threatens to deport the migrants, which would damage the gains from remittances (Piper 2006).

Migration *to* individual Asian countries has become a battleground between different Ministries. On one hand are the Ministries of Labour, which are typically more in tune with the needs of labour migrants (for example, through the creation of labour protection standards and labour complaint desks, as in Hong Kong, Japan, and Singapore), and the Ministries of the Interior, which are divorced from the concerns of migrant labour and are more interested in security issues (Piper 2004, 2006). Cooperation between Asian countries on 'irregular'/undocumented migration – again *to* Asian countries – exists, and is designed in theory to address smuggling and trafficking especially, but such agreements are not legally binding, as they often are in the European Union. Likewise, since the mid-2000s, cooperation has increased on labour migration more broadly, including the Global Forum for Migration and Development (GFMD), the 'Columbo Process' (a consultative process formed in 2003 on the protection of migrant workers overseas whose latest meeting was in 2011) (Colombo Process 2016), and the 2007 ASEAN Declaration on the Protection and Promotion of the Rights of Migrant Workers (IOM 2010). However, like the Bangkok Declaration, such agreements or 'consultations' are neither legally binding nor enforceable.

In terms of non-state actors, NGOs and voluntary organizations are proliferating across Asian countries. A number of NGOs, such as Migrant Forum in Asia based in Quezon City, Philippines, and CARAMASIA (Coordination of Action Research on Aids and Mobility) in Kuala Lumpur, have established extensive international networks from Japan to Jordan, for assisting migrants (especially women)

Continued

with their problems of welfare, education, documentation and legal needs, both in the countries of origin and destination. Likewise, migrants themselves have formed religious or ethnic-based mutual aid associations, and this is very common among the large numbers of Filipinos overseas. Depending on the country, home-and-abroad NGOs linking the entire migration chain are emerging in countries such as the Philippines, but are weaker in countries such as Cambodia and Vietnam. The point is that these NGOs are a bottom-up and often grassroots response to elite-driven migration policies (Piper 2004, 2006).

National and international instruments such as the 1990 UN Convention, mentioned earlier in the chapter, has been ratified by only Indonesia, East Timor (Timor-Leste), Philippines and Sri Lanka. The convention could offer a minimum level of protection for migrant workers, but given the low levels of ratification in Asian countries and the rest of the world, the fate of migrant workers is left to the generosity of NGOs and national governments' labour laws overseas, and even then there is a problem of enforcement. As one of the members of a pro-migrant worker NGO stated, "There are all those laws out there, but then when it comes to reality, there is nothing" (cited in Piper 2004: 224).

CONCLUSIONS

Like migration itself, the answers to why states pursue the particular migration policies that they do and the character of these policies is complex. In this chapter, we reviewed a number of approaches to migration politics and policy as a means of illustrating the often contradictory and competing objectives which are condensed into what might be broadly called 'migration management' policies, practices, and strategies. In the last part of the chapter, we moved away from theoretical perspectives in order to emphasize actual policies and practices in poorer countries. Some of the theories that have been devised in exploring richer countries may certainly apply to poorer ones, but there are limits as well, and this should be acknowledged. As we saw in Chapters 2 and 3, such theoretical approaches are often incompatible, some are complementary and some overlap. The important point to be gathered here is that migration is not a 'natural' phenomenon decided upon by magical economic laws or just the individual decisions of migrants. National states (including in the countries of origin) both control and are shaped by migrants, markets, non-state actors, international institutions such as in the EU or NAFTA, and sub-national or sub-state territories. It is difficult to say in any case which territorial forms of governance might matter more than another, though clearly national control has not disappeared, whatever the demands for the end of 'methodological nationalism' in the study of migration and immigration policies. It would seem in any case that neither supra-nationalizing nor conversely localizing migration policies necessarily address the plight of less fortunate migrants. Rather, it is the actual *content* of migration policies, and not *necessarily* the precise territorial scale from which they are conceived or emanate that matters. The content of national migration and immigration policies seem to privilege the highly skilled and the wealthy as a general competitive strategy, and exclude the legal migration of those who are not deemed to have the right 'skills'. Asylum-seekers and refugees are accepted in certain conditions and with certain limits, but protection as idealized and codified in the Geneva Convention and related conventions, whether in richer or poorer countries, fail chronically, and fail considerably short of their initial and stated intentions, whether in quantitative or qualitative terms. Yet disadvantaged and racialized/ethnicized migrants are not simply and always everywhere, the passive victims of criminalization, exclusion, oppression, and securitization; they themselves change the nature of migration policies (though often in ways which might not benefit them), and continue to organize and clamour for greater rights, and it is to these rights and citizenship that we will devote our attention in Chapter 6. In the meantime, we move to Chapter 5, in which we will spend some time on the survival of migrants once they break through the barrage of migration and immigration policies and actually settle in their country of destination. In other words, it is to the question of work and employment that we turn our attention in the next chapter.

FURTHER READING

There are countless studies of the state and migration. For broad theoretical or empirical surveys, Boswell (2007a), Geddes (2015); Hampshire (2013);

Hollifield (1992), Massey (1999), Meyers (2000) and Samers (2003a) provide reviews of theories of migration policy from Marxist, nationalist, and other liberal political economy approaches. For a sweeping review of Foucauldian-inspired approaches, one might refer to Fassin (2011). For rather conventional studies of migration and immigration politics based on national case studies around the world, including Korea and Japan, see Hollifield et al. (2014). Chapter 8 in Castles et al. (2014) offers excellent overviews of legalization programmes, refugee politics, and the intersection of migration and security-related issues. Boswell and Geddes (2011) in turn provide a thorough study of migration and mobility within Europe, as well as the importance of European-wide institutions. Gabriel and Pellerin (2010) have produced an accessible edited collection on the governance of labour migration specifically. Massey et al. (2002) have written a comprehensive historical study of the character and consequences of migration control between Mexico and the United States up until about 2000, and this is updated in Delano (2009) and Samers (2015b). Nevins (2008) provides a more recent and 'human' account of the US–Mexico border. Castles (2004) and Zolberg (2006) offer instructive overviews of the problems that states face in managing migration. Kofman (1999, 2002, 2004), Piper (2006) and Willis and Yeoh (2000) provide very useful reviews of the relationship between gender and the politics of migration. For theoretically informed, insightful, and spatially explicit studies of migration control that are sensitive to locality, at least in the UK and the US, see Hubbard (2005a, 2005b). Darling (2010) and Darling and Squire (2012) for the UK, and Leitner et al. (2008), Smith and Winders (2008), and Varsanyi et al. (2012) for the US.

SUMMARY QUESTIONS

1. Explain why migration policies in the wealthy countries are not simply restrictive.
2. Discuss some of the ways in which migration policy has been 'up-scaled'.
3. What are the Dublin Conventions and how do they effect asylum-seeking?
4. In what ways is migration policy a local affair?
5. How are migration policies gendered?
6. Discuss some of the differences between migration policies in wealthier and poorer countries.

NOTES

1 These approaches are reviewed, and their strengths and weaknesses evaluated, in Boswell (2007a), Hampshire (2013), Hollifield (1992), Massey (1999), and Meyers (2000). Our account is far from exhaustive, and Meyers (2000) discusses a rich literature which he calls a 'domestic politics' approach, an 'institutional and bureaucratic politics' approach, and a 'realism and neorealism' literature. I only touch upon some of these ideas but I do not address them in any comprehensive fashion'.

2 The Marxist term 'capital accumulation' is the rough equivalent of 'economic growth' in conventional (neo-classical) economics.

3 For those unfamiliar with the term 'bourgeoisie', it arose to prominence in the nineteenth century and referred to a long-ascending middle and upper-middle class of business people and property owners who gradually came to supplant feudal lords and aristocratic land owners, and who, in Marxist terms, exploited the mass of workers employed in the 'dark satanic mills' of industrial capitalism.

4 It is very questionable whether migrants are uniformly expected to be 'male' everywhere. However, the point that Smith and Winders are making is that in the context of the US, and for US employers, women who are pregnant are undesirable because they provide a 'drag' on productivity.

5 'Nationhood' is a rather vague concept. For our purposes here, we mean how the history and meaning of a particular 'nation' is understood.

6 Meyers (2000) himself provides a systematic critique of this approach.

7 Studies of security and migration tend to focus on the security of the state and citizens, rather than on the 'insecurity' of migrants (Castles and Miller 2009). This is a nice way of re-conceptualizing and re-focusing the 'securitization' literature, but our concern in this section, however, is not with migrants' insecurity, since We will elaborate on the conditions that migrants face in countries of immigration in Chapter 5. Rather, we will be preoccupied with how states' obsession with the threat of terrorist violence (and the use of the word 'security') assumes different spatial forms. Castles and Miller (using Adamson 2006) are also interested in emphasizing how international migration need not necessarily always be viewed in opposition to security, since states rely on migration for 'economic security', for translators who can help during a time of war, for addressing demographic decline, and so forth. Again, this is a welcome and expanded conception of security, but it is not the focus of this section.

8 According to the US Department of Homeland Security, "Removals are the compulsory and confirmed movement of an inadmissible or deportable alien out of the United States based on an order of removal. An alien who is removed has administrative or criminal consequences placed on subsequent reentry owing to the fact of the removal. (Table 39. ALIENS REMOVED OR RETURNED: FISCAL YEARS 1892 TO 2013, n.p.) https://www.dhs.gov/sites/default/files/publications/ois_yb_2013_0.pdf

9 Our discussion below draws almost entirely from Hyndman and Mountz, but also the analysis by Amnesty International (see http://www.amnesty.org.au/refugees/comments/2247/).

10   See http://europa.eu/rapid/press-release_IP-12-166_en.htm?locale=en

11   http://ec.europa.eu/dgs/home-affairs/what-we-do/policies/legal-migration/work/index_en.htm

12   Ibid.

13   See http://ec.europa.eu/dgs/home-affairs/what-we-do/policies/legal-migration/family-reunification/index_en.htm

14   Ibid.

15   Ibid.

16   Ibid.

17   See http://ec.europa.eu/dgs/home-affairs/financing/fundings/migration-asylum-borders/refugee-fund/index_en.htm

18   http://www.ecre.org/ecre-publish-note-on-the-asylum-migration-and-integration-fund/

19   As Mountz and Hiemstra point out, the term 'crisis' is "first and foremost a discourse of states" (p. 383).

20   To begin with, we need to distinguish between restriction and criminalization which are in fact two distinct but also connected processes. The first might be defined as limiting the number and/or specific categories of migrants permitted to enter a country, and the second, to a range of laws, policies, programmes, and practices which treat those who violate asylum and migration policies as 'criminals' and creates a more general atmosphere of suspicion about all migrants' behaviour and motives.

21   The term 'blacks' is used for US-born citizens who identify as such. Here, we avoid the use of the term 'African-American' (commonly synonymous with US-born 'blacks') so as not to confuse the reader with foreign-born individuals who might be born in the Caribbean or Africa, for example.

22   See the Customs and Immigration website: http://www.ice.gov/pi/news/factsheets/070622factsheet287gprogover.htm

23   http://www.minutemanhq.com, cited in Nevins (2008: 170).

24   Absolute numbers of migrants are based on 2015 figures from the Migration Policy Institute (see http://www.migrationpolicy.org/programs/data-hub/charts/top-25-destination-countries-global-migrants-over-time). Numbers are rounded, and are notoriously difficult to compare given definitional differences between countries. For additional and similar statistics, see http://www.un.org/en/development/desa/population/migration/data/estimates2/estimatesgraphs.shtml?3g3

25   We will discuss diasporic communities in greater detail in Chapter 6, but for now we can define them as a particular national or ethnic group of migrants who emigrate and re-settle in a number of different countries throughout the world, and who regroup or re-organize as co-nationals or co-ethnics in the countries of immigration.

26   One might also look at the numerous and more recent papers by Brenda Yeoh and her colleagues in Singapore and elsewhere on domestic workers and rights in Singapore.

5

GEOGRAPHIES OF MIGRATION, WORK, AND SETTLEMENT

INTRODUCTION

The writer Mike Davis (2006) paints an eye-opening and at times frightening picture of the ever-expanding city of Dubai, in the United Arab Emirates (UAE). Somehow invisible to most people who visit the gleaming new shopping malls, extraordinary mega-projects, and vertiginously high skyscrapers of Dubai, are its migrant workers, mainly from Bangladesh, India, Nepal, Pakistan, the Philippines, and Sri Lanka, who have laboured as contract construction workers and constitute something to the tune of 25 per cent of Dubai's workforce. It was estimated in the mid-2000s that some 700,000 migrants had entered the UAE for work (Buckley 2013). While the elite and the middle classes live out their more luxurious air-conditioned daily lives, many migrant workers toil six days a week, for twelve hours a day in the heat of this desert city. Racial or religious discrimination is common, as are the close watch of security guards and spies within the workforce. Employers sometimes disappear and never pay the required wages or they hold wages from their employees for months ('wage theft'). Migrant workers live in squalid quarters, in effect barracks that sometimes house up to 12 people in a room. Working toilets and air conditioning are an unheard-of luxury; in many cases, so is running water in remote and segregated desert camps from which workers are bussed to construction sites in the centre of Dubai. The term 'contract-worker' is employed, which refers to the Gulf-wide *Kafala* system; a so-called 'sponsor system' in which the legal and economic responsibility of every foreign worker (from construction to domestic work) must be ensured by a UAE citizen for the duration of the contract. In this system, obtaining citizenship or even permanent residency is nearly impossible (e.g. Buckley 2013; Pande 2013), and as Mike Davis remarked, this sponsorship system is little more than a euphemism. Passports are often confiscated at airports by recruitment agents; visas control their movements as they are tied to a particular employer. Migrant workers are effectively banned from up-market shopping malls, golf courses, and expensive restaurants. The United Arab Emirates does not observe the International Labour Organizations' labour regulations and has refused to be signatory to the International Migrant Workers convention. Human Rights Watch estimated that perhaps more than 800 people have lost their lives in construction work, covered up by the government and unreported by companies. In fact, their lives are harsh to the point that much of migrant workers' social reproduction in terms of food, hygiene, and health is usually left to expatriate elites in the form of small charities, as well as international construction companies. The work of charities and the provision of companies have been tolerated by Dubai's municipal government and the police, precisely because they concern workers' *corporeal* (bodily) needs, rather than their political desires (Buckley 2013; Davis 2006). Davis explains the government's attitude and practices towards its migrant workers:

> Dubai's police may turn a blind eye to illicit diamond and gold imports, prostitution rings, and shady characters who buy 25 villas at a time in cash, but they are diligent in deporting Pakistani workers who complain about being cheated out of their wages by unscrupulous contractors. (Davis 2006: 66)

It however does not stop at, let us say, Pakistani construction workers; Dubai's police may even throw Filipina maids into prison on the grounds of 'adultery' if they report that their employers have raped them. At the same time, weary of Shiite unrest in neighbouring Bahrain and Saudi Arabia, the governments of Dubai and the other emirates of the UAE have preferred a non-Arab labour force from mainly South Asia. Yet as these same immigrants became politically active during the mid-2000s, these governments implemented a 'cultural diversity policy' which reversed course and turned back to Arab workers to 'dilute' the Asian work force (p. 66). Employers however have failed to find the citizen workers that have been required by the government. Citizens simply did not wish to work for the $100 to $150 a month that construction firms paid in the mid-2000s. In fact, around the same time, migrant workers began to strike, mainly to protest their low pay and wage theft, the squalid, ill-sanitized, yet expensive accommodation, and inadequate transportation to and from work (their only source of transportation). By 2007, some 30,000 to 40,000 workers were involved in a two-week strike. Paradoxically, it was precisely the segregation of urban space that served both to strengthen the protests and to quell them. On one hand, workers galvanized around common issues, word spread easily between camps, and protests would oscillate between the main routes that led to the downtown construction sites, the sites themselves, the Ministry of Labour in the centre of Dubai, and the labour camps. Migrant workers suddenly became visible to the wider population and NGOs. On the other hand, in the wake of these strikes and facing concerns by construction companies in particular, the UAE government began to tolerate strikes in the private spaces of the labour camps, rather than in downtown areas. In fact, the government used the camps as sites of mediation over pay and working conditions, so as to avoid more public forms of protests. Indeed, the labour camps proved fertile ground for co-opting labour leaders by paying them off and containing the strikes. The segregation of urban space therefore led to the containment of workers' political protests (Buckley 2013), while Dubai has continued to expand on the backs of migrant workers.

Dubai may be an egregious example of the use of migrant labour to build capitalism's cathedrals, but migrant workers, whether they are extremely low-paid as in Dubai's case or on the other end of the spectrum, highly paid Indian professionals in Hong Kong, London, New York or Singapore, are central to the functioning of economies (see Table 5.1 which provides data on the number and percentage of 'foreign workers' in OECD countries). Harald Bauder sums it up concisely in the opening paragraph of his book:

Imagine if you will, that, on the same day, all migrants and immigrants decide to return to their countries of origin. The Filipina nanny would pack her bags and leave the family in Singapore whose children she has been raising. The suburban couple in San Diego would be without their Mexican gardener who worked for less than five dollars an hour. Italian farmers would find the fruit rotting on their trees because their cheap migrant workers left the orchard. New York's manufacturing sector would collapse because a large portion of the workforce is absent. Worse, Wall Street would be closed because cleaners, security guards, office staff, and taxi drivers are unavailable. Many sectors of the economy in industrialized countries would come to an immediate standstill. The rest of the economy would follow within days, if not hours. Although not your typical doomsday scenario, this hypothetical example illustrates that our economy depends on the labour of often 'invisible' international migrants. (Bauder 2005: 3)

How and in what ways do migrants become integrated into the economic activity of other countries? To begin with, it might be useful to distinguish between explaining migration and explaining *labour* migration. As we learned in the introduction to this book and the previous chapter, people migrate for innumerable reasons. If people migrate *specifically* for work purposes, this may involve at least four channels. First, either people decide to migrate for work because they expect 'better' employment conditions, including higher wages, and these expectations are shaped by various media (the internet, print media, TV, radio, and so forth) which announce work in particular countries, regions, cities or towns. Second, people migrate because friends, relatives, and acquaintances alert them to work overseas, and may provide the necessary accommodation and food, at least for a limited period of time, while the migrant 'finds their feet' and lands that expected or promised job. Third, people may emigrate simply because an employer has recruited them directly, often with the help of friends and relatives, private labour agents, or the governments in the countries of origin, which establish state-sanctioned labour recruitment or 'labour export' agencies to organize the large-scale movement of workers (as discussed in the previous chapter). This is the case especially with the Filipino and Indonesian governments for construction workers and nurses, but also a range of jobs across the occupational spectrum. Yet recruitment may also be illegal as is also common all over the world.

Table 5.1 Labour market participation (employment and unemployment rates) 2013 for native-born/foreign-born men/women

	Employment rates				Unemployment rates			
	Native-born men	Foreign-born men	Native-born women	Foreign-born women	Native-born men	Foreign-born men	Native-born women	Foreign-born women
Australia	78.0	77.8	68.6	62.0	5.9	5.8	5.0	6.0
Austria	76.7	72.7	68.9	58.5	4.4	10.4	4.5	9.3
Belgium	67.5	60.5	59.7	45.3	6.8	18.2	6.8	16.0
Canada	74.9	76.6	71.0	65.2	7.5	7.9	6.2	8.3
Chile	71.0	83.3	46.6	66.7	6.6	4.1	8.7	3.7
Czech Republic	75.5	80.6	59.6	58.4	5.9	7.3	8.4	9.7
Denmark	76.0	67.3	71.7	59.1	6.4	11.4	6.5	13.5
France	68.1	66.4	62.2	48.7	9.2	15.9	8.9	16.4
Germany	78.1	77.2	70.8	59.8	5.1	8.3	4.5	7.9
Greece	58.0	56.3	40.0	39.5	23.2	37.3	30.7	38.9
Hungary	63.4	78.4	52.5	58.3	10.4	7.4	10.1	12.8
Italy	64.2	68.6	46.1	49.4	11.1	15.9	12.4	17.5
Korea	71.6	80.9	49.7	50.8	3.2	3.3	2.9	5.8
Mexico	78.3	68.2	45.0	39.0	5.1	6.9	5.1	6.8
Netherlands	80.1	68.8	72.4	55.4	6.3	13.2	5.4	12.2
Poland	66.6	69.5	53.4	47.7	9.8	5.7	11.2	21.1
Portugal	63.4	64.1	57.6	61.3	16.4	22.5	16.5	21.0
Spain	60.3	53.2	50.7	48.4	23.5	37.4	25.2	34.1
Sweden	78.3	67.4	75.9	58.5	6.6	17.0	6.4	15.8
Switzerland	85.2	83.3	77.0	68.5	3.2	7.2	3.0	8.3
Turkey	69.6	63.9	29.6	33.0	8.1	10.2	10.8	11.5
UK	75.2	76.7	67.1	59.0	8.2	8.1	6.7	9.8
USA	69.3	79.6	62.2	57.4	8.2	6.5	7.2	7.6

Source: OECD (2015)

In the case of highly skilled immigrants,[1] this is often accomplished through private firms or public organizations such as national health services in European countries. Fourth, and related to all of the above, migrants may be trafficked, that is, they are smuggled into another country by agents of labour brokers to work in a particular setting, usually to pay back the 'cost of smuggling'. This is common even among highly skilled immigrants who are fleeing their countries for a variety of reasons; they may claim asylum, they may not. In any case, they often end up doing low-skilled/low-paid work (more on this later in the chapter).

The purpose of the remainder of this chapter however, is less to describe and explain labour migration, and more to illustrate how and why migrants are found in certain types of economic activities and employment, and how and why they experience the working conditions they do. The chapter proceeds as follows. First, in order to explore the working practices and experiences of migrant workers, we begin by revisiting some theories that attempt to explain the employment outcomes of migrants. Three broad competing perspectives stand out: human capital theory, labour market segmentation theory, and a migrant network approach. Each of these theories or approaches have their own derivations, which we will also explore. After a critical review of these approaches and several variations on them, we move on to the concept of international labour market segmentation, and use that as a guide to understand both the outcomes and experiences of migrant workers throughout the world.

UNDERSTANDING THE RELATIONSHIP BETWEEN MIGRANTS AND WORK

The conventional view (human capital theory and its limitations)

A dominant theory in the analysis of employment is Human Capital Theory (abbreviated from this point onwards as HCT), which owes its popularity to the work of the economist Gary Becker (1964), and continues to be the principle theory used by economists, some economic geographers and sociologists of immigration, and many policy-makers across the world. HCT proclaims that labour market outcomes (or how people perform in the labour market in terms of their salary or wages, that is, the 'price of their labour') are the result of the combination of what an individual – for our purposes here a migrant – brings to the labour market (such as their skills, educational qualifications, and abilities) and/or their rational choices (concerning a mix of status, job conditions, and earnings). Thus, economists say that the price of a person's labour is the result of investments into one's human capital.

There is certainly evidence to support the theory, and HCT studies have also evolved from the cruder analyses of earlier decades to using and incorporating numerous control variables to determine labour market outcomes in terms of occupation, occupational mobility, generational mobility, and income. This work has included how human capital changes over time, the 'contexts of reception' (a combination of the government policies in the country of destination, the character of labour markets, and the nature of the ethnic communities to which migrants might belong – Portes and Rumbaut 2014: 139); related to this, the presence or not of 'ethnic enclaves' (discussed subsequently), the regional concentration of 'co-ethnics', 'social capital' (in the form of co-ethnic networks and contacts with citizens in the country of destination), age, changing immigration policies, citizenship status, disability status, gender, national and ethnic background, the levels of parental education, 'race', religious affiliation, and other characteristics. There is a huge and varied literature here, sometimes with consistent but also complex results for different national groups and different countries.[2] Other analyses in this vein examine the relationship between different national 'varieties of capitalism', different welfare regimes, or different national policies and especially those related to immigration, employment, and welfare (for more recent studies, see Demireva (2011) for the UK; and for cross-national European comparisons, Devitt (2011), Fleischmann and Dronkers (2010) and Kogan (2007)). The historical evidence over the last thirty years is contradictory. Kogan (2007) for example, finds that in 'liberal welfare societies' such as Ireland and the UK, migrants are more likely to be employed than in the 'corporatist welfare states' of Austria and Germany or the 'clientelistic/residual' welfare societies of Italy and Spain. Yet Fleischmann and Dronkers (2010) show no correlation between the type of welfare regime and the likelihood of being employed. Büchel and Frick (2005) demonstrate that immigration policy seems to have a substantial impact on labour market outcomes, while controlling for 'human capital', at least in their study of Western European countries during the 1980s and 1990s. Likewise, in the US, Portes and Rumbaut (2014) compare the labour market outcomes of Mexicans and Nicaraguans with those of Cambodians, Cubans, Laotians and Vietnamese as pre-1980 refugees from Communist regimes, while also controlling for human capital variables. The ample settlement assistance provided by the US government (precisely because the last four nationalities were fleeing Communist regimes during the 1970s and 1980s) contributed to their higher levels of economic mobility in relation to Mexicans and Nicaraguans. While Nicaraguans were also refugees from the Sandinista (Communist) regime during the 1980s, US foreign policy had shifted over time, and the same level of assistance given to Vietnamese for example, had not materialized for Nicaraguans. In contrast, after the events of Tiananmen Square in Beijing in 1989 when the Chinese government cracked down on demonstrations by pro-democracy forces, the US government responded by permitting the majority of Chinese nationals in the US to work, which likely improved the earnings and income of relatively skilled Chinese immigrants (Orrenius et al. 2012). In the context of Europe, Koopmans (2010) provides a similar analysis, and shows – perhaps disturbingly for some – that more restrictive or 'assimilationist' immigration policies rather than those that have more 'multicultural' or welfare policies (such as the Netherlands or Sweden) actually leads to higher employment levels. Yet Koopmans also points out that the macroeconomic environment (in other words, low unemployment) is one of the best predictors of higher levels of employment among immigrants. However, the macroeconomic environment of national or urban economies may not be sufficient to explain employment levels. For example, Wright and Ellis (1996) show that migrant workers managed to find work and some social mobility in New York City when the city faced a precipitous economic decline in the 1970s. The reason: US citizens were leaving NYC and creating openings for immigrant workers even in the context of urban economic decline.

This leads us to the work of urban and economic geographers who, with more concern for sub-national or localized effects, have produced very sophisticated analyses by studying the relationship between human capital, social capital, and importantly social reproduction 'variables' such as housing and residential location, to understand the socio-spatial division of labour and 'employment niching' (why immigrants concentrate in particular industries) across and within urban economies, in order to explain labour market outcomes (in the US, see for example Mattingly (1999); Wright and Ellis (2000a); Ellis et al. 2007)). Ellis et al. (2007) argue that – numerous exceptions aside – the residential location of immigrants matters to the concentration of immigrants in particular jobs, or to put it in their words, spatial accessibility to jobs may be as significant as social access to jobs. Their findings that the local geographies of home and work are important to understanding the fine-grained intra-metropolitan socio-spatial division of labour suggest strong connections between the nature of capitalist production and social reproduction – a connection which we will return to later.

These instances of a more nuanced human capital theory aside, the bulk of human capital research suffers from at least four chief problems. First, many HCT studies rely on an oversimplification of migrant experiences and characteristics (such as 'holding English skills constant' or other questionable assumptions about the homogeneous character of immigrant groups – what Portes and Rumbaut (2014) call 'culturalist theories'). For example, while most Polish workers in the UK in the late 2000s were assumed in the popular imagination to be young and well-educated, some have been homeless, 75 per cent do not have university degrees, poor English language skills are common, and some are Roma, who have suffered additional forms of marginalization (White 2011). Second, and with respect to the assumptions of homogeneity, human capital theorists are guilty of 'naturalizing' social distinctions that occur in both labour markets and the wider society in which these labour markets operate (Hanson and Pratt 1991). For instance, it has not been uncommon in some French labour market statistics to use the categories of 'North Africans' and 'Black Africans', while not being clear about how this should matter precisely for labour market outcomes. Such 'naturalized' distinctions (that is, distinctions which seem natural to us, or in this case at least maybe to French observers) do not stop at national or ethnic assumptions of homogeneity however; HCT studies also assume that the legal categories of migrants remain constant while they are employed, when in fact their hiring may change their migration status. In Italy and Spain, for example, employment or the lack of employment have plunged migrants into an endless and bizarre cycle of legality and illegality, as government 'regularization schemes' which provide legal residential status to migrants are subject to the vicissitudes of having formal employment on a continual basis (Hazán 2014; Reyneri 2001; Schuster 2005). Gender too is taken at face value, rather than acknowledging that gender is actually 'produced' in the process of employment. For example, in the south-western United States, Filipina and Latina women are hired as domestic labour or nurses in part because these jobs are seen as 'women's jobs', which then reinforces the connection between domestic labour and the perception of women as 'carers' (e.g. Hondagneu-Sotelo 2002).

Since many HCT-based studies (and countless employers) assume that individuals within these labour market categories (men, women, particular nationalities, undocumented immigrants and so forth) have certain common characteristics, it is therefore not surprising that labour market outcomes are evaluated *after* scholars or even employers have already divided them in the data. Aside from the French example cited above, this is standard practice in many HCT studies of labour market performance across the globe. In short, HCT is not concerned with how employers construct or re-construct migrant worker identities, and their position within labour market hierarchies.

There is a strange irony here though. The decision of employers to channel their workers into particular jobs is often based on prevailing stereotypes (perceptions of difference based on, let us say, nationality, ethnicity, and gender) or on the basis of their bodily and behavioural performances in particular settings (manner of dress, etc. – we discuss this further in a subsequent section). This *can* have the effect of reinforcing group identities, and it raises a very tricky problem: if group identities and practices are reinforced, then are these group identities not 'real' because they are socially constructed as such by employers? If they are indeed 'real', then to work with data based on national distinctions for example, as human capital theory often begins with, is quite legitimate. Possibly, but again this can also obscure within-group differences, or indeed the primacy of other social differences and the multiple identities of workers (McDowell *et al.* 2007) as well as simply freezing group identities and practices, both on paper (in HCT) and in the heads of employers. It therefore may contribute to racial and cultural stereotypes.

A third problem with human capital theory in the context of immigration seems to be its failure to address explicitly the issue of 'socio-professional downgrading' (Reyneri 2001), 'the devaluation of immigrant labour' (Bauder 2005) or 'brain waste' (e.g. Fossland 2013). In other words, while some migrants arrive with high levels of education, and/or a range of skills and qualifications, their education, skills and other qualifications are not recognized among employers in the receiving countries, so migrants cannot easily transfer their foreign credentials to work in such sectors as architecture, engineering, law, and medicine (e.g. Banerjee and Phan 2014; Bauder 2005; Raghuram and

Kofman 2002). Furthermore, they may be unfamiliar with host country workplace regulations, as Banerjee and Phan (2014) have demonstrated more recently in the context of Canada. Perhaps then, we need to distinguish between 'highly skilled' (having a knowledge of the host country language, technical education, etc.) and 'highly qualified' (having the requisite qualifications, including certifications required by employers), as Fossland (2013) shows in the context of Norway. In the UK, Raghuram and Kofman (2002) show how doctors from India become under-qualified to work and are required to take costly and time-consuming exams for an uncertain position (if they find one) in the British National Health Service. More generally, it is common across the wealthier countries for doctors to become hospital workers, electronic engineers to become electricians, and nurses become nursing assistants (e.g. Kelly 2010), and highly educated political activists fleeing persecution to claim asylum and work in low-paid service work, or remain unemployed. In fact, many migrants and asylum-seekers are formally educated to the post-secondary level. Human capital often describes these 'poor returns to human capital' among migrants in terms of an 'immigrant wage penalty' (Kogan 2004) – a deviation from the model of potentially perfect returns, rather than as a pervasive element of labour markets created by government policies and employer decisions. In short, we can argue that there is no one-to-one correspondence between labour market outcomes and one's education, qualification, and skills. A fourth problem with human capital theory in the context of immigration is that it is preoccupied with the socio-economic mobility of migrants (that is, an improvement in their labour market outcomes) rather than with the conditions under which migrants must scrape a living. Indeed, while HCT-based studies can and do measure wages and income, rather than simply occupational attainment, the reliance on statistical data means such studies have a hard time measuring the unofficial world of informalized workers at the bottom end of labour markets in a precarious world.

THE BEGINNINGS OF AN ALTERNATIVE VIEW: THE DUAL LABOUR MARKET HYPOTHESIS AS AN INITIAL VERSION OF LABOUR MARKET SEGMENTATION THEORY

We will recall from Chapter 2 that the dual labour market hypothesis claims there are two sectors in the labour markets of 'modern industrial societies', a primary sector with more favourable working conditions, higher pay, more stable positions, and possibilities for promotion; and a secondary sector composed of jobs with poorer working conditions, lower pay, less stable positions, and little possibility of promotion ('promotion blockages'). Migrants were argued to be mainly involved in this secondary sector, because of the nature of industrial production at the time, and the unwillingness of citizen workers to accept these positions.

This is both an elegant understanding of labour markets and of migration. It is however flawed and incomplete. First, it focuses entirely on the demand side of labour markets. It says very little about the role of the state and other institutions and says almost nothing about the (gendered) social reproduction of workers. Second, there are not only two sectors of labour markets, despite the sense that richer countries are increasingly unequal, and composed of the 'haves' and 'have-nots'. In the same vein, manufacturing (the 'industry' in modern industrial societies) has declined significantly so that manufacturing employs less than about 25 per cent of all workers in most of the wealthier countries of the European Union, Canada and the US, and some Asian countries such as Hong Kong and Singapore (World Bank 2012). Calling such countries 'modern industrial societies' then, bears some critical thought, and begs the question to what extent (if at all) Piore's analysis is relevant to societies where service-oriented employment is numerically dominant. Anderson (2010) argues that in fact Piore's analysis is still extremely relevant. In her study of workers from Eastern Europe in the UK after their accession to the EU in the mid-2000s, she describes their wages and working conditions as 'recognizably Piorean' (p. 305). In any case, a third reservation is that many supposedly secondary jobs can actually be quite stable, while many supposedly primary jobs involve fixed-term projects. Fourth, Piore completely ignored both informal ('undeclared') employment and immigrant entrepreneurship, though this is not surprising since they had faded from the academic radar at the time he completed his book. Fifth, Piore employed a binary notion of citizenship (one is either a native or foreign worker) and he neglected how job segmentation involved an assumption on the part of employers for certain 'embodied performances' (more on this a little bit later in the chapter). In short, Piore's analysis had its own limitations at that time, and these limitations and others linger in the twenty-first century. These criticisms point us towards some revisions of his analysis, and we will move on ahead to what is called labour market segmentation theory.

BEYOND THE DUAL LABOUR MARKET: LABOUR MARKET SEGMENTATION THEORY

Alongside the budding research of Piore in the early 1970s, Reich *et al.* (1973) and later Gordon *et al.* (1982) developed the concept of 'segmentation' to describe how different rules of operation within firms governed different 'cells' (a segment or a grouping of job positions) within labour markets. These different segments would have different rates of pay, unique working conditions, possibilities for promotion, and so forth. Reich *et al.* argued that workers were slotted into certain segments partly based on their education, qualifications, and skills (including language) but also on other grounds, including assumptions about their suitability for work and productivity based on the colour of their skin, their nationality, their gender, and other ascribed characteristics. This can involve 'positive' stereotyping when a certain group of workers are deemed very employable for more highly paid positions, but also negative stereotyping (prejudice) whereby a certain group is segmented into poorly paid positions with difficult working conditions. What is now known as labour market segmentation theory (LMST) arose as a reaction to the limitations of not only dual labour market theory, but more importantly of human capital theory as well, even if some have tried to bridge HCT with labour market segmentation (e.g. Kogan 2004; Mumford and Smith 2004; Sousa-Poza 2004). In other words, rather than simply focusing on the relationship between individuals and labour market outcomes, scholars wanted to investigate the 'black box' of processes that explained why some immigrants did not perform as well as their 'human capital' might predict.

Peck (1996) elaborated on existing LMST by arguing that segmentation involves the intersection of what he calls 'production imperatives' (or processes relating to the demand for workers by private firms, public agencies, etc.), forces of regulation (the regulatory role of states and other non or quasi-state institutions and organizations) and 'processes of social reproduction' (conventionally known as supply-side factors; how workers are socially re-produced through families, housing, and so forth, and therefore how they come to be workers). We now see some of these ideas reflected in updates to the HCT approach as discussed earlier. At any rate, we can extend Peck's framework by including the analysis of social networks, social and cultural capital, cultural judgements by employers and workers, and racial discrimination at the time of reviewing job applications or at the interview stage (Samers, forthcoming). Now in Peck's (1996) work, these combined processes were said to be locally constituted, but his emphasis on the *local* character of labour market segmentation neglects the *international* dimension of these labour markets, and we will return to this problem in a subsequent section. Before we do this, however, we will need to say more about how labour markets and migration intersect.

VARIATIONS ON LABOUR MARKET SEGMENTATION THEORY: CULTURAL CAPITAL, CULTURAL JUDGMENTS AND EMBODIMENT

Harald Bauder (2005) uses the sociologist Pierre Bourdieu's work to show how 'cultural judgments' and processes of 'distinction' on the part of *employers*, and 'corporeal (or bodily) performances' and one's 'embodied cultural capital' on the part of *workers*, intersect to produce particular labour market outcomes for migrants in certain cities and more rural areas in Canada, such as in Vancouver and on the farms of south central Ontario. Segmentation involves more than just stereotyping and employers have a specific set of traits they see as desirable for a given job or job situation. Such bodily performances are necessary to obtain or maintain a job, and immigrants must 'play by the rules' and 'look and dress the part' (e.g. Bauder 2005; McDowell *et al.* 2007; Waldinger and Lichter 2003). Those who do not are deemed to lack 'cultural competence', and risk unemployment or less desirable jobs. As he puts it:

> Employers and workers who insist on corporeal performances to express competence for certain occupations participate in a cultural segmentation of labor. Workers who do not possess the code for legitimate corporeal presentation and who fail to give the expected corporeal performances face the devaluation of their labor and exclusion from workplaces in which a certain embodied cultural capital is required. (Bauder 2005: 45)

Some workers have this requisite 'cultural capital' and others do not, and the 'cultural judgments' on the part of employers are rooted in notions of class, ethnicity, gender and citizenship, which serve to separate or distinguish employers (often 'white' Canadians) from racialized non-Canadian others. However, Bauder is careful to warn that 'embodied cultural capital' is not something that workers simply bring to the labour market, or as he expresses it, it is not "an exogenous pre-market variable" (p. 82). Rather, it is the question of the interplay of employers' expectations and worker's performances in specific contexts, what he refers to as 'spatial contingency'. This spatial contingency might involve a city, a neighbourhood, a firm or another organization, a school or a café. In Vancouver, he shows how many migrants of South Asian origin who apply for jobs are not accepted either because of their accent, their dress, or even the smell of their bodies. Yet Bauder also discusses how one woman of South Asian origin could not obtain a paid job in any library in Vancouver. Library organizations claimed that patrons were unlikely to understand her accent if she had to speak English over the phone. And yet, she had worked already as a librarian in New York and in the Supreme Court Library of India where she spoke daily with many English-speaking foreigners. In other economic activities however, one's accent may be largely irrelevant. For example, Bauder raises the issue of some South Asians who apparently have poor grammatical and pronunciation skills, but this does not preclude them from working as taxi drivers in Vancouver. From a different angle, he also recounts Zukin's work from New York in *The Cultures of Cities*, in which the "embodied performances of some gay men may be an asset for getting a job as a waiter in Manhattan, but in Queens [another borough of New York] it is a barrier" (p. 86). However, Bauder points out insightfully that in certain circumstances, a migrant's distinctive clothing may be an advantage. Here, he refers to Sikhs in Vancouver, many of whom work in the security business. The turban that Sikhs wear has now become a symbol of reliability and integrity, rather than a symbol of cultural otherness. If particular immigrants are concentrated in a particular industry with specific cultural practices, then the expectations of consumers change. As Bauder writes humorously:

> To use a hypothetical example, if ethnic networks channeled large numbers of traditional, lederhosen-wearing southern German men into the pizza-delivery business, then wearing lederhosen might become a legitimate practice in this occupation. If this group dominates the occupation, lederhosen may even become a trademark of the occupation which customers learn to expect from the delivery personnel. What this silly example illustrates is that the concentration of an immigrant group in a given occupation affects the corporeal conventions that dominate in that occupation. (Bauder 2005: 46)

Similarly, McDowell et al. (2007) show that in a major executive-oriented hotel in West London, (legal) migrants are selected for their embodied characteristics (smile, dress, and appearance, ways of relating to customers, language ability, and also skin colour). Yet employers associate some embodied characteristics with stereotypical views about employees based on their national origins: for example, 'he or she is Indian so they must be pleasant, well-groomed and helpful'. Using Salzinger's (2003) notion of interpellation, McDowell et al. (2007) argue that performances involve a 'dual form of interpellation' in which employees must perform for both managers' idealized conception of service, and the expectations and fantasies of customers. The enchantment of the customer is a central element of the performance of their job in the hotel they study. While these characteristics are appropriate for a corporate hotel; they may not be useful in other domains. In short, embodied cultural capital is spatially contingent; that is, some forms of 'cultural capital' and 'corporeal performances' matter in some places and not others.

In fact, argue Waldinger and Lichter (2003), hiring practices and labour market segmentation are not simply a matter of 'negative prejudice', but often the result of stereotypes based on nationality and skin colour. Rather, hiring practices are shaped by whether migrants are perceived to be subservient workers, particularly at the low end of the employment spectrum where working conditions are tough and employers assume they will encounter resistance. Employers hardly wish to hire employees who might 'give a lot of lip'. If they have hired an employee from Guatemala who is deferent to the boss, then they may hire the employee's sister who may also be perceived to be deferent, and the employer does not have to spend too much money looking for another worker. This brings us to the question of migrant networks, which I explore below.

IS LABOUR MARKET SEGMENTATION THEORY (STILL) RELEVANT? THE LABOUR MARKET LITERATURE AND THE MIGRATION LITERATURE

In the 'super-diverse' (Vertovec 2007) or 'mongrel' cities (Sandercock 2003) of Asia, Australia, Europe and North America, where the employers may not be citizens, and/or of the dominant ethnicity, the issue of segmentation becomes more complex than simply an issue of prejudice based around the dominant and the less dominant ethnicity; more complex than the distinction between citizen and non-citizen; more complex than simply a distinction between 'white' employers and 'black' or 'brown' employees. Rather, for Waldinger and Lichter (2003), the debate around segmentation seems to concern those who perceive firms and organizations, whether they are large and bureaucratic or small and family-owned, to segment workers along discriminatory lines (the labour market literature) and those who privilege migrant networks as shaping hiring practices (the migration literature).

Concerning the migration literature, Waldinger and Lichter (2003) argue that employers do not simply advertise jobs and hope workers will show up at the door, although that is still common. Rather, 'social capital' in the form of 'social networks' play a role in allocating workers to certain jobs. Indeed, employers rely on social networks among migrants to recruit other migrant workers, particularly for jobs that are shunned by local citizens. Recruitment of migrants at the bottom end of the labour market where proficiency in the dominant language is often not required is facilitated by the gravitation of migrants with poor language skills to such jobs. However, in terms of being chosen by an employer for a job, their lack of fluency in the dominant language – let us say English – may not always be a disadvantage, especially if all the other workers with which they are likely to interact also speak a language other than English. Language requirements are a function of the nature of the job, including how the worker is likely to interact with others inside and outside the firm or organization. For example, if the majority of the customers are Arabic, Spanish, or Urdu-speaking, would it make sense to hire a person who could only speak English? What happens if most of the workers do not speak English but the customers do? In this case, some workers may need to be bi- or multi-lingual, including in the dominant language. This is required if workers and/or customers are of diverse backgrounds and communication is facilitated with the knowledge of several languages. For example, this appears to be the case in many hospitals in the Los Angeles region. Likewise, employers do not always have a choice in who they hire; they must draw on the available pool of workers, which is dependent upon the nature of local labour markets. As we discussed above, employers' racial prejudices and negative stereotypes about workers from the 'third world' may be pervasive, but reading off employer decisions on the basis of skin colour or national origin alone is too simple. In fact, American employers may have a preference for 'Latinos' over 'white workers' at the bottom of the labour market as the former are perceived to be more docile, but also hard-working – a requirement of tough jobs, for which 'whites' or 'African-Americans'[3] are often not perceived to be suitable. This is abundantly clear in the context of ethnic enclaves where the employers are themselves migrants or of migrant origin, and hiring those of similar ethnicity or at least of migrant origin or migrant status is quite common (more on this in a subsequent section). In fact, argue Waldinger and Lichter (2003), in the context of both ethnic enclaves and other labour markets, "network hiring [among immigrants] can allow a linguistic minority to establish monopoly control over a set of jobs and, when the characteristics of the job permit, to use language to exclude those who only speak another tongue" (p. 78). Migrant networks and job search can thus lead to 'social closure' where certain sectors are dominated by certain nationalities or ethnicities, as well as by men or by women. For example, in the deindustrialized city of Dayton, Ohio, the number of Turkish and other Middle Eastern entrepreneurs has expanded rapidly since the early 2000s. Yet, many African-American leaders in a city where 43 per cent of the population is African-American, were wondering whether they were being hired by these entrepreneurs in the context of high unemployment (New York Times 2013).

The above discussion is presented as a complication of labour market segmentation but it is not designed to deny personal or institutional forms of racism or discrimination in the work place, especially by 'white' citizens towards 'people of colour', or indeed other variants of racism. In fact, Batnitzky and McDowell (2011) offer a telling study of just how pernicious institutional racism can be in the hiring and promotion process in the UK's National Health Service. What it is designed to do is to hint at how important the relationship is between particular kinds of spaces, migrant/ethnic communities, and employment. We visit this further in the next section.

Complicating labour market segmentation theory: the question of ethnic immigrant enclaves and ethnic/immigrant dispersal

Thus far, we have said very little about the geography of labour markets and migration. Both geographers and sociologists have contributed enormously to showing how 'space' and 'place' matter for the labour market outcomes of migrants. In this section, we will briefly explore three prominent spatial phenomena which might have a distinct effect on migrant workers: ethnic/immigrant enclaves, ethnoburbs, and the process of geographical dispersal. The point of such a discussion is to highlight two issues.

First, that stereotypes and expectations of migrants may be different in areas where the majority or at least a large proportion of the employers and employees are migrants than when the local labour market is dominated by 'white' and/or citizen employers. Second, that the migrant networks involved in hiring may also be different in certain areas dominated by migrant employers and employees. This does not mean that the concept of LMS is invalid in 'ethnic enclaves' – only different.

Continuing in the vein of the Chicago School of the 1920s and 1930s (which, broadly speaking, argued that the socio-economic mobility of 'white' European immigrants in Chicago led to their residential mobility outwards from the city to early suburban areas), sociologists resumed their study of the spatial manifestation of ethnic/immigrant enclaves and economies in the 1980s onwards, this time with a far broader range of nationalities. This body of work assumed added importance in the context of industrial decline in the richer countries as the supply of industrial jobs withered for migrants. In fact, racism in the wider labour markets of the countries of immigration, the lack of employment opportunities as a consequence, and the possibilities for higher income are among the primary reasons that drive immigrants to either find work in 'employment (or ethnic) niches, immigrant/ethnic economies and enclaves, or entrepreneurial (self-employment) niches (Logan et al. 2003). Yet what are the differences between these seemingly overlapping terms? Employment niches "are economic sectors [. . .] where group members are disproportionately represented in the labour force, either in public sector jobs or in private businesses that are typically owned and managed by whites or members of another ethnic group" (Logan et al. 2003: 346). For example, Vietnamese women in Canada and the US, but also in Belgium, France, and the Netherlands have both developed and dominate nail care – a standardized form of personal service which Eckstein and Nguyen (2011) refer to as 'McNails'. They have come to preside over this 'industry' through both formal and informal networks that have excluded other immigrant workers. However, even if certain immigrants may eventually dominate particular workplaces or sectors, they may only be able to shape their labour market outcomes to a limited degree, since labour market demands, immigration and other policies, their human capital and segmentation processes also bear on their fortunes.

'Ethnic economies' (Bonacich and Modell 1980) involve the self-employed and their co-ethnic employees; they are economies that are spatially dispersed across cities, regions and countries, but whose networks of interactions are specifically 'ethnic' in character (their ethnic character may be defined by a researcher, or self-defined by migrants themselves, and it commonly involves only a single 'ethnicity') (Light et al. 1994). Such ethnic economies could therefore be transnational or more translocal in character, and more formal or informal. Light et al. (1999) draw a further distinction between ethnic economies and immigrant economies; in the latter, immigrant entrepreneurs hire other immigrants, but not co-ethnics, such as Korean entrepreneurs hiring Mexicans and Ecuadorians in New York's garment industry (Kim 1999).

Note that the ethnic or immigrant economy idea involves both the self-employed (entrepreneurs) and their employees (or workers), and thus the literature on these economies is different from dual labour market theory insofar as the latter is not concerned with self- employment or entrepreneurship. The main thrust of the discussion here though, will not be about entrepreneurship (see Box 5.1), so as to focus on the question of low-wage labour in and outside ethnic/immigrant economies, whether enclaved or not.

Box 5.1 ETHNIC OR IMMIGRANT ENTREPRENEURSHIP?

Self-employed entrepreneurs and their businesses account for a substantial amount of the employment of migrant wage-earners. Rates of entrepreneurship vary by country, region, nationality, gender and so forth. Data from the OECD/SOPEMI (2015) suggest that the percentage of those self-employed among the foreign-born in European countries, for example, range from 2.5 per cent in Hungary to 50.5 per cent in Luxembourg. Yet one of the neglected questions of the entire 'ethnic entrepreneurship' literature is whether the entrepreneurs and firms involved are actually different from the firms owned and run by citizens and/or those of the dominant ethnicity (Light 2005). The answer is not entirely clear.

Questioning the significance of ethnicity to entrepreneurship then, a group of scholars based in the Netherlands now use the term 'immigrant entrepreneurship' to denote how the issue of citizenship in particular shapes the fortunes of non-citizen entrepreneurs (Kloosterman et al. 1999; Kloosterman and Rath 2015). This has much logic; after all, we are all 'ethnic' (Samers 1998a), and there is no reason to assume a priori that businesses owned by citizens are any 'less ethnic', or that ethnically specific social networks (social capital) support either citizen or immigrant businesses. This does however ignore the long established problem of racism in the wider labour market and the obstacles that immigrants and ethnic minorities face in not being of the dominant ethnicity ('labour market disadvantage') as well as having fewer resources ('resource disadvantage'). Thus, migrants who face both labour market disadvantage and resource disadvantage suffer from 'double disadvantage' and are more likely to establish a more informal than formal business. From this perspective, ethnic entrepreneurship seems to be quite the correct term. For those migrants who do not suffer from either of these, they are more likely to establish formal businesses (Light 2005), and what may distinguish them is only their citizenship. From this perspective, immigrant entrepreneurship seems the correct term.

In the US literature, 'interactionism' has been the preferred conceptualization of immigrant/ethnic entrepreneurship, which refers to the relationship between the self-employed and their customers in an ethnic economy or ethnic enclave (Light 2005). This perspective persisted until Kloosterman et al. (1999) – writing in the context of the Netherlands – argued that the issue of regulation has been neglected by American scholars, in part because the European regulation of entrepreneurs and business in general is more stringent than in the United States (Light 2005). The result is that Turkish bakers or butchers in Amsterdam, for example, will evade a whole range of regulations, and immigrant entrepreneurship in the Netherlands therefore exists along a continuum of the formal and informal. Given then the extent of regulation of small firms in

Continued

European countries, Kloosterman *et al.* (1999) developed the idea of 'mixed embeddedness', by which they mean the interconnections between economic, institutional, and social contexts. Thus, the rise of immigrant entrepreneurship is, in theory, a product of changes in urban economies on one hand, and changes in 'socio-cultural frameworks' on the other. "The interplay between these two different sets of changes takes place within a larger, dynamic framework of institutions on neighbourhood, city, national or economic sector level" (p. 257).

Wilson and Portes (1980) were the first to speak of the 'immigrant enclave' – a spatial concept involving the geographical concentration of, and the relationship between self-employed or entrepreneurial migrants, and their co-ethnic employees, although the emphasis remained on employees. Whether or not the customers were citizens or immigrants did not matter in their original formulation. The concept of immigrant and later 'ethnic enclave economy' has been stretched in a number of different directions (Light et al. 1994), including the incorporation of co-ethnic residence within the enclave (Sanders and Nee 1987, 1992). A debate ensued on the pages of the *American Sociological Review* (1987, Vol. 52, No. 6) about what exactly constituted an ethnic enclave economy, and as Light et al. (1994) quipped, the term has become something a 'rubber yardstick' (p. 69). In any case, what came to be called the 'ethnic enclave hypothesis' generated a number of debates. First, a question emerged over whether wages were higher among employees in the 'ethnic enclave' than they were for co-ethnics in the 'general labour market' (the issue of 'relative wages') and to what degree, the self-employed and employers rather than employees were responsible for some of this difference (Sanders and Nee 1987). This rested on three propositions; that employers in the enclave benefit from a co-ethnic labour force by trading training for loyalty and low labour costs (Bailey and Waldinger 1991), that migrants could find jobs in the ethnic enclave, despite their labour market weaknesses (e.g. language); and that those with skill would not suffer from socio-professional downgrading, find work more suitable to their skill level, and earn more (Wilson and Portes 1980). The overall evidence for the benefits of the ethnic enclave seem to be inconclusive, and as Logan *et al.* (2003) write, "The ethnic strategy is not a magic bullet, but neither is it a poison pill" (p. 381).

A second and parallel debate emerged over whether migrants were 'trapped' in the ethnic enclave economy which inhibited their 'acculturation' and socio-economic mobility (Bonacich and Modell 1980). A third debate concerned the question of social reproduction. For example, in her widely received book *Chinatown* (1992), Zhou notes how Chinese employers allowed Chinese women to take care of their own children *at work*. This was unlikely to be permissible outside the 'ethnic enclave' of Chinatown, and thus employment for Chinese women in the wider labour market would not be possible.

Ethnoburbs, heterolocalism, and geographical dispersal: implications for labour market segmentation

In these immigrant or ethnic enclaves, the concept of labour market segmentation does not lose its validity but the process and patterns of stereotypes, and expectations of 'others', that we learned about earlier may take different forms than when the local labour market is dominated by citizen employers. Migrants however, do not reside exclusively in enclaves. Yes, they may be disproportionately concentrated in the world's 'global cities' (Ellis *et al.* 2014; Hu 2015; Sanderson *et al.* 2015; Sassen 2006a) or 'gateway cities' (e.g. Price and Benton-Short 2008), for example, about 75 per cent of all immigrants in the US reside in urban areas of one million or more compared to less than 50 per cent of US citizens, but immigrants are not confined to the central city's limits (Ellis *et al.* 2014). As Hardwick (2008) summarized, "new incoming groups of immigrants are just as likely to settle in the suburbs upon their arrival in the United States as they are to reside in downtown neighbourhoods" (p. 164) (see also Hardwick 2015; Singer 2004). While such a statement requires far more nuance, the same may be true broadly speaking for any number of urban areas in the UK, continental Europe, Australia, Canada, and New Zealand. In this regard, Li (1998a, 1998b) coined the term 'ethnoburb' in referring to suburban clusters of residential areas and business districts with a dominant, but not exclusive ethnic group, such as those of Chinese or Vietnamese immigrants in Orange County outside Los Angeles (Portes and Rumbaut 2014).

The idea of ethnoburbs however suggests that migrants have simply moved from the central core of large cities to the suburbs, but migrants' dispersal has extended well beyond the largest metropolitan areas of richer countries – what some have called 'new immigrant destinations' (e.g. Winders 2012) or 'new immigrant gateways' (Singer 2004) in the context of the US at least. Yet immigrants have always settled beyond the largest metropolitan areas in the wealthier countries, and certainly these ideas have earlier conceptual antecedents. Indeed, situating their analysis in the context of 'globalization' and in contrast to the 1920s and 1930s Chicago School of urban sociology, Zelinsky and Lee (1998) coined the term 'heterolocalism' to refer to the dispersion of migrants in the late twentieth century, from central cities to metropolitan or non-metropolitan areas based on declining transport and communication costs (see Figure 5.1).

Businesses and residential spaces no longer needed to exist in spatial proximity because of these technological innovations associated with 'globalization'. Migrants could maintain contacts across physical distances, not only on a 'metropolitan scale', but internationally as well. Light (2004) for example, identifies a process of settlement whereby as the number of

Figure 5.1 'Water, electricity and gas included': 'for rent' sign in Spanish on the West Side of Lexington, KY.

migrants exceeds the number of jobs (migration saturation), the 'buffer' of informal economic activity expands (buffer expansion). This buffer itself reaches some sort of saturation (buffer saturation), and then migrants leave a particular city for less expensive housing, higher wages and a police force more tolerant of poverty. As we saw in Chapter 2, there may be other reasons why migrants move, such as the promise of plentiful jobs (not just higher wages), other familial reasons (to be close to relatives, etc.), or simply the refugee settlement policies of governments. So whether Light's idea is correct, and there is no doubt evidence to suggest it is, migrants are certainly dispersed throughout the European Union and North America. In what is sometimes called 'secondary migration', many Dominicans have moved out of costly New York City to the much cheaper, deindustrialized city of Reading, Pennsylvania to take up jobs in chicken processing plants, hotels and restaurants (New York Times 2006). Refugee settlement policies have led to the development of Somali-owned small businesses in Columbus, Ohio and Minneapolis, Minnesota, while direct recruitment in Mexico has led to legal and undocumented Mexicans and other Central Americans working in the slaughterhouses of south-western Nebraska, the hog farms of Iowa, or the horse farms of Lexington, Kentucky. Algerians have for a long time settled in Roubaix, in northern France, initially to work in the now forlorn factories and mines, and from the 1980s onwards, in temporary construction work (Samers and Snider 2015; Samers 2017), while in England Pakistanis in Bradford and Leicester initially came to work in the textile mills in the 1950s (e.g. Castles 1984). As the case of Reading above suggests, the availability of decent work is not the only element that attracts migrants; they also seek cheaper housing, which is often found in the smaller deindustrialized (and inevitably poorer) cities of the rich world. There is more than a little irony here, since in many cases, it has been an earlier cohort of immigrants who built up these now deindustrialized cities in the first place, from the Welsh coalminers of Pennsylvania to the Belgian, Polish, and Italian workers who worked in the factories of north-eastern France (Noiriel 1984). In short, there is little reason to assume that migrants only find work in the largest or even the most economically dynamic cities.

Having thus far explored more nuanced readings of labour market segmentation, including certain geographical dimensions from the transnational to the trans-local, we take these insights to develop what Samers (2010d) calls international labour market segmentation.

INTERNATIONAL LABOUR MARKET SEGMENTATION

'International labour market segmentation' (ILMS) (Samers 2010d) can be understood as having three dimensions. First, it entails the dividing up of labour across the world by national and supra-national governments (such as the EU) and their respective immigration policies. Here we could add other non-state or quasi-state bodies as well. By 'international', we are referring to international *actors* – that is national and supra-national governments and other international organizations that sort or segment labour on *international grounds*. A second dimension of ILMS is that these national and supra-national immigration policies also divide up labour within national economies, in such a way that migrants are segmented into particular kinds of sectors and jobs according to national origin or other characteristics deemed desirable by national states, such as occupational skills deemed necessary by national governments, firms, or NGOs, and so forth. This is reflected in a battery of stratified rules that governments use to recruit certain kinds of migrant labour as discussed in Chapter 4. A third dimension of ILMS is that it involves the segmentation of migrant workers within both private firms (the common understanding of labour market segmentation theory) and organizations. We include organizations because large numbers of migrants, and especially from China, the Dominican Republic, Ghana, India, Ireland, Jamaica, the Philippines, and South Africa work in the hospitals, clinics and other medical sites of international health care systems in North America and Western Europe. This segmentation within firms has social and spatial dimensions. Socially, the precariousness of employment created through visa restrictions forces certain nationalities into sectors or occupations where migrants work routinely without papers, such as agriculture and farm work in Europe and North America.

ILMS is distinctive from existing LMS approaches because it assumes that it is not simply the colour of a migrant's skin or their 'embodied cultural capital' that matters, but how *citizenship* also becomes a source of division. Indeed, Van Parijs (1992) spoke of 'citizenship exploitation', in other words, the ways in which the citizenship of a person matters to the conditions they face in labour markets. Certainly, some HCT studies now take this into account but they assume that citizenship remains constant before and during the period of employment. Second, the idea of ILMS assumes that social reproduction is not just 'local' in character (e.g. the nature of housing, transportation, welfare policies, and so forth), but that it is shaped by international financial flows between sending and receiving countries and back again. As was suggested in Chapter 2, these financial flows serve to mediate the relationship between household/ family decision-making, migration, the level of vulnerability in the work place (in short, how desperate someone is to work), and thus the willingness to accept certain types of jobs (e.g. Kofman 2005a, Kofman and Raghuram 2006; Yeates 2004). And in terms of remittances for example, many extremely low-paid undocumented immigrants from Pakistan working in typically small 'take-way' (take-out) Pakistani-owned fast-food restaurants in east London are supported financially by their relatives in Pakistan, not the reverse. This enables them to carry on in the miserable conditions under which they either 'voluntarily' work, or are forced to because they have been trafficked (Ahmad 2008a).

A MIGRANT NETWORK APPROACH IN 'SUPER-DIVERSE' SOCIETIES

Rather than an emphasis on employers and their actions, in the migrant network approach, the emphasis is on the practice of immigrants, and especially the networks of 'co-ethnics' or 'co-nationals' for finding jobs and related resources. Put differently, such networks help 'social capital' and the formation of an 'economy of favours' within groups (Ledeneva 1998). Co-national and co-ethnic ties are sometimes referred to as 'bonding capital', while 'bridging capital' refers to contacts outside a migrant's national or ethnic group, and in particular ties with citizens (Putnam 2000; Lancee 2010). Both 'bonding' and 'bridging capital' may entail 'weak' or strong ties'. As Granovetter (1973) argues, weak ties may be just as significant for obtaining resources (jobs, housing, and so forth) as 'strong ties', and certainly not all co-ethnic ties are strong, divided as ethnic groups may be by class, gender, generation, and other inter-sectionalities. In fact, Reyneri and Fullin (2011) show that 'weak ties' among immigrants may actually be more effective in landing a job, obtaining greater resources and achieving higher incomes. In Kanas et al.'s (2012) study of Germany and Lancee's work in the Netherlands (2010), they find that social contacts are positively correlated with higher occupational status, and that ties with German and Dutch citizens ('bridging capital') also led to higher occupational status. However, Kanas et al. do not observe a correlation with higher incomes, while Lancee (2010) finds that incomes actually increased. Beyond employment, other research shows that financial resources can prove to be especially useful. For example, among undocumented Ghanaians in London, Vasta and Kandilige (2010) report that for those who were unable to open a bank account, they could 'rent' one from a fellow and *legally resident* Ghanaian. In some cases, a payment for this 'renting' is expected, in others, a favour in return is expected (in other words, 'mutual aid'). In contrast, Grzymala-Kazlowska (2005) demonstrates in her study of undocumented Polish migrant workers in Brussels, that there is considerable suspicion among many Polish workers and that cooperation has given way to competition and the undermining of mutual aid. Instead of the social capital that one might assume would develop among Polish workers in London, she refers rather to a 'quasi-community'. Gill and Bialski (2010b) nuance this however, by exploring how the reliance on co-ethnic social networks is shaped by 'socio-economic status', so that surprisingly, higher income Polish migrants relied on 'stronger ties', while lower income individuals relied on 'weaker ties'.

Having reviewed more or less seven approaches to understanding the incorporation of migrants in labour markets, we now turn to accepting that they may be combined; that is, migrant workers' labour market outcomes may be related at once to their 'human capital', to segmentation processes, to social or spatial formations that involve enclaves, niches, dispersals, and ethnoburbs, to their 'cultural capital', to migrant networks, and to *international* segmentation processes that involve both regulatory and social reproduction processes.

A DEMONSTRATION OF A MIXED APPROACH IN THE RICHER COUNTRIES

In this section, we demonstrate the usefulness of a mixed approach, that is, by combining the insights from the various approaches discussed earlier. However, we emphasize ILMS and the evidence we present is analysed through the lens of Peck's (1996) categories of 'production imperatives', 'forces of regulation', and 'processes of social reproduction' (or again in more simple terms: demand, supply and regulation). We mainly focus our discussion on wealthier countries, but turn to a brief account of poorer countries as well. This distinction may again seem both 'methodologically nationalist' and crude for at least two reasons. There is significant diversity within the rich world, such as the enduring acceptance of family migration in the US, and the official absence of family reunification in most Asian countries (Seol and Skrentny 2009). Many of the jobs that migrants perform in the 'global north' and 'global south' are often very similar, and after all there are wealthier and poorer regions, cities, and so forth, in both the north and south. The economic structure, labour demand, and labour migration are equally diverse in southern countries and at any rate they are connected to the north through various labour migration systems (Kofman and Raghuram 2012). Nonetheless, a few rough distinctions can be drawn. First, in contrast with richer countries, much agricultural and domestic work is still performed by citizens; second, informal employment (explained later) is far higher in poorer countries than richer ones; third, immigration policies in wealthier countries entail a highly stratified set of immigration policies ('point systems' or other tiered systems which are designed to address skill shortages) which many poorer countries have simply not established for complex reasons. Fourth, the conditions of social reproduction are on the whole considerably more grave than they are in wealthier countries, although as we shall see, the social reproduction of migrants in many, but not all, wealthier countries can also be somewhere between dismal and horrific (e.g. Darling 2009; Wilkinson and Craig 2012).

Labour demand and the segmentation of migrant workers

In terms of 'production imperatives' (labour demand), it is difficult to paint a uniform picture of the economies of the wealthiest countries, and again, it may be less wise to analyse these matters through the lens of national states, than through regions, cities, other intersecting territorialities, and the diversity of employers. As Ruhs and Anderson (2013) note, "A key consideration in assessment of employer demand for migrant workers is that 'what employers want' (i.e. the skills, competencies and attributes required of employees) is critically influenced by what employers 'think they can get' from the available pools of labour" (p. 72). No doubt the 'available pools of labour' are at least partly shaped by national regulations and societal contexts, so we will proceed (caveats on our shoulders) with both a certain methodological nationalism, and an attention to the sub-national character of demand. Below we sketch out a handful of these characteristics of the economies of wealthier countries. These include:

- The dominance of service-oriented employment in terms of the numbers of workers employed, and a decline in the number of those employed in manufacturing; what has come to be called 'post-Fordist', 'post-industrial', or 'knowledge economies'.
- A significant demand for highly educated/'highly skilled' (or 'skilled') employees (computer and other engineers, business executives, and scientists, to name a few). As above, this demand leads some to describe such economies as 'post-Fordist/post-industrial economies', 'knowledge-intensive' or 'knowledge economies' or even 'knowledge societies'. However, there is often a concentration of such jobs in so-called urban or regional 'islands of innovation' within countries (think of 'Silicon Valley' in California, for example), that attract the highly educated and highly skilled. The demand for highly skilled workers often occurs on a project-basis. Project-oriented employment (e.g. Grabher 2002) refers to short-term projects that are associated with research and development, and other innovative activities in hardware and software design, bio-informatics, pharmaceuticals, high-end consumer products and so forth. In many cases, this has led to the 'resurrection' of temporary worker programmes' to suit such demand (Castles 2006; Samers 2016; Surak 2015).
- A demand for personnel in 'welfare sectors', anyone from doctors and nurses, to social workers and teachers, many of whom may be classified as 'highly skilled' by various governments.
- Despite the development of so-called knowledge-intensive jobs, there is a consistent demand for low-skilled/low-paid labour to perform jobs across the occupational spectrum, but particularly in agriculture, construction, manufacturing, services, especially the three 'Cs' (care, cleaning, and catering), but also tourism.
- The proliferation of 'precarious', 'flexible', 'atypical', 'contingent', insecure, part-time, and temporary/short-term low-paid employment (e.g. Anderson 2010; Fudge and Strauss 2014; Kalleberg 2009; Theodore et al. 2015) which has led to the term the 'precariat' (Standing 2011). Some of this work will also involve informal(ized)/undeclared[4] employment, and as with more highly skilled employment, may also occur through the channel of temporary worker programmes (at the low end, this is often called guest-worker policies). All of this 'atypical' employment may sometimes eliminate processes of rigid segmentation, or is combined with it (Peck 1996). Yet outside

of work in private and public enterprises, precarious work extends to domestic work as well in homes and elsewhere, which is strongly gendered across the world (Kofman 2013).

- The 'vertical disintegration' of firms that involve the 'outsourcing' (sometimes overseas; i.e. 'offshore outsourcing') of some element of the production or development of a final product or service of a parent firm, to other firms. These other – often smaller firms – then rely on even smaller firms in an endless chain of sub-contracting in which the firms engage often, but not always, in informal economic activity and which frequently employ undocumented/irregular immigrants, but also legal immigrants and citizens, in an informal fashion. However, there is considerable variation in the percentage of workers who are informally employed across various countries. For example, southern European countries (especially Greece and Italy) have much higher rates of informal employment[5] than do Scandinavian countries, and other northern European countries (Williams and Nadin 2014).

- Since especially the global and financial economic crisis of 2007–9, and elevated unemployment in the richer countries, there has been a purported increase in the size and extent of 'sharing economies' and 'gig economies' that entail not only a supposed increase in informal employment, but the growth of 'mutual aid'/mutual assistance within and between groups of people, whether more legal or illegal in character, and an increase in both formalized and informalized, off-line and online trading systems for especially the trade in low-end goods and services.

These are argued to be basic attributes of twenty-first-century economies, and some of the characteristics are more relevant to particular countries, regions, cities, rural areas, and so forth, than others. For example, in most of the wealthier countries, what are deemed to be highly skilled migrants far exceed the number of those admitted as low-skilled workers, but the reverse is true in South Korea, even if government policy encouraged highly skilled migration and heavily regulated less-skilled migration (Lee 2007). At the same time, none of these features cited above *necessarily* point to the use of migrant labour, as employers have a number of other strategies they can deploy to demand or not demand migrant workers, including, as Ruhs and Anderson point out:

increasing wages and/or improving working conditions to attract more citizens who are either inactive, unemployed, or employed in other sectors, and/or to increase the working hours of the existing workforce; this may require a change in recruitment processes and greater investment in training and up-skilling; (ii) changing the production process to make it less labour intensive by, for example, increasing the capital and/or technology intensity; (iii) relocating to countries where labour costs are lower; (iv) switching to production (provision) of less labour-intensive commodities and services; and (v) employing migrant workers. (Ruhs and Anderson 2013: 74)

With respect to these geographies of labour demand for immigrants in the richer countries, one of the most novel and path-breaking discussions can be found in Sassen's 'global city hypothesis' and later revisions (1991, 1996b, 2006), and we will use her analysis as a starting point to explore the geographies of migrants' labour market segmentation and their experiences across the wealthier countries.

Global cities, urban labour demand, and migration

What is particularly useful about Sassen's analysis is how she connects the economic structure and character of labour demand with labour migration ('labour supply'). In the global city hypothesis (the GCH), certain cities emerged during the 1970s to become centres of 'command and control' of the world economy, involving the rapid development of financial markets and the rise of 'producer' or 'specialized business services' (accountancy, law, management, financial consulting, and related activities) that have contributed to burgeoning property markets, the gentrification of large cities, the creation of luxury-oriented consumption, and a demand for 'highly skilled' workers, both citizens and international migrants alike. At the same time, the growing presence of a highly paid, highly skilled class of workers 'requires' vast numbers of low-skilled/low-paid workers to 'service' them. Again, these workers may be citizens or immigrants. Sciortino and Finotelli (2015) add here that the demand for low-paid service work can be considered to be the result of what Baumol (2012) calls the 'cost-disease' problem – that is, the inability to increase productivity in the service sector relative to other sectors which leads to the employment of extremely low-paid workers, especially domestic workers. Now back to Sassen: she devotes considerable time to the 'informalization' of economic activity (in other words, the development of more informal economic activity). She argues that 'third world' immigration does not produce informal economic activity; rather informal economic activity has remained an age-old feature of these cities. For Sassen then, the important question to ask is how immigration from poorer countries contributes to this informalization. What she argues is that increasing inequality in terms of earnings together with the inability of some production-oriented firms to compete for the necessary and unusually expensive resources in global cities such as commercial space, business inputs, related services, and labour, have led to informalization, often involving 'sweatshop conditions'. This she refers to as 'down-graded manufacturing'. Yet low-paid migrants and other ethnic minorities cannot afford the luxury goods on offer in global cities. As a consequence, they rely on 'co-ethnic' producers, and/or other low-cost immigrant-run shops for their required goods – what she refers to as 'down-graded mass consumer services'. At the same time, niche-market small-batch goods that target more affluent consumers– what she refers to as upgraded 'non-mass consumer services', leads to the development of labour-intensive, small-scale sub-contracting in these cities which itself is dominated by migrant labour. However, in a later discussion (Sassen 1991), her argument assumes a more explicitly 'supply-side' perspective. In other words, she claims that the growth in the number of migrants (the 'supply') in global cities has led to the expansion of small-scale producers that can compete with large chain stores and supermarkets, although competition is fierce, profits are slim, and this in turn creates a demand for ever cheaper labour.

Criticisms of Sassen's arguments abound, but we will simply point out three. First, all of these supposed 'global cities' (let us say London, Hong Kong, New York, Paris, Seoul, Singapore, Sydney, and Toronto) are somewhat different in terms of their economic, political, and social characteristics (e.g. White 1998) and thus the 'production imperatives' and labour markets associated with them are shaped through different forms of governance. Second, what actually distinguishes global cities from other cities is unclear (McCann 2002; Samers 2002), and every city in the world is shaped by flows that circulate around the world (Robinson 2002). As Taylor (2004) put it so tersely, "There is no such thing as a non-global city" (cited in Robinson 2002: 42). To take but one striking example, evidence of ethnic diversity across the world's largest cities suggests that the city of Mecca is more diverse than Los Angeles

(Benton-Short *et al.* 2005), the former hardly to any scholar's mind a 'global city'. This has two implications; either the nature of 'production imperatives' are similar between supposedly 'global' and more 'ordinary' cities, or these 'ordinary' cities have different geographies of work and migration that demand our attention. Before we explore some of these other geographies however, a third point should be raised about Sassen's argument, and this concerns the growth of, and relationship between, undocumented immigration and informal employment. So let us evaluate some of the tenets and limitations of the GCH further, by first looking briefly at 'highly skilled' migration and then to informal employment and 'less-skilled' migration.

Constructing highly skilled migration

We will recall from above that highly skilled migration is argued to be a central element of so-called global cities. By 'highly skilled', it usually means a combination of experience and tertiary degrees, though the notion of what exactly constitutes 'highly skilled' is vague, or determined by governments and employers in specific times and places, and subject to change. Ruhs and Anderson explain:

> Some 'skills' are credentialized (e.g. National Vocational Qualifications, professional qualifications, and apprenticeships), but what is and is not credentialized changes and jobs can shift from being classified as 'low-skilled' to 'skilled' and vice versa without necessarily changing in their content. The limitation of formal qualifications as a measure of skills becomes most apparent when one considers 'soft' skills not captured through formal qualifications. They cover a broad range of competencies, transferable across occupations (rather than being specialized) from 'problem solving' to 'teamworking' and 'customer-handling skills'. Soft skills are often said to be particularly important in sectors where social relations with customers, clients, and/or service users are important to the delivery and quality of the work. Certain 'skills' may be necessary to make sure the job is done in a way that contributes to a good service experience, rather than simply to complete the task. For example, the quality of care delivered in both health and social care sectors is affected by the soft skills of those providing care, with some service users actively expressing a preference for

personal qualities over formal qualifications. At the same time, 'skills' can also be used to refer to attributes and characteristics that are related to employer control over the workforce. A demand for soft skills can easily shade into a demand for employees with specific personal characteristics and behaviour. (Ruhs and Anderson 2013: 71–2)

Skills are also gendered, so that many 'care workers' with quite varied domestic and health care skills may not be considered as 'highly skilled' (e.g. Iredale 2005; Kofman 2014). In any case, those deemed 'highly skilled migrants' are a diverse range of individuals working in quite varied contexts. Kofman (2014) argues that most 'skilled migrants' are women in Europe, but that does not preclude specific sectors from being dominated numerically by men. Mahroun (1999, cited in Williams et al. 2004: 30) offers a five-fold categorization of highly skilled migrants: managers and executives, engineers and technicians (oddly, those working in health sectors, such as doctors and nurses are included in this category); academics and scientists, entrepreneur migration, and student mobility. We could also add media and telecommunications workers (Hu 2015) as well as school teachers, as Iredale et al. (2015) do in an unusual analysis of the Pacific Islands. However, evidence for these last two categories are not presented here, nor is data for academics and scientists, but we do discuss these other categories very briefly in turn.

In terms of highly skilled migrants and executives then, Beaverstock (2002, 2005) and Beaverstock and Boardwell (2000) conducted research on British accountant and banker inter-company transferees in New York, London, Hong Kong, Singapore, and Zurich. Most of these were men. Their argument is that highly skilled migrants reinforce capital accumulation (economic growth) in these international financial centres, which in turn strengthens the demand for highly skilled migrants. Similarly, Bauer and Kunze (2004) argue that it is often foreign-owned multinational corporations concentrated in the wealthiest regions that demand a multinational labour force. Evidence certainly bears this out. For example, in 2002, some 91 per cent of all the work permits issued to migrants in financial services were concentrated in London and its surrounding region (Home Office/DTI 2002, cited in Williams et al. 2004), and roughly 70 per cent of H1B visas (highly skilled visas in the US) were concentrated in 20 of the largest metropolitan areas in the US (Ruiz et al. 2012). Bernard et al. (2014) report even greater concentration for the

Czech Republic. In terms of engineers and technicians, Williams *et al.* (2004) point out that they are more spatially dispersed than the executives and managers within global cities. No doubt this is the case, although there tends to be a concentration of foreign engineers and technicians in 'high-technology' cities and urban regions, from Austin, Texas, to Stockholm, Sweden, and to Munich and southern Germany in general (Bauer and Kunze 2004). In fact, about 70 per cent of Indian-born 'computer professionals' in Sweden were working in Stockholm between 2008 and 2010, and of these, some 87 per cent were men (Hedberg *et al.* 2014). Silicon Valley offers yet another well-known example of recruiting such individuals, where 53 per cent of all scientists and engineers were born overseas, and approximately a quarter were Indian and Chinese around 2000 (Saxenian 2005).[6] In the UK, engineers and technicians live within a triangle that stretches from Bristol through Oxford, to Cambridge and London. Most of these are men, too. In contrast, only 12 per cent of all highly skilled migrants in Britain's National Health Service (NHS) were located in London and the south-east of England (Williams *et al.* 2004). The matter of foreign doctors in the NHS is instructive in terms of what we have referred to as 'socio-professional downgrading'. The number of doctors qualified and migrating to the UK from outside the European Economic Area[7] (EEA) increased from 23 to 26 per cent between 1995 and 2000, yet in contrast to those from within the EEA, they were concentrated in less remunerated 'non-consultant career grades', in part a function of how they migrated to the UK in the first place, that is as students (Raghuram and Kofman 2002) or sometimes as asylum-seekers and refugees (Stewart 2008). If we narrow our analysis to the movement of nurses, and of teachers, then this migration is overwhelmingly composed of women (Iredale 2005).

With respect to immigrant/ethnic entrepreneurship, the range and types of businesses, and the sectors and niches into which entrepreneurs insert themselves, is geographically diverse (see Kloosterman and Rath 2003, 2014). One of the more studied cases is of Hong Kong-born 'astronauts', that is highly skilled, and especially high-income businessmen who 'buy' their citizenship from the United States or Canada, and fly between Hong Kong and the west coast of North America to conduct their often transnational businesses (Kobayashi and Preston 2007; Ley 2003, 2006; Ong 1999). Although it has remained a strategy to recruit such entrepreneurs by the Canadian government, they are far from always economically successful, owing in part to their unfamiliarity with the regulations and practices associated with operating businesses in, let us say, Vancouver (Ley 2003, 2006). Yet immigrant and ethnic entrepreneurship extends way beyond the typically cited global cities. Rather, the more numerically modest immigrant or ethnic 'communities' in smaller cities and towns facilitate the sustenance of entrepreneurship far beyond the bright lights of the big city.

Students, as we noted in Chapters 1 and 2, constitute a significant proportion of migrants across the world, and they are emblematic of the troubled distinction between temporary and permanent migration, as well as between other categories such as asylum-seekers and refugees (King and Raghuram 2013). Countless surveys show that at least some students stay in the country where they pursue their higher education, while others return to their 'home' countries (Hazen and Alberts 2006). Like other forms of highly skilled migration, student migration is distinct in terms of its geographies. In 2001, some 59 per cent of the world's foreign students were concentrated in just five countries (Australia, France, Germany, the US, and the UK)[8] and although that percentage fell to 55 per cent in 2014 (given the increase in student mobility within Asian countries, and particularly China) there is still a relative concentration of international students in about ten countries[9] (IIE/Open Doors 2015a), and these geographies are complicated by differences in the ease of mobility between macro-regions such as the EU and countries such as the United States. In the EU, policies of unfettered student movement between EU countries (and programmes to encourage it, such as the ERASMUS plus programme for 2014–2020, and its predecessors) have greatly enhanced student migration. Yet even within the EU, there is a concentration of foreign students in the UK in 2014 (about 11 per cent of the world's total compared to the next two EU countries: France and Germany – both 7 per cent). Some of this concentration is attributed to the desire of students to learn English (in order to acquire 'language capital'). At the Masters and PhD levels, foreign students tend to concentrate in the most prestigious universities, which have an urban inflection to them (Williams et al. 2004; IIE 2015a). In the US for example, although urban universities seem to capture the largest number of foreign students, at the undergraduate level, students are also found in universities in smaller cities and rural areas alike (IIE/Open Doors 2015b).

The events of 11 September 2001 curtailed student migration to the US from 2001 to about 2005 by complicating the procedures and checks for obtaining a student visa. In 2002/3, the number of foreign students enrolled in US universities decreased from a little over 586,300 in 2000–01, to approximately 564,700 in 2005–6, and showed year on year declines between 2002 and 2005. Not until 2006/7 did the foreign student population increase again (by 3.2 per cent) (Open Doors 2008). The increase in 2006/7 reflected at least in part an effort by US immigration authorities to simplify visa procedures for student migration, which may also affect their propensity to stay in the US and reside as highly skilled permanent residents (Hazen and Alberts 2006). While the economic and financial crisis of 2007–9 certainly slowed the growth of the international student population to about 3 per cent per year, it has since increased rather steeply from 690,000 in 2009 to about 970,000 in 2015 to nearly 5 per cent of the total student population in the US (IIE/Open Doors 2015b).

Undocumented migration and informal employment: the intersection of production imperatives and forces of regulation

A second element of the GCH that we mentioned concerns low-skilled/low-paid migrants working informally, but this is hardly confined to so-called global cities and extends from agriculture to mining and countless other paid and unpaid economic activities. One argument is that migrant workers have become essential to enabling the development of low-skilled, low-paid work, and so the economic structure of wealthier countries is thus intimately tied up with the availability of a 'willing' migrant labour force (Castles and Kosack 1973; Castells 1975). Migrants throughout the wealthy countries do in fact accept jobs well below their level of education and qualifications (Ruhs and Anderson 2013), and there is a clear logic here in explaining the dependency on migrant workers. However, as Waldinger and Lichter (2003) and many other observers have insisted, citizens (and in the US case, especially US-born 'blacks') would accept these so-called 'migrant jobs' to one degree or another if employers offered such jobs to them. After all, wages might increase in sectors where immigrants were unavailable which might then raise wages and then attract citizens (e.g. Cohen 1987; Gomberg-Muñoz 2011). In short, this 'dependency' argument is problematic.

In any case, countless numbers of low-paid and unpaid undocumented and legal migrants work informally (e.g. Marcelli et al. 2010). Informal employment has historically been part and parcel of global capitalism, but it does seem to be an increasing, if geographically variable component of many, but not all richer countries, at least from the 1980s to the late 2000s (Schneider 2012). But before we plunge any further into the relationship between undocumented immigration and informal employment, we need to at least provide some definitions and then explore the various legal or illegal(ized) scenarios of undocumented immigration and informal(ized) employment. In terms of 'undocumented', 'irregular', 'unauthorized', or 'illegal immigration', Walters (2008) notes insightfully that states do not actually have a stable definition of any of these, hence the now common focus on 'illegality' over time and space (De Genova 2002). Perhaps it is easier then, to define 'informal employment'.

The first point we wish to emphasize is that rather than speak of a formal and informal sector (e.g. Hart 1973), it is more appropriate to think of the formalized and informalized character of jobs on a spectrum or a continuum (Sassen 1998; Williams and Lansky 2013; Williams and Nadin 2014). Jobs or livelihoods may have both formal or informal moments, and some positions or tasks in a firm might be declared and others not (Smith and Stenning 2006; Williams and Nadin 2014). Following from the earlier definition of Castells and Portes (1989), Williams and Windebank (1998) define the 'informal sphere' as 'all productive' or 'work activities' that are "unregistered by or hidden from the state and/ or tax, social security and/or labour law purposes, but which are legal in all other respects" (p. 4). Yet they build on this definition by distinguishing further between informal economic activities that are paid – they call this informal employment – and two other related forms of informal economic activity: unpaid informal work (that is 'mutual aid' between individuals and groups), and illegal employment. The latter refers to the production of a product that would be illegal under any conditions (think of sweaters produced informally; the product is legal, but its production is not. This is in contrast to let us say cocaine which most citizens cannot really 'produce' legally under any conditions, at least in most countries).[10] In this chapter, we will be mainly concerned with paid informal employment, unpaid informal work, and 'illegal' (or illegalized) employment (especially trafficking) but not mutual aid or illegal (illegalized) employment, even if they are often intertwined. Now, with these definitions behind us, let us turn to some possible scenarios of informal employment and immigration that will help us understand these relationships (adapted from Samers 2005):

1. Legal migrants (including asylum-seekers and refugees) with legal residence and work authorization employed illegally producing licit (generally legal) goods and services. The process of production or service provision is illegal, although the final product or service, if produced or performed under legal conditions, would not be illegal – e.g. the 'sweatshop' production of garments.
2. Legal migrants (including asylum-seekers and refugees) with legal residence and no work authorization employed illegally producing licit goods (as in 1 above).
3. Legal Migrants (including asylum-seekers and refugees) with legal residence and work authorization employed illegally producing illicit goods ('sweatshop' production of banned items, drug trafficking, sex/sexual labour) (see Box 5.2).
4. Legal migrants (including asylum-seekers and refugees) with legal residence and no work authorization employed illegally producing illicit goods (as in 3 above).
5. Undocumented migrants (that is, they have neither legal residence nor the authorization to work) employed illegally producing licit goods and services (as in 1 above).
6. Undocumented migrants employed illegally producing illicit goods and services (as in 3 and 4 above).
7. Undocumented migrants employed legally producing licit goods and services (this is less common, and although it is illegal from the standpoint of residence and work authorization, the employer has hired the person legally based on falsified documents) (see Box 5.3).

Box 5.2 TRAFFICKING AND SEXUAL TRAFFICKING

We will recall from Chapter 1 that trafficking usually refers to a combination of an illegal mode of entry tied with informal or illegal employment, often involving 'exploitative conditions' (a 'slave-importing operation'). For the United Nations (UN 2000) "Trafficking in persons" shall mean "the recruitment, transportation, transfer, harbouring or receipt of persons, by means of the threat or use of force or other forms of coercion, of abduction, of fraud, of deception, of the abuse of power or of a position of vulnerability or of the giving or receiving of payments or benefits to achieve the consent of a person having control over another person, for the purpose of exploitation. Exploitation shall include, at a minimum, the exploitation of the prostitution of others or other forms of sexual exploitation, forced labour or services, slavery or practices similar to slavery, servitude or the removal of organs" (UN 2000, Article 3a, p. 3). Such a definition is inadequate though to cover the type of situations involved. In fact, trafficking might include actual physical violence, physical confinement to the work place, and threats of denouncing a migrant to the police or immigration authorities if the migrant does not comply with the traffickers (Anderson and Hancilová 2011). Furthermore, de Lange (2007) points out in the context of Burkina Faso, that child migration and child trafficking within the country are not so easily distinguished for a whole range of reasons, and thus what exactly constitutes 'exploitative' and 'forced' is unclear. Likewise, the relationship between smuggling and trafficking is more 'grey' than is asserted by the UN, since for example, smuggling can also involve the repayment of debts through difficult work with extremely low wages (e.g. Ahmad 2008b; Kyle and Koslowski 2001; Salt 2000). Nevertheless, the UN's conceptualization of trafficking above is widely seen as a benchmark definition, including by the IOM (International Organization for Migration), NGOs, and a range of scholars concerned with 'trafficking'. As with most illegalized flows, precise figures for trafficking are far from available, but it seems to be a widespread phenomenon, involving perhaps hundreds of thousands of people across the entire world. The reasons and conditions of trafficking vary from country to country and from one sub-national region to another. What seems clear from surveys is that trafficking, whether sexual or otherwise, entails a spectrum between a certain degree of voluntarism on one hand, involving a combination of cajoling and seduction about future opportunities (the lure of a 'western' lifestyle and consumer goods for example), and on the other hand, kidnapping, coercion, threats, fraud, false promises of acceptable work, forced labour,

Continued

and pseudo-imprisonment (Askola 2007; Ahmad 2008b; Salt, 2000). Anderson and Hancilová (2011) are far more critical of the idea of a spectrum, questioning the degree to which 'consent' is really possible when the conditions which migrants leave are so deplorable that they are more afraid of forced *non-work* in the countries of origin than they are of forced *work* in the country to which they have been trafficked.

Sexual trafficking seems to have received the greatest scholarly and media attention (Askola 2007). Agustín (2006) and Kempadoo (2007) have argued that sexual trafficking should not necessarily be viewed as forced migration, as women are aware that some level of sexual work will be involved (which they may not necessary dislike). In fact, from experience in their own countries, some women may prefer prostitution over other work. Thus, for Agustín (2006) and Kempadoo (2007), it is not incorrect to characterize at least some of the supposedly 'sexually trafficked' as simply migrant women who are prostitutes, and there is a need to separate migration for the purposes of sex/sexual work, from sexual trafficking. Thus, reference by governments and the popular press to 'victimization discourse' and the misreading of migration for sex/sexual work as only sexual *trafficking* rests in part on moral expectations of women's proper sexual behaviour and simply leads to tighter migration control over 'Third World populations' and a policing of women's 'sexual agency' (Kempadoo 2007: 82; FitzGerald 2016). However, as Agustín (2006) recognizes, many women often do not expect the working conditions they end up experiencing, and exposés by newspapers and magazines paint lurid and grim pictures of the phenomenon.

Box 5.3 HOW CAN UNDOCUMENTED IMMIGRANTS WORK LEGALLY?

It may seem impossible for undocumented immigrants to work legally. However, given the necessary social networks and the ability to pay, obtaining a 'social insurance' (SI) number is relatively easy. Indeed, in the UK a huge market exists for stolen SI numbers, and undocumented immigrants may find themselves working alongside legal migrants (Ahmad 2008a). Similarly, in the US, there seems to be an ever-expanding market for very convincing identification cards (*mica*) and other necessary documents that can be purchased in Mexican cities and certain US cities with large Mexican populations (Cornelius 2005; Meissner and Rosenblum 2009; Nevins 2008).

Points 1–7 above tells us that the categories we may have assumed to be quite black and white are actually much more complex. As noted in the introductory chapter, because migrants live in and between these categories, they sometimes find themselves in a grey area between legality and illegality with respect to immigration and employment laws – what Anderson *et al.* (2006) refer to as 'semi-compliance' (see also in the Introduction to this volume). That is, some may have a residential permit, but not a working permit. The working permit itself might entitle a migrant to work for only so many hours, which they may or may not violate. In sum, knowledge of this legal complexity allows us to critically engage with often sensationalized media representations, especially about the apparent growth, characteristics, and pervasiveness of informal employment.

It is equally common to hear that undocumented immigration is increasing across the rich world. Notwithstanding the problems of a stable, trans-state definition of undocumented immigration, there is considerable anecdotal and some statistical evidence to suggest that the number of undocumented immigrants (as defined by individual states) had in fact grown from at least the early 1990s until the mid-2000s in many wealthier countries. Yet, evidence from 2007–10, at least in the EU, suggests that the numbers of undocumented immigrants may be lower than before the mid-2000s (Morehouse and Blomfield 2011). Similar estimations have been found for Mexicans in the US, although not necessarily for many Central Americans (Massey *et al.* 2014). In terms of the relationship between these two phenomena (undocumented immigration and informal employment) one would assume that employers wish to hire the cheapest, most docile, reliable and productive workers and who have what Ross (2003) calls 'zero drag' (having no 'baggage' whatsoever, including children).[11] If we assume this, it might seem logical for employers to prefer undocumented migrants over legal migrants, yet this may not always be the case. For example, Iskander (2000) notes how employers in the Paris garment industry began to substitute legal migrants for undocumented migrant workers during the 1990s because only legal migrants, and not undocumented migrants, could operate across the different phases of production in the industry, some of which were illegal (or illicit) and some legal (or licit) in character.

Where do legal and undocumented migrants work?

Whatever the patchy evidence for the size and varied geographies of undocumented immigration and informal employment, it has become common throughout Asia, Europe and North America for undocumented and legal migrants who are deemed to be 'unskilled' for whatever reason, to be employed across the occupational spectrum, sometimes side-by-side, and in diverse locations. It is common for migrants to perform work both inside and outside ethnic/immigrant niches such as in agriculture, clothing/textiles and other down-graded and light manufacturing, construction, food and poultry processing, in small groceries, landscape gardening, mining, retail, street peddling and the 'three Cs' (care, cleaning and catering). Across some of these sectors at least, one might characterize the work that migrants perform in terms of the 'three Ds' (dirty, dangerous, and difficult). We might add 'demeaning' to that as well, perhaps when it relates to sex work, or wherever and whenever migrants are treated in ways which they themselves see as unacceptable. In certain countries however, there are exceptions to the sectors in which migrant workers are found, such as in South Korea, where migrants have been employed less in agriculture than elsewhere (Seol and Skrentny 2004). In other wealthier countries and regions, agriculture remains one of the most consistent employers of both legal and undocumented immigrants, from Polish and other eastern European workers picking vegetables in southern and eastern England (Anderson and Rogaly 2005; Rogaly 2008, 2009), to Jamaican and Mexican men gathering strawberries in Ontario and California (Bauder 2005). For example, Rogaly's (2008, 2009) research is exemplary of ILMS. He explores the demand for migrant workers by supermarket retailers through the 'intensification of British horticultural workplace regimes' and the increased and strong preference for migrant workers since the 1990s because they were believed to be "reliant, flexible, and pliant" (2008: 500) and "'crucial' to their businesses" (ibid.: 499). In terms of the 'supply of workers' or the 'processes of social reproduction', the 'willingness' of migrant workers (mainly young men, students, and from Eastern Europe) to accept such jobs were two-fold; they tolerated the low wages because they looked rather favourable when converted into Eastern European currencies at the time, and because seasonal work provided extra income for life projects such as furthering their education, and renovating or even building a house back in their country of origin. They were often housed in cheap, on-site accommodation which lowers the cost of their social reproduction and therefore the need to pay higher wages. Yet, the demand itself is generated through the power that food retailers have over 'growers' and packers, including in terms of requiring high quality produce at an ever competitive price. This has in turn driven suppliers (the growers/packers) to rely on legal 'gangmasters' (labour contractors) who then search and employ migrant workers through the 'Seasonal Agricultural Workers Scheme' (SAWS). Yet with respect to 'forces of regulation', the SAWS must be viewed both in the context of the licensing of 'gangmasters' which in turn drove out smaller growers, consolidated retail power over growers, and placed pressure on wages, as well as EU policy that involved the partial international integration of Eastern European migrant workers within the EU during the 2000s.

Elsewhere, migrants have worked in clothing and textile production, from predominantly Turkish men in Amsterdam's garment industry (Raes et al. 2002), to Central American men and women in Los Angeles and New York City (Light et al. 1999; Kim 1999), and to Vietnamese women (among others) in Malaysia (Crinis 2010). In construction, we discussed earlier how an 'army' of Indian and Pakistani migrants have for some time now laboured in the terrible construction jobs of Dubai, or Latino migrants in burgeoning Texas cities (Torres et al. 2013), but during the 1990s, both legal and undocumented Polish workers built the 'new Berlin' in similarly tough conditions before Poland's 2004 integration into the EU (Wilpert 1998). At the same time, some 21 per cent of the Lithuanians in the UK were concentrated in London's construction industry in the first few years of the 2000s (Spence 2005). Other migrant construction workers, such as Algerians and Moroccans are segmented into the least desirable and least paid jobs in the French building industry (Jouin 2006). The hundreds of thousands of sandwiches that are consumed each day in the UK would not be possible without the thousands of African and South Asian migrants working in food processing and sandwich-making factories in the London area (Holgate 2004) (see Box 5.4 for another example of food processing).

Box 5.4 ARE ALBERTA AND IOWA THAT DIFFERENT FROM DUBAI?

Slaughterhouses and meatpacking have become significant employers of both migrants and refugees, particularly in Canada and the US, although the percentage of Latino migrant workers declined from 57 per cent in 2005 to 48 per cent in 2010, after well-publicized raids on employers took place in 2006–7 in the US, and because the hourly wage (US$12/hr around 2010) in meatpacking jobs is high enough to attract legal migrants (Martin 2015). Nevertheless, in Canada, by 2006, 60 per cent of the workers in one plant in Brooks, Alberta were migrants and refugees, and in fact, 90 per cent of these same workers were refugees from Afghanistan, Ethiopia, Pakistan, Sudan, and Somalia, but also from other southern and eastern African countries. Exhausted and frustrated with their working conditions, they staged a 3-week strike in 2005, ultimately ending with the plant managers changing their recruitment strategies and hiring workers from China, the Philippines, El Salvador and the Ukraine (Broadway 2007). In the US, Immigration and Customs Enforcement raided the Crider, Inc. chicken processing plant in Stillmore, Georgia in 2006, finding that some 20 per cent of the 7,000 workers were undocumented (Martin 2015). On 12 May 2008, US immigration officers raided a Kosher meatpacking plant owned and operated by Agriprocessors, Inc., in the town of Postville, Iowa. The plant had already been cited previously for a whole host of violations, in part at the behest of Jewish community leaders elsewhere in the US who were dissatisfied with conditions in the plant. The raid found 389 undocumented immigrants mainly from Guatemala, working in a plant that routinely violated labour and safety violations, including hiring workers under the age of 18 (the state-defined legal working age in the meatpacking industry). Managers provided little training, they harassed workers – sometimes in a derogatory manner – and employees were forced to work overly long hours with little rest. At least one Guatemalan, named Elmer, worked 17 hours a day, 6 days a week, and the *New York Times* reported on July 27 that "he was constantly tired and did not have time to do anything but work and sleep". In the article, Elmer claimed that "I was very sad", and "I felt like I was a slave." This led to protests by, on one hand, Rabbis not associated with the plant, pro-immigrant supporters, and legal and undocumented Latino workers themselves (many of whom face arrest and/or deportation). On the other hand, there were locals and anti-illegal immigration counter-protestors from the Federation for Immigration Reform, content that immigration authorities had rounded up 'illegals'. The firm itself claimed that they did not willingly hire workers under the age of 18, and that the documents of workers were forged. The immigrants claimed that managers knew they were under age although many acknowledged using false papers to find work (*New York Times* 2008b). Does all of this sound that different from the story of south Asian workers in Dubai?

The employment of migrants in landscape gardening is particularly important in the United States, where the owners of often large, multi-acre executive homes demand pristine lawns and gardens. In this 'industry', commonly Mexican and other Central American migrants of different nationalities wait near hardware stores or similar informal sites to work as 'day labourers' in teams or as individuals. Those who choose them may be small business people with landscape firms or simply private home-owners (Valenzuela 1998). While day labouring extends across the employment spectrum in the US, and often involves both legal and more shadowy temporary firms (Valenzuela 1998; Theodore 2003; Theodore et al. 2015), this sort of day labouring is however, not widespread in EU countries, and may even be increasingly curtailed in the US in part because of municipal ordinances outlawing such sites from Hazleton, Pennsylvania to Carpentersville, Illinois and Danbury, Connecticut (*New York Times* 2007; *New York Times Magazine* 2007).

Employment in the 'three Cs' is heavily gendered, with so-called 'caring' jobs or 'emotional labour' (Hochschild 1983) increasingly dominated by migrant women, and particularly 'women of colour' from poorer countries.[12] But it is more than that. Hochschild (2000) sees this movement of domestic labour as part of a 'care chain', that is, "a series of personal links between people across the globe based on the paid or unpaid work of caring" (2000: 131). Typically, a care chain involves "an older daughter from a poor family who cares for her siblings while her mother works as a nanny caring for the children of a migrating nanny who, in turn, cares for the child of a family in a rich country" (ibid.). Such caring jobs include childcare, domestic work in the form of both live-in and live-out cleaners/maids, nursing and general health care work in hospitals, clinics, elderly/retirement homes, and private residences. There is a vast amount of research here from all of the wealthier countries going back to at least the 1990s (but more recently, see e.g. Ford and Kawashima (2013) on Japan; Lutz and Palenga-Möllenbeck (2010) on Germany; Silvey (2004b) and Pande (2013) on the Gulf States; Yeoh and Soco (2014) on Singapore; Lee (2006) and Song (2016) on South Korea; Anderson (2007) on the UK; and Hondagneu-Sotelo (2002) and Parreñas (2001) on the United States, among many others).[13] More sweeping contributions and critiques are offered especially by Ehrenreich and Hochschild (2004), Kofman (2012, 2013), Kofman and Raghuram (2012) and Yeates (2009, 2012). However, the care chain metaphor remains popular, and in terms of nursing specifically, Yeates claims that

> Countries at the top of the 'global nursing chain' (Yeates 2009) are supplied by those lower down the chain. For example, the United States draws nurses from Canada; Canada draws nurses from the United Kingdom to make up for its losses to the United States; the United Kingdom draws nurses from South Africa to fill its own vacancies; and South Africa draws on Swaziland. (Yeates 2010: 426)

Aside from the 'care chain metaphor', we addressed some of the reasons for the demand for migrant women as domestics in Chapters 2 and 3. Kofman (2013) and Raghuram (2012) are hardly dismissive of the care chain metaphor but they wish to reorient some of the debate back to the idea of 'social reproduction' (see Box 2.4), some of which is addressed through the 'global house-holding' literature (see Chapter 3) and to diversify our explanations of 'care work' across different countries (e.g. India) and differently 'classed' households. For example, Kofman (2013) notes (in drawing on Wang 2007) that wealthy and middle-class households in Asian countries may hire a foreign domestic worker for child-care and household chores, while in poorer households, a man might bring in a 'foreign bride' to address domestic needs.

Though working conditions vary enormously, dozens of studies point to the difficult, demeaning, and sometimes downright repressive conditions under which migrant women work. In the context of domestic labour in particular, we also discussed in Chapters 2 and 3, the 'always on' (Anderson 2001b) condition of migrant women that makes them especially attractive for employers. Migrant women typically work 60–75 hours a week with little rest and the line between public and the private life of the domestic worker becomes blurred. Domestic migrant workers may work in expensive but isolated homes with few amenities and little or no means of transport. Some women are restricted from bringing home boyfriends during the work week or entertaining guests (Hondagneu-Sotelo 2002). In extreme cases, they are sexually abused and/or their passports are taken (Anderson 2001b), and nationality and citizenship status are important determinants of what working conditions a migrant may accept. These work difficulties do not stop at domestic workers, as migrant nurses also face a host of problems in the hospitals and clinics of the rich world, from low pay to long hours (e.g. Yeates 2004, 2009, 2010).

Catering and cleaning on the other hand employ both men and women, and this involves a whole range of activities. Among them, Ahmad (2008a) and Wahlbeck (2007) document the difficult lives of Pakistani and Turkish migrants in London and Finland's take-away (take-out) shops – what Wahlbeck (2007) calls work in the 'kebab economy'. Here, the all-too-familiar working conditions reign. The undocumented 'Dead men working' (the title of Ahmad's paper) tells the story of long hours, low pay, but also a poor diet composed of the greasy and unhealthy food that the workers are usually provided free of charge by their employers. Sure, some migrants prefer working for co-ethnic bosses and may even report quite fair treatment, but many regret the confinement to the 'tiny physical space' of work and home in an 'endless cycle of work', the chronic instability that freezes them "in a vacuous present fraught with anxiety and question marks about tomorrow" (Ahmad 2008a: 315). That is a story of dismal lives in an 'ethnic enclave', but there is labour market segmentation and frustration even in the up-scale hotels of West London, where in the food and beverage division studied by McDowell et al. (2007), there were few people who were not white in the 'front' while 'people of colour' were relegated to back of the shop operations.

The cleaning sector in London and the UK is not remarkably different. Not associated with a particular ethnic enclave, there is little doubt that cleaning has become an immigrant (and not an ethnic niche) in cities across the wealthier countries. Recruiting migrant workers as cleaners has coincided with the privatization and contracting out of cleaning. Typically, in this cost-conscious sector, migrant workers are recruited because they are simply cheaper (May et al. 2007; Pai 2004). Whether it's cleaners in hospitals, schools, supermarkets, other large retailers, or London's Underground (metro), more than 90 per cent surveyed by May et al. were foreign-born. Some migrated to the UK for tourism purposes, others as asylum-seekers, students, au pairs, and so forth. In the study by May et al., they reported mainly legal workers from principally Nigeria and Ghana. There seemed to be a gender division of labour across the industry. Women were numerically dominant in semi-private spaces (e.g. hotels), and men in semi-public spaces (office cleaning, the Underground). Some 90 per cent were earning below the 'London Living Wage', working for about £5.45 an hour (about US$10), and £10,200 a year (about US$20,000), and this was before US taxes! Most were working as legal workers with written work contracts, and they paid taxes and social contributions.

Pai (2004) found both legal and undocumented migrants working for a firm that had been contracted out by a supermarket chain in London's Canary Wharf towers. The migrants were from Ghana, Nigeria, Ethiopia and South America. The supermarket chain had 'opted out' of the British Working Time Directive, and therefore cleaners could be requested by the contracted firm to work for more than 48 hours a week. Working for more than 48 hours entitled one to four weeks of holiday (vacation) pay, but this regulation had been routinely violated. Some migrants were paid differently for the same work that other migrants performed. The contracts did not confirm hourly pay, and there seemed to be the all too typical divide in terms of working conditions between those who were legal and those who were not. That points to the importance of regulation in shaping the working experiences of migration.

Forces of regulation

Entry and settlement policies as well as other economic and social policies, from labour market regulations to planning, zoning and housing initiatives, matter to the likelihood of migrants finding employment, and their position and experiences in workplaces. Such policies matter as much for legal immigrants as they do for undocumented ones. As Ruhs and Anderson (2009) put it, immigration controls produce illegality, and illegality shapes labour market outcomes. Since such interrelated policies are geographically diverse and national regulations overlap with supra-national forms of regulation, notably in the EU (Geddes 2014), it is impossible here to explore in any depth the myriad 'forces of regulation' that intersect with migrant workers' lives. Such a task would require at least another chapter in itself. Nonetheless, a few broad points should be either reiterated or emphasized at this stage, namely the role of entry and settlement policies in shaping labour market outcomes and experiences. These policies and regulations are aimed at both migrants and employers.

Migrants' labour market outcomes or segmentation are shaped by supra-national (i.e. EU) governance in at least three principal ways. First, EU authorities have sought to increase the recruitment of highly skilled workers through the EU 'Blue Card' as discussed in the previous chapter. Yet, it is now known that the EU's 'Blue Card' to attract highly skilled migrant workers to Germany is hardly viewed as a 'success', and that the Common Asylum and Migration Policy creates enormous obstacles to the recruitment of such workers and for asylum-seekers. Access to EU member states' labour markets is not only highly varied but often delayed, and with innumerable restrictions (Cerna and Chou 2014; Weber 2016). Second, the EU's Common and Asylum migration policy creates rules that determine where and when asylum-seekers or refugees settle (if only temporarily) in an EU country. Since asylum-seekers do not often have a choice about where they seek asylum, their language skills figure prominently in whether they will find work. Third, because of the restrictive nature of the CAMP, and the difficulty of claiming asylum, let alone being accorded refugee status, this can lead to undocumented status, which has its own consequences for employment as we have seen.

National governments use formal citizenship and visa regulations to control access to, and the duration of, different kinds of work. This might include tying work permits to proof of residential status, and tying residential status to the proof of employment, demanding evidence of adequate housing and evidence of sufficient financial resources, in order that migrants are not a 'burden on the public purse' (a phrase uttered to virtually all foreign residents in the UK by customs officials). Quite clearly, these policies are stratified so that different working regulations exist for different categories of migrants, and as we will see in the subsequent section, immigration policies together with economic and other social programmes can contribute to the weakened position of migrants vis-à-vis their employers – described earlier as 'citizenship exploitation'. Asylum-seekers and refugees in particular often have stipulations on the number of hours they can work or if they can work at all. In the UK for example, most asylum-seekers are prevented from working until they are granted 'leave to remain' (that is residential or refugee status). Furthermore, even if they are considered refugees, they only have the right to remain in the UK for five years. Two refugees that Stewart and Mulvey (2014) interviewed explained the implications of this for employment:

> actually it was quite difficult for me because while I study I did found job but with someone looking at my paper it's for five years, they don't want to take me because it's training, they have to spend time and money for training but they think maybe she's not forever here and we don't want to pay and just it was so difficult for me to find job. (Sophia)

> I can't find a job because I'm a refugee for five years . . . and I understand for employers why they don't take me for this . . . I have a lot of rights, I mean permission to work blah blah blah but the problem is that employers not very happy with my situation. (Gina)

As is common elsewhere, in Canada and Sweden, the propensity to be employed (and working conditions that one accepts) is shaped by the mode of entry of the asylum-seeker/refugee. In a quantitative analysis, Bevelander and Pendakur (2012) compare differences between asylum-seekers and refugees in terms of whether they receive the full or partial range of services (such as in Sweden), including paid or unpaid language training. In Sweden, both asylum-seekers and refugees receive paid language and employment training, while this is not the case in Canada, where only government-settled refugees receive language training with a stipend. While their analysis is far more nuanced than can be detailed here, they essentially conclude that the wider difference between the policy treatment of asylum-seekers and refugees in Canada in terms of paid or unpaid language training may explain the wider differences in employment levels between these two 'intake categories' in Canada.

In contrast, in Sweden, where treatment is more or less identical in terms of asylum-seekers and refugees (with the exception of the option of where they are allowed to settle in the country), the difference in employment levels between these categories is narrower.

Beyond asylum-seekers and refugees, governments deploy employer sanctions (fines and criminal charges) to dissuade employers from using either undocumented migrants, or employing legal migrants in an informal or illegal manner. These however are only sporadically enforced (see Box 5.5).

Box 5.5 EMPLOYER SANCTIONS – DO THEY MATTER?

Since 1986, the US government has imposed employer sanctions on those employers who 'knowingly' hire undocumented immigrants. However, between 1986 and 2003, the US Immigration and Naturalization Service only devoted 2 per cent of its funding to enforcing sanctions. This involved only 124 immigration officials for the whole country compared to 9,500 at the border, and the number of hours devoted to investigations fell by 50 per cent between 1999 and 2003 (Cornelius 2005). While the border became increasingly 'militarized' (Andreas 2000), the enforcement of violations fell by 70 per cent from 1992 to 2002. And in 2002, only 53 employers in the entire United States were fined for violations and only four employers were prosecuted in 2003. The average fine was about $9,729, which, for most large employers anyway, was very little money. Clearly, employer sanctions were not a priority for the US Congress by the early 2000s, even if they were for the Immigration and Customs Enforcement Agency. There have been attempts by the US Department of Justice to prosecute large corporations such as Tyson Foods (a food and especially chicken processing company) and Walmart for hiring undocumented workers. Tyson Foods, for example, were accused of hiring undocumented immigrants in 2001, but were never convicted. Part of the problem was that the clause 'knowingly' in the 1986 Immigration and Reform Control Act

---

Continued

meant that employers did not have to verify if the documents presented by potential workers were authentic, only that they checked for the presence of such documents and stated this on an 'I-9 form'. On some occasions, this also led to discrimination on the part of employers against Latino and other 'minority' workers who were *suspected* of being undocumented. In 2003, the US government developed the E-Verify system which is similar to the previous system, except that the same information is filtered electronically through various organizations, including the Social Security Administration and the Department of Homeland Security in order to verify that the worker has the legal right to reside and work in the US. If a worker cannot be authorized in databases, both employers and workers need to verify if there is a software problem, a data problem, or if indeed the worker is unauthorized to work. However, data and software problems impact on potential workers through no fault of their own; it is often costly and time-consuming to rectify this, especially for potential workers, and it even affects legal immigrants and foreign-born citizens, among others (Meissner and Rosenblum 2009; Rosenblum 2011). Despite the E-verify system (and despite the rapid up-take of the E-Verify system), between 2005 and 2010, enrolment has been voluntary (with the exception of the US state of Arizona which has made the system mandatory, in the hopes of pushing undocumented workers to seek work elsewhere), and non-compliance (that is, not submitting a query to the system) seems to be widespread. Furthermore, some states such as California and Illinois restrict the ability of employers to use E-Verify. Together, these continue to be the principal limitations of the system (Martin 2014; Rosenblum 2011). Similarly, there are no laws in the United States against private home-owners hiring undocumented immigrants, and we have seen how significant this sort of hiring is, from domestic work to landscape gardening (e.g. Cornelius 2005). *Perhaps more importantly, there has been a shift in the US from an emphasis on employer sanctions (that is, criminalizing employers) to criminalizing immigrants* (Jones-Correa 2012) as we saw in Chapter 4.

In the UK, employer fines have also proved weak and ineffective (Layton-Henry 2004), though fines rose sharply from around

£5,000 (about US$10,000) in 2001 to £20,000 (US$30,000) in 2013 for each 'illegal migrant' employed (UK government website, 2016).[14] The government of Spain increased inspections of employers during the financial and economic crisis, though it did not necessarily increase fines, and in any case, the recession and higher unemployment would prove far more effective at dissuading undocumented migration than any inspections did (Hazán 2014). In Japan, whose employer sanctions are modelled on those of the US and Europe, they too have proved inadequate (Tsuda and Cornelius 2004). The only country where employer sanctions seems to have any significant recorded effect is in France during the late 1990s, perhaps because of the comprehensive and wide-ranging cooperation between different agencies (Marie 2000).

In China, labour market segmentation through regulatory policies operates as much on internal migrants (i.e. Chinese citizens) as they do on international migrants. That is, during the 1950s, China developed the hukou (or household registration) system that prevented rural people from migrating easily to the coastal cities. The hukou entitles a resident to certain state subsidies and benefits for urban residents, but not for rural migrants. The Chinese government relaxed this system in the last years of the twentieth century to allow migrant workers to reside in the large coastal cities especially, but the most desirable and prestigious jobs are reserved for those with a hukou, and migrants without are forced into the least paid, and most unpleasant jobs. In short, resident status continues to shape the job prospects of the vast number of rural-to-urban internal migrants in the 2010s, despite moves to modify its restrictive character (e.g. Wang and Fan 2012).

With respect to employer regulations, as we mentioned in Chapter 4, governments create rules and incentives to attract certain kinds of workers, whether highly skilled/high-income or otherwise. Some have noticed a 'sectoral turn' (Caviedes 2010) in this form of recruitment. Yet earlier in the 1990s, Canada pioneered the idea of a 'points system' (Hiebert 2002) that we referred to previously, in which a combination of a migrant's 'adaptability', age, educational qualifications, investible income (including the willingness to purchase and operate a farm in Canada), and language ability provides differential access to citizenship. The various national policies to recruit highly skilled and entrepreneurial individuals entail variations on the theme of the Canadian point system in countries such as Australia, France, Germany, New Zealand, the UK, and the US. Koslowski (2014) distinguishes between three different 'models': the Canadian 'human capital' model using a points system to recruit permanent immigrants (not that the majority of immigrants in Canada are recruited for permanent residency!) (Siemiatycki 2015), an Australian 'neo-corporatist' model based on state selection using a points system based on criteria established by businesses and workers; and a market-oriented, demand-driven model based primarily on employer selection of migrants (in the US). In 2008, the UK introduced its own points system, building upon a similar and earlier system of 'tiers'. A decade later, the UK's 'points system' continues to be a complicated system of various tiers including entry for highly skilled migrants, entrepreneurs, sponsored skilled workers, temporary workers, European Economic Area (including Swiss workers), and 'other workers' (UK government website).[15] In this brave new world of points, some people are instantly disqualified. That is, if a migrant is categorized as low-skilled and low-paid (e.g. care workers), and from outside the EU and the European Economic Area, it is almost impossible to gain entry *as a worker*, though they be able to enter on other grounds. Likewise, as Kofman (2014) maintains, these points systems, although they appear 'gender neutral', are in fact gendered, given how they privilege certain sectors and occupations, which have their own gendered characteristics that range from sexism to the de-skilling of women over time through motherhood (Fossland 2013). The Australian government has had strong policies to recruit highly skilled migrants, but it chooses from students and temporary skilled migrants for permanent residence, as they are deemed to have more 'Australian experience' than migrants from outside the country (Hugo 2014).

Processes of social reproduction

Forces of regulation intersect with process of social reproduction to produce differential employment opportunities and experiences for migrant workers. The discussion that follows suggests how legality and social support, or indeed the lack of it, may create varying degrees of docility among migrants in the workplace. The equation we have in mind is simple enough; the less social support and the fewer rights migrants have, the more vulnerable and docile they are likely to be. This may not always be the case in practice, but this assumption will nonetheless motivate the following discussion. Again, it is impossible to produce more than a sketch of social reproduction across the richer countries, as the territoriality of social reproduction is exceedingly complex. Below we focus on two dimensions of social reproduction, a general picture of welfare transformations and how housing issues shape labour market participation and experiences.

In all their diversity, these social reproduction processes commonly include state-provided welfare (social and subsidized housing, food programmes, health care, etc.) and other 'integration' policies from free schooling and English language lessons to job training (e.g. Sainsbury 2012). Beyond the state, the availability of rentable housing, family and kinship dynamics, including 'social capital', social support provided by worldwide NGOs such as the Red Cross, CBOs (Community-based Organizations such as 'one-stop shops', drop-in legal clinics, and the like), hometown or homeland associations, houses of worship or religious organizations, and trade unions in both the countries of origin and destination fill in the holes where states have formally retreated from the international, national, regional, and localized landscapes of social reproduction.

One certainly broad and pervasive argument that encapsulates this retreat is that the neoliberalization of social reproduction has entailed a reduction in state-provided social assistance for citizens across the wealthier countries over the last 30 years or so and by extension for migrants. This has led to what Hondagneu-Sotelo (2002) calls the 'commodification of social reproduction' and thus the development of 'care chains'. By this she means that if social welfare is reduced and there is an increase in the proportion of dual-earner couples, then someone has to be paid (i.e. commodification) to perform the task of social reproduction (caring for children, cooking, cleaning, etc.). Migrant domestic workers are that someone.

Since state and non-state policies, including those related to domestic migrant labour, vary over space, we can say that social reproduction is shaped by complex territorialities. One certainly broad and pervasive argument is that 'neoliberalism'[16] has reduced state-provided social assistance for citizens across the wealthier countries over the last thirty years or so, and by extension for migrants. If this is true for migrants, it has certainly not been uniform (Bommes and Geddes 2000; Sainsbury 2012; Schuster 2000), and some individuals who would be classified as 'undeserving' years ago are now considered 'deserving', although the reverse is true as well (Sainsbury 2012; Sciortino and Finotelli 2015). While many observers have argued that in most wealthy countries, legal migrants and refugees are provided with social assistance on a par with citizens (Bevelander and Pendakur 2012; Bommes and Geddes 2000; Geddes 2000; Portes and Rumbaut 2014), Sainsbury (2012) argues that "major discrepancies exist between the social rights of immigrants and those of native citizens" (p. 280); moreover, they are dependent on time of residence, and increasingly of diminished value (Dwyer 2005). Furthermore, in most EU countries for example, what seems clear is that state policies have been tightened to reduce state provision for *asylum-seekers and undocumented immigrants specifically* (e.g. Darling 2009; Dwyer 2005; Sainsbury 2012; Schuster 2000). This has cemented a divide between those who do not have EU citizenship or the right to remain in EU, with those who are EU citizens and/or those who have longer-term residential status. In Spain for example, the Programa Greco (*Programa Global de Regulacion y Coordinacion de la Extranjeria y la Immigracion en Espana*) implemented in 2000, ruled that only immigrants who pay social security and taxes can have access to 'integration services'. This has had the effect of excluding both undocumented immigrants and legal immigrants working in informal employment from some of these services. On the other hand, *all* immigrants, *regardless of legality*, have in theory access to free medical care and school enrolment. Other organizations such as the Red Cross, Caritas, and SOS Racismo have tried to supplement governmental funding by providing services to migrants through *centros de acogida* (social service centres) that are managed by NGOs but financially supported (if inadequately to meet the demand) by the government and the Catholic Church (Cornelius 2004). However, we should note that the level and quality of social services varies by region and municipality, and has been undermined by the 2008 financial and economic crisis, to the point that the town of Vic in the region of Catalonia began to deny medical care and education to undocumented migrants (Hazán 2014). The services have included child-care, legal aid and Spanish language instruction. The situation for the children of undocumented immigrants has proved more difficult. Whereas few children are likely to be rejected from schools because they are not of legal status, they are not entitled to scholarships or vocational training, they cannot be given a diploma at the end of their elementary school studies, they cannot be provided with a work permit which in turn limits them to informal employment. As such, it is rare to see the older child of an undocumented immigrant working in more formal jobs (Cornelius 2004).

In the United States, the welfare reform bill of 1996 (the Personal Responsibility and Work Opportunity Reconciliation Act or PRWORA) created a new complex territoriality of welfare policies involving in large measure a shift from the Federal level to greater discretion by individual US states over the provision of welfare. On the whole, this legislation stipulated that 'non-citizens' could no longer access welfare funds in the first five years after their arrival, and could no longer use welfare for more than five years in their lifetime. They could not access the food stamp programme (a system in which clients purchase food with vouchers) or the SSI (Supplemental Security Income or 'old age assistance'). The only exceptions were refugees and asylum-seekers, and those who could prove they had lived and worked continuously in the United States for ten years. Though some of these benefits (food stamps and SSI) were partially restored by 2002, especially for children, most legal immigrants and many elderly and disabled immigrants either could not access, or had difficulty accessing, these benefits, often in the context of racialized welfare delivery (Marchevsky and Theoharis 2006). In its place, the 'roll-back' of welfare provision has been accompanied by the 'roll out' of workfare, in which legal citizens are either required to work in order to access benefits or simply work and not receive benefits (Peck 2001; Peck and Tickell 2002). The result of this 'roll out' and 'roll forward' neoliberalism is an often docile and flexible labour force of migrants (Marchevsky and Theoharis 2006) who frequently work two jobs a day at poverty-level wages. Furthermore, the curtailing of non-citizens' access to the food stamp programme has led to food insecurity among older and immigrant children in particular (Nam and Jung 2008; Van Hook and Balistreri 2006), as well as undocumented immigrants of any age (Hadley et al. 2008). Like welfare policies in the EU, it has created a divide between citizens and non-citizens (Hero and Preuhs 2007; Marchevsky and Theoharis 2006; Zimmerman and Tumlin 1999), and produces different attitudes and weaknesses with respect to employers.

Housing is a central element of social reproduction that shapes the employability of migrants and the conditions of their working lives. While finding suitable housing among high-income immigrants is unlikely to be difficult, for low-paid migrants, asylum-seekers and refugees, access to, and the availability and cost of, social or rental housing are likely to preoccupy their initial settlement, and may continue to weigh on their lives in subsequent years. Proximity to jobs is vital given the costs of commuting. In Philadelphia, for example, Iskander *et al.* (2013), show that for the immigrants they studied, the 'physical proximity' between the restaurants and construction sites in which they worked enabled them to use low-paid work in restaurants to subsidize them while they apprenticed in construction projects in the hopes of eventually obtain higher wages. The availability of relatively low-cost rental housing nearby in south central Philadelphia proved key to this relationship.

For social housing, access varies according to citizenship. In many countries, refugees are accorded social housing or vouchers for the rental market (although the standard obtainable is very poor), but for asylum-seekers and undocumented immigrants, access to social housing is generally not provided. The UK has a larger proportion of social housing compared to many EU countries, especially southern European countries, but there are sub-national differences. Waiting lists for such housing have ranged from seven years in the south-east (in and around London) to several months in more northern towns and cities. Overall in the UK, the amount of social housing has decreased, consistent with arguments concerning the neoliberalization of social reproduction (Dell'Olio 2004).

Private rental housing is not a simple alternative to social housing, and in many countries the supply is very constrained, although again with substantial intra-country differences. Indeed, the brutally high property prices of so-called global cities restrict the amount of affordable private rental housing, combined with an inadequate stock of social housing, leaving migrants to struggle for a place to live (Drever and Blue 2011). Yet this may be equally true in other cities or rural areas with a limited social or rental housing sector, as in Belgium, Italy, Finland, Ireland, and Luxembourg. The Netherlands, Sweden, and the UK have a comparatively larger rental market than other countries in the EU, but there is a sub-national geography to this supply of housing. In fact, asylum-seekers in the UK were dispersed to northern England, precisely because of the lack of affordable housing in the south-east of England (Audit Commission 2000). This in turn limits their employment possibilities in cities that have generally weaker labour demand (Phillimore and Goodson 2006). Discrimination against migrants 'of colour' and charging 'above-market' rents exacerbate the shortage. The result is 'overcrowding'[17] which presents innumerable problems for migrant workers, from lack of sleep and inadequate cooking facilities, to the problems of accessing jobs based on their residential location (Ahmad 2008a; Blumenberg 2008; Dell'Olio 2004; Dwyer 2005; Ozuekren and Van Kempen 2002). The problem of housing and even homelessness is vividly portrayed in Juan Carlos Frey's film *Invisible Mexicans of Deer Canyon*. In this film, he portrays the lives of undocumented Mexican immigrants living in isolated shacks without electricity or running water in the mountain canyons that surround the urban region of San Diego, California, often only a stone's throw from the multi-million-dollar homes that crown the surrounding ridges. These Mexicans move between their mini-shanty towns and mainly the sporadic informal 'day labouring' they can manage in San Diego's suburbs. It is difficult then to develop one's 'human capital' living and working under such conditions.

LABOUR MARKET SEGMENTATION AND MIGRATION IN POORER COUNTRIES

In terms of 'production imperatives' in poorer countries, it is difficult to create a universal picture out of their indisputable diversity. In fact, from the now modestly growing industrial heartlands of Asia and Latin America, such as Argentina, China, India, Mexico, the Philippines, Thailand, and Vietnam, to the declining fortunes of Brazil in the mid-2010s, and to the robustly growing but persistently poor sub-Saharan African countries, the image is one of uneven economic and social development (social and spatial inequality) both within and across poorer and more middle-income countries (e.g. *Economist* 2016b). Equally diverse are the forms of labour migration associated with this uneven development (Kofman and Raghuram 2012; Castles et al. 2014). In this very brief section, we will focus mainly on South Africa as vaguely representative of at least some poorer or even middle-income countries, since the gleaming offices in parts of downtown Cape Town and Johannesburg scrape the sky not far from immigrant-dominated shopping districts and the most abject urban neighbourhoods and townships. In the latter, unemployment may be close to the overall national formal unemployment rate in South Africa of around 40 per cent (Grant and Thompson 2015).

Indeed, on one hand, a limited number of relatively wealthy urban regions in poorer countries have witnessed rising living standards for some people, from Mumbai and Nairobi to Rio de Janeiro and Shanghai. On the other hand, a pervasive argument among social scientists is that three decades of what many have called structural adjustment, 'neoliberalism', or even 'post-neoliberalism' have created severe problems of joblessness, food insecurity, and chronically poor health for both the citizens and migrants of countless poorer countries. In this context, some scholars in the early 2000s claimed that because of structural adjustment and other forms of neoliberalism, informal employment had done nothing but expand in the face of widespread formal unemployment and minimal social entitlements in the cities, towns and rural areas of the global south (Beneria 2001; Davis 2004; Harriss-White 2003). Such claims are confirmed by both more recent data (although without the attribution to neoliberalism) (Schneider 2012), and more national or localized qualitative studies of South Africa for undocumented migrant workers and entrepreneurs alike (e.g. Bloch 2010; Grant and Thompson 2015). Indeed, work in agriculture (see Box 5.6), domestic and other services, manufacturing, and mining have all created considerable demand for extremely low-paid migrant workers, both internal and international, trafficked and not trafficked, from Argentina to Kazakhstan to India and South Africa (e.g. Anderson and Hancilová 2011; Bloch 2010; Crush 2011; Kofman and Raghuram 2012). Yet such demand is accompanied by a vast range of formal and informal entrepreneurs, from especially Habesha (Ethiopian/Eritrean) retailers of handbags, clothing and textiles, among countless other goods and services, to Mozambican food vendors and other Somalian traders in Johannesburg and in South African townships (Grant and Thompson 2015).

How do these sets of production imperatives merge with 'forces of regulation'? Here let us briefly mention two examples. The first is immigration policies, which are clearly tighter and more strictly enforced in some countries than in others, depending on economic conditions and political will, as well as crucially, the resources made available for enforcement. The border between South Africa and Zimbabwe is in fact heavily policed, although such control is routinely flouted by undocumented immigrants who either evade control individually, are smuggled and/or trafficked, or who bribe border guards and other officials. Such evasion is common elsewhere, and rampant where borders are even more porous.

In terms of labour standards, which have, in many instances, been curtailed under structural adjustment anyway, residual regulations are routinely ignored or violated (e.g. Hughes 1999), especially with respect to work-related immigration rules. The lack of resources or the unwillingness to enforce work regulations negates the importance of such regulations on paper. In South Africa, Bloch (2010) notes that the Corporate Work Permit Scheme under the 2002 Immigration Act allowed private firms to hire workers of *any status*, provided that they register them and that the Department of Labour verify that their pay and benefits are identical to all workers. Yet, Zimbabwean farm workers were being paid below the minimum wage, illegal deductions were common, and overtime work might be declared at the last minute, and far from voluntary. This describes, unfortunately, much of the employer-migrant relationships in the labour markets of the global south.

Box 5.6 AGRICULTURAL PRODUCTION AND THE DEMAND FOR MIGRANT FARM WORKERS IN SOUTH AFRICA

(From Johnston 2007)

In the country's predominantly rural Free State during the early 1990s, 'white' farmers began to switch from hiring 'black' South Africans to hiring mostly Basotho women (citizens of neighbouring Lesotho) to work in harvesting and processing a diverse range of crops at an approximate wage of a meagre $1.65 *per day*, for a gruelling 10 hours a day, 6.5 days a week (see Map 5.1). A number of 'forces of regulation' (or more appropriately de- or re-regulation) on the eve of the fall of Apartheid conspired against white farmers; these included a de-regulation of marketing and processing of many agricultural products; lower crop-specific subsidies from the South African government; a decline in other financial subsidies including for cheaper credit; a worsening macro-economic environment that involved the deteriorating exchange rate of the South African currency (the Rand), which led in turn to a rise in the cost of imported agricultural inputs; and drought, which exacerbated farmer indebtedness. All of these together placed enormous cost pressures on most white South African farmers.

As a consequence, farmers began to plant more asparagus and other horticultural products in the Free State, mainly because it proved more profitable than cultivating other staples such as wheat and maize. While certainly, labour costs remained a significant

---

Continued

component of costs for employers, that alone did not explain the hiring of Basotho workers and women in particular, since many men offered to work at the same wage. Rather, the answer lies partly in the nature of the product and its 'production imperatives', but mainly in gendered stereotypes. Asparagus farming, which is extremely delicate and time-sensitive, is heavily mechanized and labour costs were actually a declining percentage of total production costs for employers. However, given how time-sensitive the harvesting of asparagus is, employers depended upon a night-shift-ready flexible labour force that could be called to the fields and canning factories at precisely the right moment. At the same time, employers were worried as much about labour stoppages or outright strikes as they were about cost, anticipating that 'black' South Africans would become more demanding after the official fall of Apartheid in 1994. In short, labour costs figured in the production equation, but so too did flexibility and docility. Thus, using labour contractors, employers turned especially, but not exclusively, to Basotho women with the legal right to work in South Africa. Women were *considered to be* appropriate for what employers perceived to be 'women's work'. In Basotho women, employers saw a more docile, harder working, and more dexterous workforce than they did in South African or even Basotho men (by the way, dexterity and docility are also common stereotypes of Asian women).[18] This might explain why some 60–75 per cent of the farm workers were women, and why on at least one farm surveyed by Johnston, the percentage of Basotho seasonal migrant workers increased from 10 per cent in 1985 to 82 per cent in 1992. Employers preferred to recruit women with older children (their average age seemed to be 40), rather than those with infants or very young children. They believed that Basotho women with older children were saddled with considerable household responsibilities, and therefore would be more desperate for the money they sought to earn. As Johnston (2007) points out astutely, while white South African farmers claimed this would improve the lives of these poor women, ironically, these same farmers protested against earlier South African legislation that would have improved the pay and conditions of farm work.

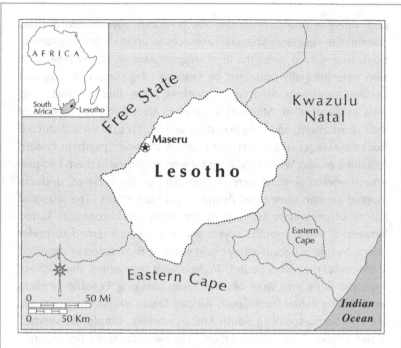

Map 5.1 Lesotho and the Free State

This contract labour system appears to be an ideal system for employers but it ran into the problem of losing skilled canning labour throughout the year, and thus missing out on maximizing their investment in the on-site canning factories. Farmers began to plant other vegetables outside the asparagus growing season, lengthening the contract of some workers and issuing promises of employment in subsequent seasons. According to Johnston, employers' efforts in this regard were emblematic of a nascent 'migration system' between the Free State and Lesotho in order to maximize worker productivity and thus returns on farmers' investment.

Legal restrictions on hiring undocumented workers included considerable fines (for both employers and employees) and a five-year prison sentence for each migrant worker employed. Fines however,

Continued

were rarely imposed, since the South African government did not commit the necessary financial resources to upholding these regulations. In any case, even the legal migrant women in the Free State who were allegedly protected by Lesotho's legislation on pay and working conditions for migrant workers, often had contracts that were easily violated. Migrants were often forced to pay the cost of their recruitment, and transportation and health care were deducted from their wages in a haphazard manner; the wage payment system remained erratic. While legal migrant women (some of them had previously worked illegally) were predominant on many farms, undocumented women were hired despite legal restrictions. The seasonal nature of asparagus farming meant most work contracts lasted between 4 and 6 months a year and employers seemed to prefer migrant workers because they could easily be repatriated or deported at the end of their work period. Workers were segmented along ethno-linguistic lines into work teams; Sotho-speaking Lesotho workers were distinguished from South African Xhosa or Tswana speakers, but also Sotho-speaking South African workers. Employers seemed to use ethnic-based segmentation to prevent workers from organizing, which is an all too common strategy among employers in poorer and richer countries alike.

The same ethnic distinctions mentioned above spilled over into the sphere of reproduction, where dormitories near the farms were also segregated along ethnic lines. Yet the abysmal conditions of social reproduction in Lesotho as well, were also crucial to this putative migration system, and as we discussed in Chapter 2, may be an exemplary case of 'super-exploitation'. Employment opportunities – especially more formal employment – were scarce, and wages extremely low in Lesotho. Basotho migrant workers originated from particularly poor families with little education and poor health care. Many were single mothers with large numbers of children, and if they were married, remittances from male family members were unlikely. As with so many other labour migrations, this created a desperate labour force willing to work under conditions that would allow white South African farmers to remain profitable (Johnston 2007).

The case study discussed above represents a set of processes which are by no means unique to other poorer countries or to richer ones, but nonetheless in its timing and territoriality, creates a specific constellation of production imperatives, forces of regulation and processes of social reproduction which converged to create a geographically specific experience for migrant workers.

CONCLUSIONS

In this chapter, we briefly discussed the reasons for labour migration, and then outlined three prominent theories that concern the relationship between labour markets and migration, namely human capital theory, dual labour market theory, and labour market segmentation theory (LMST). We expanded LMST further by examining some variations including (for lack of a better expression) 'a cultural capital, cultural judgments and embodiment approach', the significance of what Waldinger and Lichter (2003) called the 'migration literature', as opposed to 'the labour market literature', as well as what Samers (2010d) calls 'international labour market segmentation'. Without completely discounting human capital theory and even less the idea of dual labour markets, we chose to harness LMS and ILMS in order to focus on how policies, the practices of firms, organizations and households, ethnic and immigrant communities, and spaces, rather than simply one's human capital, shape outcomes in twenty-first-century labour markets. Yet, we should not ignore 'human capital variables' including the more sophisticated ways in which human capital approaches have been modified (with their attention to age, gender, 'contexts of reception', and so forth). Yet unemployment and the 'devaluation of immigrant labour' through problems of unrecognized skills or qualifications that are shaped by immigration policies, employment policies, employers, organizations, their practices of negative or even positive discrimination; and their relationship with social reproduction processes, suggest that a more critical approach along the lines of ILMS might be more fruitful for understanding the labour market position and experiences of migrant workers. That is, this approach typically implied that citizen employers segmented migrants from poorer countries into – generally speaking – low-paid jobs based on migrants' immigration status, nationality and other ascribed (stereotyped) ethnic, gender, or racial characteristics. Thus, skill is *constructed*, not just an attribute of workers. So we began with some general 'production imperatives' that were roughly universal among the wealthier countries, but acknowledged that such a 'spaceless' perspective is inadequate. As a consequence, we turned initially to the global city hypothesis (GCH) which serves as a framework for examining both 'highly skilled/highly paid' employment, but also the relationship between undocumented immigration and informal employment, which we maintained were central features of migration and work in both wealthier and poorer countries. The GCH is insufficient however. For the last twenty years or so in richer countries, migrants have either increasingly settled in less expensive cities ('new gateway cities') and/or in the suburbs (sometimes 'ethnoburbs') of the largest metropolises. Within this variety of cities, migrants are often found in ethnic 'enclaves' or similar spaces, and frequently working in immigrant or 'ethnic' economies where the employers are themselves migrants. In these sectors or occupations, migrant workers or small-scale entrepreneurs constitute such a large proportion of the labour force (that is, they involve ethnic or immigrant 'niches') that migrants rather than citizens are recruited for jobs. This micro-geography of labour markets matters for how workers find jobs, what jobs they might perform and perhaps the experiences they have of work, which are often awful, and *sometimes* promising socio-economic mobility over time. This, we argued, is a useful means to understand the position of migrants through different territorialities, not least those of cities, but also rural areas and small towns, national states, and macro-regions such as the European Union. We finished the discussion of wealthy countries in particular, by exploring in more detail some of the 'niches' in which migrants work, and then outlining some basic points concerning the reproduction and regulation of labour migrants. The chapter came to conclusion by exploring some examples of labour market segmentation in poorer countries, which seemed to mirror many of the processes associated with segmentation in richer countries.

Since economic expansion and job creation is such a crucial goal for the majority of ruling governments in the world, labour migration is deemed to be a pivotal strategy to meet this objective. Wealthy country governments are especially preoccupied with what is viewed as more highly skilled labour and poorer countries with retaining highly skilled labour, but most governments recognize the need to create channels for all different levels of skill for a diverse range of sectors and occupations. Thus work and its geographies feature at the heart of governmental responses to migration and immigration in countries around the world, as well as the survival of immigrants. Their experiences of work and employment both shape and are shaped by citizenship and senses of belonging, and it is to these latter ideas that we now turn in Chapter 6.

FURTHER READING

Readers interested in labour migration in general can start off with some general texts already mentioned in Chapter 2, including Castles and Miller's *The Age of Migration* (4th edition). Some of the major volumes and articles that influenced this chapter include Harald Bauder's (2005) *Labor Movement: How migration shapes labor markets*, or Waldinger and Lichter's (2003) *How the Other Half Works: Immigration and the social organization of labor*. On informal employment, Williams' (2010) chapter in Marcelli, Williams, and Joassart's (eds) *Informal Work in Developed Nations* provides a more recent assessment of informal employment. Chapters 6 to 15 in the 2016 *Routledge Handbook of Immigration and Refugee Studies* has several chapters relative to work. Kofman and Raghuram (2012) conduct a useful survey of domestic labour in poorer countries, as well as some of the theoretical debate on domestic labour. SOPEMI provides data on OECD countries in its annual publication *International Migration Outlook*, and the IOM (International Organization for Migration) provides data for the entire world, but thankfully as well for the vast number of countries beyond the OECD.

NOTES

1 Later in the chapter, we discuss the understanding of skill in more detail.
2 For more recent evidence from the 2010s, see e.g. Bradatan and Sandu (2012) on Spain; Chiswick and Miller (2010) for Australia; Creese and Wiebe (2012), Rajkumar et al. (2012), Zuberi and Ptashnick (2012), and Preibisch, (2010) for Canada; Connor and Koenig (2013), Koopmans (2010), and Reyneri and Fullin (2011) for comparisons of various European countries; Fleischmann and Dronkers (2010) for a European comparison in terms of unemployment; and Lessard-Phillips et al. (2013) for the 'second generation' in Europe; Kanas et al. (2012) and Kanas and Van Tubergen (2011) for Germany; Waldinger and Luthra (2010) and Portes and Rumbaut (2014) on the United States.

3 The term 'African-Americans' refers to US-born 'blacks', rather than to those born in Africa who migrated to the US.

4 There are various terms for informal employment, such as 'informalized' or 'undeclared' employment, 'shadow economy work', and so on, e.g. Schneider (2012), Williams and Nadin (2014). For the sake of simplicity in this chapter and throughout the book, we will simply refer to it as informal employment (this is common usage), recognizing the limitations of this term, and the diversity of conditions involved.

5 At least until the 1990s in Sweden, the low rate of the employment of undocumented immigrants in informal employment could be explained by the regulation and power of trade unions and collective agreements in regulating labour markets, Hjarnø (2003), Schierup et al. (2006).

6 The term 'ethnic niching' is generally not used to refer to the concentration of highly skilled migrants in such sectors, but it would not be incorrect to use this term.

7 In 2008, the European Economic Area consisted of Liechtenstein, Iceland, and Norway and these countries benefit from preferential trading and other agreements.

8 Tremblay (2002, in Iredale (2005)) provides a figure of 80 per cent. The comparative historical data that we use is from the International Institute of Education.

9 The British newspaper The Guardian has an extremely useful interactive map on the geography of international students. See http://www.theguardian.com/higher-education-network/blog/ng-interactive/2014/jul/17/international-students-where-do-they-go-to-study-interactive?view=desktop

10 Again, for the sake of space, we have kept this discussion rather simple. For more on the debate around defining informal employment, see Williams and Lansky (2013) and Williams and Nadin (2014).

11 Cited in Smith and Winders (2008).

12 Raghuram (2012) has questioned the political value of the persistent emphasis in the migration studies literature on marginalized women from the global south. That is, she is concerned that scholars might be reproducing a certain stereotype in the process of trying to tackle inequalities and marginalization.

13 See also Chapter 3, footnote 6.

14 Refer to https://www.gov.uk/government/news/employers-exploiting-illegal-immigrants-face-tougher-sanctions

15 See https://www.gov.uk/browse/visas-immigration/work-visas

16 In Chapter 2, we acknowledged some difficulties with this term.

17 'Overcrowding' is always relative to prevailing citizen and state-defined norms, practices, and laws.

18 The stereotype of expected docility is not easily dismissed since critical researchers have pointed this out about undocumented migrants in general, who, faced with the threat of deportation, acquiesce to employer demands. It is true, few Basotho workers had the institutional channels to complain, and often few did. However, according to Johnston's study, Basotho women and other workers did organize themselves by ethnic group or job type to protest frequently employment conditions although no strikes longer than about half a day took place.

6

GEOGRAPHIES OF MIGRATION, CITIZENSHIP, AND BELONGING

INTRODUCTION

Consider the lives of one Ecuadorian family in New York City, as reported in the New York Times in April 2009.[1] The parents have settled as undocumented immigrants in Queens, a borough in the city. The mother, who had a difficult career as a computer systems analyst in Ecuador now does babysitting. The father, who had a series of low-paid engineering jobs in Ecuador, and had already studied for an unfinished engineering degree in New York in the 1980s, now works as a draftsman for a Chinese immigrant-owned construction company. The father arrived clandestinely via Texas in 2001, while the mother and daughter arrived in New York on tourist visas, but overstayed in the same year. The son was born in Miami and is an American citizen. They are therefore a 'mixed status family', where the son is legal and the daughter is undocumented. While the son can travel freely back and forth between Ecuador and the United States, and pursue a formal job in the US, he wishes to return to Ecuador and the family feels that he has taken his citizenship for granted. Meanwhile, the mother has tried to use her son's citizenship status to obtain a 'green card' (meaning a permanent residency card) for herself.

The daughter, who loves living in the United States with all her friends, and has graduated with honours from a local high school and university,[2] fears leaving the city because of the danger of being apprehended by immigration authorities. She cannot take a high-paying job with the accounting degree she holds, obtain a driving license or a social security number. The diligent daughter worries about her future, but has found a reasonable, if low-paying job working legally as a bookkeeper for a small company that provides immigrants with information about immigration policy and visas. She works legally because the US government provides tax identification numbers for people without social security numbers. The mother wants her daughter to find an American husband to obtain citizenship status. She is reluctant, though perhaps slowly changing her mind, but nevertheless works with her undocumented boyfriend for an NGO that is lobbying the American government to push the 'Dream Act' which would give legal status to undocumented high school graduates brought to the US by their parents.

The above anecdote points to the centrality of citizenship and belonging to migrants and migrant families. This chapter has two aims then, to continue the discussion as to how citizenship figures in immigration policies and the relationship between forms of citizenship and migrant belonging. The study of this second relationship, particularly with respect to nationality law is surprisingly young. It was widely assumed by academics during the nineteenth and most of the twentieth century, that migrants would gradually assimilate over time (especially in the United States) or that they would be restricted to the status of 'guest-workers' (in Europe, for example) who would eventually return 'home'. Not until the 1980s did the relationship between migration, citizenship and belonging begin to generate vigorous debate among academics in Europe and North America, as it became clear that neither assimilation nor 'guest-workers' returning home were simple or inevitable outcomes (Bauböck 2006; Hanson and Weil 2002b). While spatial ideas and arguments have permeated academic debates, these debates still remain spatially 'immature', having on occasion hastily thrown out the nation-state in favour of either transnational citizenship or local participation and belonging. This chapter is designed to understand these complex geographies of citizenship, migration and belonging.

After an explanation of the chapter's geographical premise, we draw on some basic distinctions now used in the literature on citizenship and migration, namely the differences between citizenship as legal status; citizenship as rights; citizenship as belonging and identity, and citizenship as political participation (e.g. Bloemraad et al. 2008; Bosniak 2000; Leitner and Ehrkamp 2006; Joppke 2007b; Lister and Pia 2008), and we use these academic and social distinctions to divide the chapter's discussion into four separate, but related sections.

SPACE, MIGRATION, AND CITIZENSHIP

A central debate in the migration studies literature around citizenship and belonging involves the continual salience of nation-states as the determinant or locus of citizenship and belonging in the face of international law, the emergence of an international human rights regime, transnational belonging and the apparent 'denationalization' of citizenship (Sassen 2006b). This is sometimes referred to as the debate between 'nationally oriented' forms of citizenship on one hand, and 'globalist', 'postnational', 'denationalized' or 'transnational' forms of citizenship on the other (e.g. Joppke 1998a, 1998b; Sassen 2006b). To add to this are the calls by Favell (2001, 2008) to move away from national states as a lens on citizenship, in favour of viewing citizenship from the perspective of cities and their wider urban areas. Indeed, in light of neoliberalism and new forms of political participation, we may need to stretch our understanding of citizenship beyond states altogether, in order to examine new forms of trans-local economic, political and social participation and resistance (for a review, see Ehrkamp and Jacobsen 2015). It may even be necessary to lower further the resolution of any analysis down to households. The opening anecdote of this chapter tells us this, but Elias (2008) also argues that the household or 'home scale' is crucial. Drawing on work in Malaysia, Elias insists on this because the discourse and delivery of migrants' rights is distinctly gendered in Malaysia; that is, Indonesian and Filipino women constitute the bulk of domestic workers in the country. Rights associated with the public sphere are often meaningless in the private worlds of households in which domestic workers circulate.

Despite the promising corrective of analysing citizenship both in terms of supranational and sub-national entities, this obscures how nation-states as forms of territories still matter to the creation and contours of citizenship. In that sense, it is still too easy to dismiss the significance of nationally centred forms of citizenship and belonging amidst the fervour and haze of transnationalism or globalism that so pervaded debates about citizenship during the early 2000s (e.g. see most recently, Isin (2012)), or on the other a seemingly new penchant for urban or local citizenships given the supposed dissolution of national states. Thus, we maintain that interlocking supranational, national, and sub-national territories matter for the production of differential, 'stratified', or 'segmented' forms of citizenship, belonging, and participation. This does not mean however, that immigrants (or even citizens for that matter) are passive receptors of territorially rooted citizenship; they produce this territoriality as well. Despite this complexity, for many migrants around the world, citizenship (and even better dual citizenship) in the country of immigration is an ultimate goal, though permanent residency may in some cases be equally desirable. In that sense, we might think of citizenship as a 'strategy' too (see Box 6.1).

Box 6.1 CITIZENSHIP AS STRATEGY

Many studies on migration and legal status discuss migration as either a 'waiting room' where migrants 'do their time' and finally achieve full legal citizenship, or they show how migrants have actively protested for regularization, residence permits, and other social rights. While the transnational citizenship literature certainly explores the obtainment of citizenship as a (coping) strategy, this is still a less common perspective. In contrast, Ong's well-received book *Flexible Citizenship* (1999) speaks of citizenship as a 'flexible strategy' ('flexible citizenship') in order to survive and even prosper economically. Specifically, she discusses the lives of the so-called astronauts, that is, highly skilled, and especially high-income businessmen who 'buy' their citizenship from the United States, and fly between Hong Kong and the west coast of the United States to conduct businesses. Meanwhile, their wives and children, and sometimes only their children, live out their lives in the suburbs of San Francisco for example, taking advantage of educational and other opportunities in the United States. In the process, they transform the urban and suburban economic, political, and social landscapes of the west coast of the United States (see also the research in Canada by Kobayashi and Ley (2005), Preston *et al.* (2006), and Waters (2003)). Contrary to Ong's (1999) research of 'flexible citizenship' for economic ends, Preston *et al.*'s (2006) study of Hong Kong men and women living in Toronto and Vancouver found different motivations and strategies. To begin with, many of the men were not 'astronauts' at all, and some had relatively low incomes in these two Canadian cities. Few had remaining *economic* ties or owned property in Hong Kong. Most women had migrated either because their husbands wished to settle in Canada or because they thought a Canadian education would be better for their children. In addition, for men, family reasons and the fear of political repression after Hong Kong's hand-over from the UK to China in 1997 were the source of seeking citizenship in Canada. Overall then, for both men and women, citizenship seems to be a goal for the sake of their families.

From a completely different context, Mavroudi (2008) speaks of 'pragmatic citizenship' among Palestinian refugees in Athens, Greece. Her study reveals to us the profound feeling of exile and marginalization that Palestinians feel in Greece and the wider world. At the same time, many dream of returning once again to a Palestinian state, while others are ambivalent about returning, or not as hopeful about the prospects for peace. In both cases, however, Palestinians use the complex and difficult reality of their largely useless Palestinian passports or statelessness, the obstacles to obtaining Greek citizenship, and their low income to forge a 'pragmatic citizenship'. This practical citizenship entails either waiting for Greek citizenship – for themselves or for their children – or uses friends and relatives in, for example, Canada or the UAE, in order to gain new forms of mobility with the object of a better life, perhaps with the aim of ultimately returning to a Palestinian state, or perhaps not.

Migrants' diverse origins and destinations, and their different levels of formal citizenship affect, but do not determine, the type, quality, and scope of rights that they receive. Yet it is not states that unilaterally provide these rights in some generous top-down manner; rather, as we noted above, migrants themselves clamour for specific rights to which states respond out of appeals to humanitarianism, matters of (economic) practicality, or the fear of large-scale protest. Indeed, as we saw in Chapter 4, protest for citizenship and greater rights are common among immigrants throughout the world. Like other forms of citizenship, such contestation on the part of migrants is inflected by different places, spaces and territories, and these geographies of citizenship will prove central to the perspective adopted in the remainder of this chapter.

CITIZENSHIP AS LEGAL STATUS (FORMAL CITIZENSHIP, NATIONALITY AND NATURALIZATION)

The idea of citizenship is as old as the Greek *polis*, but if we can speak of the 'modern form' (nation-state-based citizenship), then its origins may be traced to the French Revolution and the formation and consolidation of nation-states in the nineteenth century (e.g. Brubaker 1992; Hansen and Weil 2002a), or even further back into feudal Europe (Bauder 2014b). This modern form of citizenship, which gave birth to such elements of control as the 'passport' (Torpey 2000), concerns bounded populations with rights and responsibilities and it excludes others by virtue of nationality. We should distinguish between citizenship and nationality however. While both involve the national state, each of them, as Sassen (2006b) explains, "reflects a different legal framework. Both identify the legal status of an individual in terms of state membership. But citizenship is largely confined to the national dimension, while nationality refers to the international legal dimension in the context of an interstate system" (Sassen 2006b: 281). Thus, it is important to emphasize that while nationality may be another word for citizenship, they are not synonymous, but "two sides of the same coin" (Bauböck 2006: 17).

This section of the chapter is in essence about the attribution or acquisition of nationality and from that, full citizenship. This can occur through citizenship by birth, by descent, by marital status, or by residence (Hansen and Weil 2002b). Depending on the country in question, full citizenship might involve a set of rights and responsibilities from the right to vote in national elections to holding public office and required military service. While the idea of even formal citizenship might seem complex, the point to be gleaned from the above discussion is that formal citizenship and nationality are by definition exclusionary – they exclude others – even if they are not bounded by a single territory such as a nation-state.

In the introduction to this book, we highlighted how certain migration categories used to define migrant statuses have been the subject of increasing criticisms. Such criticisms are warranted, but they are also exaggerated. Certainly, individuals who migrate as students for example, may convert their status to permanent residency, and eventually obtain nationality over time on the grounds of, let us say, family reunification or employment. This is particularly common among students in Australia, for example (Hugo 2014). Similarly, migrants from outside the European Union and living in a European country may be entitled to a range of social benefits that are accorded to citizens, either immediately or over time. However, undocumented migrants may remain undocumented for decades, depending on the availability and timing of amnesty or regularization programmes, the migrant's age, time of entry, national background, and gender. In this case, it may be somewhat far-fetched to label their status as 'fluid'. This suggests that nationality and formal citizenship are not to be dismissed as simple artefacts of a more nationally oriented age.

National 'models' of formal citizenship?

For most of the twentieth century, researchers typically couched the relationship between migrants and citizenship in terms of national 'models' of citizenship. These models of citizenship were argued to shape rates of naturalization, as well as notions of 'integration' and belonging (Brubaker 1992; Favell 1998 [2001]). For example, Brubaker (1992) contrasts France's 'model' of jus soli[3] (citizenship by birth in a territory or 'law of soil') with Germany's jus sanguinis (citizenship by ethnic descent, or 'law of blood'). In France, jus soli essentially granted citizenship automatically to a child of foreign parents if the child was born on French territory. The idea rested on a 'statist', republican and universalist ideal. In other words, the French government entertained an expansionist conception of citizenship that welcomed those who wished to be part of French political culture (the statist and republican dimension), regardless of their ethnic origins (the universalist dimension).[4] It has been argued that French universalism explains in part why the legal, political, and even social recognition of 'race' and 'ethnic minorities' have been taboo since the inception of the French Republic (e.g. Feldblum 1993). And it may explain in part the angst about the wearing of veils (and other visibly religious symbols) within public institutions in France. In contrast, Brubaker (1992) argues,

Germany's *jus sanguinis* is founded on the notion of an 'ethnic community' (*volksgemeinschaft*) or 'community of descent', a sense of 'Germanness', in which citizenship would only be granted sparsely to those who could demonstrate their ethnic German background, usually through parental heritage. Brubaker (1992) claimed that this explained why it proved more difficult to obtain legal citizenship (measured in rates of naturalization) in Germany than in France.

Using such a typology of states and citizenship, scholars held that most European countries and the UK lay somewhere in between the poles of territory and descent, with the exception of Austria, Greece and Switzerland, which were rooted in a *jus sanguinis* notion of citizenship. Australia, Canada and the United States were argued to lean more towards *jus soli*. These 'national models' were never pure however (e.g. Kastoryano 2002). For example, during the 1950s and 1960s, the German government seemed to combine *jus sanguinis* with quite liberal refugee policies, while the French government, and various other public and private actors within French society have scrutinized African and Muslim immigrants on the basis of a 'cultural problem' of integration, regardless of their territorial birth right and willingness to participate in the political life of France (Bloemraad et al. 2008; Laurence and Vaisse 2006). If nation-states did harbour comparatively different conceptions of citizenship during the twentieth century, then they seem to have converged over the last decade and a half (e.g. Geddes 2003), or as Vink and Bauböck (2013) admonish, "the common association of *jus sanguinis* with ethnic conceptions of nationhood and of *jus soli* with civic ones is in several ways misleading. It does not reflect the historic origins of both principles and wrongly accuses them as polar opposites" (p. 629).[5]

In fact, some 15 years ago, Faist (2000) argued that advocating the existence of such 'models' is discredited. Furthermore, Ehrkamp and Jacobsen (2015) maintain, by asserting such top-down models of citizenship, other alternative forms of citizenship that might involve more grassroots forms of democracy and power are eclipsed.

Dual and plural nationality

Dual and plural nationality concerns the acquisition or eventual attribution of citizenship through *jus sanguinis*, *jus soli*, or marriage for a person holding nationality in another country or countries. However, it is not simply a matter of a country of immigration granting naturalization to a migrant who has another nationality, the country of emigration must also accept that one of its nationals is adopting nationality elsewhere. The country of origin may be more lenient if the second nationality is acquired through jus soli, that is involuntarily, rather than by purposeful naturalization (Schuck 2002). Though little comparative quantitative evidence is available on the extent of dual nationality across the world, case studies of certain countries suggests that it is growing, and growing rapidly. It is estimated that about half of the world's countries now have dual nationality or dual citizenship provisions (Bauböck 2006; Kraler 2006; Sejersen 2008; Vink and de Groot 2010), although some countries such as the Netherlands and Spain (and Germany for non-EU nationals) do not allow dual citizenship (Green 2012; Yanasmayan 2015). It is especially significant for countries such as Switzerland, where it was estimated in the late 1990s that some 60 per cent of its citizens were living abroad as dual nationals (Koslowski 1997, in Schuck 2002: 67). Dual (and even triple) nationality emerges because, let us say, "a German marrying a Turk and giving birth in the US will give birth to a Turk and German by descent and an American by birth" (Hansen and Weil 2002b: 3). Thus multiple citizenships will tend to increase with increasing international migration (Hansen and Weil 2002b).

Yet despite the apparent proliferation of dual or multiple nationalities, many migrants still find it difficult to adopt another nationality without their home country granting this wish, such as for Haitians living in the US, Japanese living in Brazil (the Haitian and Japanese governments do not allow dual nationality), and, let us say, Syrians living in Argentina (the process of renunciation in Syria is extremely arduous and therefore uncommon) (Escobar 2007; Forcese 2006; Schuck 2002; Surak 2008). Argentina, in fact, represents an extraordinary case of the problems of dual nationality, as the state does not explicitly allow dual or multiple nationalities, and will not allow for the renunciation of citizenship. Thus, its citizens abroad have dual nationality anyway because they adopt it without the Argentine government being aware of this (Escobar 2007). One might distinguish in this sense between 'open', 'tolerant', and 'restrictive' dual nationality regimes (Aleinikoff and Klusmeyer 2001 in Escobar 2007: 47). For example, let us look at India, the UK and the US, which represent different shades of openness, toleration and restriction (see Boxes 6.2 and 6.3).

Box 6.2 LIBERALIZING DUAL CITIZENSHIP IN INDIA, FOR SOME

(From Dickenson and Bailey 2007)

In contrast to Argentina's very restrictive stance, India's 2003 Dual Citizenship (Amendment) Bill and its 2005 revision liberalized dual citizenship for many, but not all Indians overseas. There are two categories: Non-Resident Indians (NRIs), or those who live outside India for more than 183 days, and A Person of Indian Origin (PIO), that is, anyone who once held an Indian passport, or whose parents or grandparents were Indian citizens, or who has married an Indian citizen. The liberalization of dual citizenship has its motivation in the government's renewed interest in attracting back to India those who are perceived to be the successful artistic, business, intellectual, and professional 'heroes' among Indians overseas. After all, there are about 20 million Indians living outside India. For the government, attracting back Indians from abroad hopefully serves two purposes: it stimulates national economic development by encouraging foreign direct investment and the transfer of technology; and it serves to reinforce an imagined nationalism centred on a moderate and tolerant form of Hinduism. Concerning economic development, dual nationality becomes an element of the general move towards the liberalization of the Indian economy. In fact, the government developed a two-pronged strategy which entailed The Federation of Indian Chambers of Commerce and Industry (FICCI) organizing the *Pravasi Bhartiya Divas* (Overseas Indian Day, which really lasted three days and involved meetings, performances, parties and speeches) in order to celebrate the contribution of NRIs and PIOs to build closer links with India. A second feature of this strategy was to provide dual citizenship to some of these NRIs and PIOs. Ultimately, dual citizenship has been offered to Indians in 16 countries in Australasia, Europe and North America but not in Africa, Southeast Asia, and the Middle East, which contain much larger Indian populations and which are excluded from the list of countries (two exceptions are Nigeria and Lebanon). Those who are ultimately accorded dual citizenship can travel to India without a visa. They can stay without registering, they can invest in various economic activities, they can buy land and property, and they can enrol their children in Indian universities.

It is important to recognize that India distinguishes between its pre- and post-independence diaspora, in which the pre-independence diaspora is associated with the 'old India' of British colonialism and oppression. In that way, this liberal attitude of dual citizenship is not extended, for example, to South African Indians, who arrived in South Africa as indentured workers or traders between 1860 and 1911, and who appear not to conform to the Indian government's ideal of the perfectly integrated citizen overseas or desirable member of the diaspora. That is, Indians in South Africa are perceived by the Indian government to have rejected multiculturalism, aloof from the black majority and living behind 'gilded cages' (Dickenson and Bailey 2007: 768), and ultimately too far removed from a 'modern' India. In contrast, India's post-independence diaspora is viewed as able to prosper economically through transnational networks and able to contribute to the 'new India'. As Dickenson and Bailey argue, the Indian state is constructing a notion of diaspora around "professional success, ecumenical Hinduism, and multiculturalism" (ibid.: 765). For these authors, there is a class dimension to dual citizenship, since it is only offered to those who sought educational and professional activities in 'the west', while denied to those who left as indentured servants in what are now poorer countries. In short, the Indian government distinguishes between who is and who is not worthy of dual nationality based on their geographic and social origins, which is itself infused with ideas of 'Indianness'.

Box 6.3 DUAL CITIZENSHIP IN THE UK AND THE UNITED STATES COMPARED

Perhaps the most effective way of documenting the nature and significance of dual nationality is to compare its actual contours in the UK (a 'tolerant regime') and the United States (by law restrictive,

Continued

but in practice tolerant). In contrast to Germany or the United States for example, the UK government has maintained a liberal or indifferent approach to dual nationality, neither pushing migrants to nationalize, nor encouraging them to renounce their citizenship.[6] This indifference has been mirrored by the lack of protest among the majority of the British polity. For Hansen (2002), tolerance for dual nationality in the twenty-first century reflects first, the creation of 'British citizenship' in 1948 as 'plural citizenship' (that is citizenship with a sense of a United Kingdom and a greater British Commonwealth); second, that dual nationality is seen quite simply by the government as a means by which migrants may be better 'integrated' into 'British society'; and third, that it seems to create few if any practical problems for the British government. There are virtually no limits concerning dual nationality, and the British government does not maintain statistics on its occurrence. British citizens abroad can obtain citizenship in another country while immigrants in the UK do not have to renounce their former citizenship on obtaining British nationality. In fact, the UK Home Office does not query the intentions of the individual wishing to naturalize in the UK, and does not inform or divulge the process of naturalization to the government of the individual's original nationality. However, two caveats are necessary here. First, in what Lewis and Neal (2005) call 'neo-assimilation', or 'civic integration policies' Joppke (2007a), citizenship tests are proliferating across the rich countries, including the UK. This will ensure that while dual nationality may be tolerated, it will probably be more difficult to obtain, and particularly for migrants originating from Islamic countries. Second, dual nationals are not accorded exactly the same rights as nationals. For example, in a situation in which a dual national is naturalized as a UK citizen, and is accused of 'high treason' and/or a threat to 'the security of the British people', a government-appointed *ad hoc* committee may revoke a person's status as a national, unless that will lead to the individual's statelessness. British nationality cannot be revoked under any circumstances for an individual born with British nationality (Hansen 2002). One of the significant implications of Hansen's analysis is that there is no evidence of a relationship between dual nationality and loyalty to the British state. This is illustrated in the notable decision by the British government not to detain British-German dual nationals during the Second World War, unlike in the United States where Japanese-Americans were imprisoned in camps. The above discussion, however, is not to suggest that somehow the British government entertains a general liberal set of policies vis-à-vis migration and settlement. After all, as Hansen (2002) points out, "a senior Conservative politician suggested as late as 1990 that West Indians take a 'cricket test' as proof of their loyalty (which side do they cheer for? England or Jamaica?)".

In the United States, the Federal government remained for a long time suspicious of dual nationality since the signing of the US Constitution, although since at least the 1990s, the government has increasingly tolerated but not encouraged it (Schuck 2002), particularly for the problems it is argued to raise for political allegiance and loyalty to the United States (whatever these ideas mean exactly?), but also for perhaps less obvious reasons such as the complexities of offering diplomatic protection to dual nationals in the context of a 'war on terror' (Forcese 2006). Thus, dual nationality is constitutionally forbidden, and a requirement for naturalization is the renunciation of one's original citizenship. However, it is increasingly tolerated partly because first, it is difficult or impossible to verify, since many states do not divulge this information to the US government. Second, the US government does not have a legal requirement for this renunciation of citizenship to take place, and thus it is hardly enforceable. However, the record number of naturalizations requested along with the liberalization of dual nationality in Latin American countries suggests that it is growing rapidly in the United States as well. The US government has now dramatically increased the amount of resources it devotes to naturalization, paradoxically during a decade in which it also escalated its spending on border enforcement. For a migrant to obtain naturalization in the US, she or he must have had a certain period of residence (Schuck 2002). The toleration of dual nationality in the United States shifted some political

Continued

burden on Latin American governments and other governments with large numbers of its emigrants living in the US, including Canada, India, and the Philippines to address the issue of dual nationality (Escobar 2007; Portes and Rumbaut 2006; Schuck 2002). Not surprisingly, dual nationality and naturalizations became increasingly common among Latino migrants, for example. Martin (2002) identifies at least five reasons for this:

1) The number of migrants eligible for naturalization rose sharply during the 1990s.
2) 'Green cards' replaced easily counterfeited long-term residence cards. These were as expensive as the cost of naturalization and many migrants simply chose to naturalize rather than obtain a 'green card'.
3) The Immigration and Naturalization Service moved towards emphasizing 'naturalization' and established a programme to handle the increasing number of naturalizations.
4) The liberalization of Mexican dual nationality encouraged naturalization among Mexicans in the US, as discussed above.
5) Welfare restructuring encouraged (impelled?) migrants to naturalize in order not to lose social benefits.

To Martin's list, we can add the effects of 9/11 and its manifestation in the 2001 Patriot Act, which facilitated the deportation of 'non-resident aliens'. Quite simply, Latino and other migrants viewed naturalization as protection against deportation, and at the same time, it would allow them to cement ties to their 'home' countries, especially the ability to vote in home elections (Escobar 2007; Levitt 2002).

From national models to denizenship and *jus domicili*: the convergence of national models?

France and Germany's nationality laws illustrate the question of both the convergence of naturalization policies in 'western liberal democracies' and the rise in the importance of length of residence as a determinant of rights.

Changes in France's nationality code

In France, Articles 44 of the Nationality Code of 1889 concerned the acquisition of citizenship and Article 23 concerned the attribution of citizenship. Article 44 granted citizenship (upon request and at the age of majority) to children born on French soil, regardless of whether the parents were themselves immigrants of foreign origin. Article 23 concerned what was called double jus soli – the attribution of citizenship to children of Algerian immigrant parents especially (Algeria was a Département or region of France until 1962). In late 1993, France passed a Nationality Law (effective January 1, 1994) that restricted Article 23, by requiring a minimum number of years of residence for immigrant parents from former French colonies (including Algeria) before children born in France could be attributed French nationality. The same law also reversed and disavowed Article 44, and children would no longer be entitled to citizenship if their parents were of immigrant and foreign origin, but lived on French soil. As Feldblum (1999) writes: "The 1993 reform inverted this logic [. . .] Children born in France of immigrant parents were to be defined by their birth status – as foreigners – until they achieved their birthright (the right to be French). They were to be foreign until they could prove themselves French" (p. 149).

Yet the 1993 reform did not abolish jus soli, it simply became more difficult to obtain French nationality. In somewhat of a reaction against the previous conservative administration's decision to tighten citizenship policies, the Guigou Bill of 1998 under the French Socialists re-implemented the quasi-automatic attainment of citizenship for immigrants at age 18, provided they maintained consistent residency in France for at least five years from the age of 11 (young immigrants could also decline the acquisition of nationality if they wished). It also reintroduced double jus soli rights for the children of Algerian immigrants, which had been eliminated during the 1993 reforms. However, this revision did not apply to other Francophone migrants of colonial origin, and certainly not to those immigrants from other countries beyond the European Union (Feldblum 1999). Finally, the 1998 Guigou law allowed for the regularization of some undocumented immigrants, particularly those with links to families in France, or conversely single people, and these were defined in relation to Article 8 of the European Convention of Human Rights, suggesting that the supranational legislation had some impact on determining national laws (Kofman 2002).

Changes in Germany's Nationality Law

Many scholars argue that until the 1990s, Germany's citizenship policies were dominated by jus sanguinis which made naturalization comparatively difficult (e.g. Ersanilli and Koopmans 2010; Green 2001, 2012; Klusmeyer 2001; Köppe 2003; Rotte 2000). Following on from the 1990 Aliens Law, there were further calls to liberalize citizenship in Germany during the late 1990s. On 1 January 2000, the German government implemented the German Law on Nationality, whose provisions and revisions (in 2005 and 2007) actually involved both a liberalization and a tightening simultaneously. The tenets of the 2000 Law included a reduction in the minimum required residency for naturalization from 15 to 8 years. At the same time, the government tripled the naturalization fees from €51 to €255 (about US$280) and it stipulated (as in many other countries) that any potential citizens could not have served a prison sentence for 6 months or more (or the equivalent in fines) which was reduced to a cumulative of no more than 3 months in 2007 (Green 2012) and they could not have relied on social security or social welfare 'for any reasons of their own making' (Köppe 2003: 440). Furthermore, candidates for naturalization must pronounce their faith in the German political (constitutional and democratic) system, and demonstrate 'a sound knowledge' of German. Furthermore, the German government introduced elements of jus soli for the first time, only accepting dual citizenship with naturalization until the age of 23. In other words, 'second generation' immigrants (i.e. the children of migrants) automatically received German nationality by birth (provided one of their parents had lived in Germany for at least 8 years or they had an 'unlimited permit of residence') but they had to choose either German nationality or their original nationality between the ages of 18 and 23 (Green 2001, 2012; Klusmeyer 2001; Köppe 2003). Thus, in principle, young people of the second generation would have to give up their foreign nationality in order to remain German citizens, but in practice this has not necessarily proved the case. For example, the Turkish government developed a 'Blue Card' (not the same one as in the EU) that allows Turkish citizens to 'retain a privileged form of membership to Turkey' in which they are not officially full citizens (Yanasmayan 2015: 2) but accords them the right to live and work in Turkey, the right to vote, the right to own land or the right to inherit. At any rate, this Optionsmodell ('option model') of choosing citizenship at the age of majority in Germany drew its inspiration from the

French nationality reform of 1994 (Green 2001, 2012; Klusmeyer 2001; Köppe 2003; Rotte 2000), and it appeared that France and Germany converged to one degree or another in terms of the importance of length of residence (Joppke 2007).

Denizenship and jus domicili

In an influential book, Hammar argued in *Democracy and the Nation State* (1990) that long-term resident immigrants in various European countries enjoyed a number of social and political rights, including access to social welfare and sometimes the right to vote in local or regional elections. He referred to this situation as 'denizenship'; that is a spectrum of partial and stratified citizenship depending on a combination of a migrant's country of origin, migration category, and length of residence. This state of differential rights has also been usefully referred to as 'civic stratification' in the context of the EU (Morris 2001; Kofman 2002) or 'differential inclusion' (e.g. Carmel *et al.* 2012; Mezzadra and Nielson 2012). It is length of residence that prompted scholars such as Faist (1995) to argue that access to social services in countries such as the UK, Germany, Sweden, and France increasingly involve legal recognition based on a migrant's *length of residence in the country* as much as they might involve either place of birth or ethno-national belonging. He in turn called this *jus domicili* (loosely translated as law of residence, or more figuratively as citizenship and rights based on length of residence). As Bauder (2014b) explains succinctly: "Domicile-based citizenship is granted to people independently of the place and community of birth, and applies to migrants after they entered a territory and established a residence in this territory" (p. 93). In countries of the EU, migrants are subject to a sort of tiered citizenship, whereby citizenship is based on eight categories: citizens of a particular national state, citizens from an EU country residing in another EU country, those migrating to, or residing in the EU from a non-EU country, dual citizens, those with bilateral agreements, asylum-seekers, refugees and undocumented immigrants. In short, migrants in Europe are stratified or segmented. While Hammar's book focused on EU, the citizenship policies of countries around the world certainly involve civic stratification and denizenship based on *jus domicili* to one degree or another. In fact, they combine all three elements: *jus soli, sanguinis,* and *domicili* (Bauder 2014b). Even beyond these three, we are also witnessing the growth of a 'neoliberalized' citizenship or 'investor citizenship' (e.g. Carrera 2014) (see also Box 6.1). As Carrera recalls: in 2013, the government of Malta announced that it would accord citizenship to foreigners who donated money to Malta, *regardless of length of residence*. This 'citizenship-for-sale' scandal as it came to be known quickly drew the ire of the EU Parliament (EP) which protested that there should be a "'a genuine link' or 'genuine connection' between the applicant and the country or its citizens" (European Commission 2014c, cited in Maas 2016: 541). After negotiations with the EP, the Maltese government did capitulate and introduced some residency requirements. Yet, investor citizenship hardly began in 2013 in Malta. Indeed, Austria, Bulgaria, and Cyprus, and Ireland had all begun similar programmes with no length of residency requirement, while France, the Netherlands and the UK required a 'genuine link' (Maas 2016). Thus, jus *domicili* and investor citizenship can also be combined.

CITIZENSHIP AS RIGHTS

Citizenship, as we noted in the introduction to this chapter, is more than just nationality. Substantive or *de facto* citizenship also involves rights. Certainly, they intersect, but our focus in this section is on the differential landscape of economic, political, and social *rights* accorded to migrants, especially in *relation to* citizens. One of the starting points for much of the literature on citizenship and rights with respect to migration is T.H. Marshall's now widely read book *Citizenship and Social Class* (1950). In it, he argued that people first obtain civil rights (e.g. the right to a fair trial, 'free speech', 'free movement', etc.), then political rights (e.g. the right to vote) then social rights (e.g. access to social welfare). This linear sequence has been criticized precisely for its linear-evolutionary character, its Anglocentrism, and because Marshall neglected the struggles between migrants and governments over the very acquisition and content of citizenship (e.g. Bloemraad *et al.* 2008; Hampshire 2013; Isin and Wood 1999). In the first decade of the twenty-first century, this sequence seems to have been re-ordered, so social rights may be accorded initially, but political rights such as voting in local elections for example, may be highly contested (e.g. Guiraudon 2000). Even for undocumented immigrants without legal status, social rights are not necessarily eliminated, such as in Spain or the United States. It is not simply a matter of a new order replacing a previous sequence, but rather a complexity of orders that vary geographically, both across and within countries. However, rather than document endlessly the actual accumulated rights of migrants in different countries, sub-national regions, or localities, it will be perhaps a better route to review some general arguments and concepts concerning the acquisition of rights for migrants across the richer countries, and introduce the actual landscape of rights through these different arguments.

Towards post-national forms of citizenship?

If national forms of citizenship were either converging during the 1990s, and/or based increasingly on other criteria such as *jus domicili* in Europe, they were also experiencing other processes of change, which concerned both legal status and rights. These changing forms of citizenship are often referred to as post-national, transnational or global. Unfortunately, these processes involve less a theory than an approach, framework of analysis, or research agenda, and the differences or similarities between these terms are sometimes unclear (Lister and Pia 2008).

Drawing on an analysis of European countries in particular, Soysal (1994) argues in *Limits of Citizenship*, that national forms of citizenship are moving towards universal forms of 'personhood', whereby supranational or international charters, codes, conventions, and laws increasingly attach universal rights and privileges to individuals regardless of their membership status in a nation-state. She speaks of 'post-national membership' and calls this new 'model' *post-national citizenship* (or as Jacobson (1996) puts it: 'rights across borders'), and it is predicated or legitimated on the idea of international or global human rights. On these grounds, not just individuals but migrant 'groups' are accorded protection on the basis of rights to their identity, including for example language and religion (Soysal 1997). Thus for Soysal (1994, 1997) national forms of citizenship are deteriorating, accompanied by a 'de-coupling' of rights from national territories. As she insists, a 'guest worker' in Sweden need not have a knowledge of Swedish history or even the Swedish language to access a bundle of rights in that country. However, Soysal recognizes that paradoxically, post-national rights are organized by national states.

Criticisms of Soysal's idea of 'post-national citizenship'

Soysal's argument set an agenda for research, and her claim that post-national rights were in fact 'organized' by national governments in the EU, proved to hold much validity (Hampshire 2013: 111). However, a number of significant criticisms have been levelled at her work over the last 20 years. First, 'post-national' is not quite the same as 'denationalized' (Bosniak 2000) or what Sassen (2006b) calls the 'denationalization' of citizenship. For Sassen, post-national citizenship is something beyond national states, but 'denationalization' refers to change within national states and the way in which they have absorbed international and post-national norms. Thus, 'de-nationalization' means that an international or global human rights regime is slowly at work, but this involves national courts using international human rights instruments for jurisprudence (the interpretation of the law) and legal decision-making. For Sassen then, the issue is not simply a matter of something happening beyond the national state, but that 'the national' is itself changing and absorbing international and post-national norms (e.g. the acknowledgement of dual citizenship). This is why it is crucial to distinguish between denationalization and post-national citizenship, which are different but deeply intertwined processes.

A second group of reservations focuses on the relationship between migrants and higher (often European) courts. Among the questions raised are whether migrants' legal status and rights are still meted out by national states, and whether individual migrants have access to higher (European) courts to claim rights and privileges that they are not accorded by national states. Similarly, a question remains as to the degree to which migrants actually make these claims to higher authorities, rather than to national governments, and the extent to which the decisions of international institutions have the 'legal teeth' to render decisions over and above the rights and responsibilities offered and dictated by national states, never mind the enforcement of these decisions (e.g. Joppke 1998a; Kofman 2005b; Surak 2008). In terms of these third group of criticisms, it is worth spending more than just a few moments to discuss the actual territorial progression of rights in the European Union, since the EU is held up as the region of the world in which the 'supra-nationalization' or 'post-nationalization' of rights has progressed the furthest. In this respect, some important legal decisions emanating from both the Luxembourg-based European Court of Justice (the supreme judiciary of the European Union) and the Strasbourg-based European Court of Human Rights (or the Single Court)[7] during the 1990s and 2000s have in fact come to supra-nationalize or post-nationalize citizenship and the extension of rights to a degree (e.g. Geddes 2003, 2014; Guiraudon 2000; Kostakopolou 2002).

Indeed, some crucial caveats are necessary. To begin with, a citizen of any European country is a citizen of the European Union,[8] but it is national governments that determine the citizenship of 'Third Country Nationals' (or TCNs – those migrants from outside the European Union) whether they are asylum-seekers, family members, or labour migrants. Thus, it should be emphasized that any reality of post-national rights is partly dependent on an individual's denizenship in a national state, a denizenship which seems to be increasingly determined by *jus domicili* (e.g. Bauder 2014b). For example, TCNs who have lived in a European country for less than five years do not possess the same rights as those who have resided in an EU country for a longer period of time. Yet even those nationals of an EU country living in another EU country, who are by law citizens of the European Union, are entitled by European law to similar economic and social rights, but political rights (such as holding public office) is still not permitted in many EU countries (Perchinig 2006; Maas 2016). For asylum-seekers in the EU, the situation is varied but overall grim, where most states now provide the most meagre social support and isolated accommodation as to avoid alleged 'asylum-shopping' among would-be asylum-seekers. In the UK for example, benefits for asylum-seekers were progressively reduced during the 2000s, and for the most part asylum-seekers can only receive benefits if they are deemed 'destitute' (Darling 2009; Dwyer 2005).

The result is different levels of citizenship in the EU, based on differences between European citizens living in their own country, European citizens living outside their country of nationality, legally residing TCNs, asylum-seekers, and undocumented migrants (e.g. Kofman 2005b). Thus in terms of economic, social, and political rights, the situation is sometimes unfathomably complex, dependent as it is upon not only the *jus domicili* of member states, but also European directives and other legal rulings concerning the provision of rights, and bilateral labour and European association and cooperation agreements between European countries and between European countries and those outside the EU (e.g. Maas 2016). While there has been some post-nationalization of rights, the evidence for a convergence of national rights is not convincing. For example, Koopmans *et al.* (2012) point out that for the ten European countries they investigated between 1980 and 2008, they found no evidence for cross-national convergence in terms of rights. As they put it, "Rights tended to become more inclusive until 2002, but stagnated afterward" (p. 1202).

In light of the difficulties associated with Soysal's formulation, and given our spatially sensitive approach in this book, we can instead think of migrant rights as shaped by multiple territorialities that are well-captured by Feldblum's (1998) idea of 'neo-national membership'. By this she means the "developments whose effects are to reconfigure cultural, national, and transnational boundaries to ensure closure" (p. 232). Yet, we also need to think of *local* forms of rights, some of which may be officially inscribed in municipal policies, and some of which may simply involve the everyday practices of citizens and migrants alike. In sum, multiple territorialities may be at work in the acquisition and provision of social rights, producing in this case, European, national and local denizenships. In sum, if Soysal's arguments concerning the post-nationalization of rights are problematic, this says nothing about the actual *content* of rights, so let us examine this a bit further.

The 'neoliberalization' of migrants' rights?

Oddly, so much of the research on the intersection of citizenship, migration, and belonging is divorced from arguments about neoliberalism/neoliberalization or at least broad political-economic changes (for exceptions, see e.g. Mitchell 2003; Ehrkamp and Jacobsen 2015; Schierup *et al.* 2006). This is not entirely surprising, given the emphasis on the *increasing* acquisition of economic, political and social rights for at least *legal* migrants around the world. However, the actual *financial support* associated with these newly acquired rights – particularly economic and social rights in North America and for most EU countries for most categories of migrants – has deteriorated, consistent with arguments concerning neoliberalism or 'commodification' (the latter refers to the withdrawal of, reduction in, or privatization of government-financed social entitlements so that individuals become increasingly responsible for their own 'social reproduction') (e.g. Andersson and Nilsson 2011; Carmel *et al.* 2012; Sainsbury 2012; Schierup *et al.* 2006). In this respect, the US has always been assumed to be a weak welfare state, but one that also appears to have increased the 'commodification' of migrants further under 'roll back neoliberalism' (see Box 6.4). By contrast, Scandinavian countries are still assumed to be models of 'de-commodification' or strong welfare states, but let us evaluate whether this remains the case (see Box 6.5).

Box 6.4 MIGRANTS AND SOCIAL RIGHTS IN A 'WEAK' WELFARE STATE: THE UNITED STATES

Here we re-visit the US case in a little more detail, having already touched upon changes in social entitlements in Chapter 5. In contrast to European countries, there is no supranational authority in the United States, and the delivery of rights in the US is shaped by the relationship between Federal, state, and local governments. Prior to the 1996 Personal Responsibility Work Opportunity Reconciliation Act (PRWORA), or what came to be known as 'welfare reform' in the US, legal migrants enjoyed most of the same social rights as US citizens, and even undocumented immigrants were entitled to some social services prior to the 1990s. For legal migrants, this included access to Aid to Families with Dependent Children (AFDC), Supplementary Security Income (SSI) for elderly and disabled people, and Food Stamps. Not everyone was immediately eligible however, and a stipulation called 'deeming' required that the 'sponsors' of an immigrant (most family category immigrants required sponsors) have their income estimated ('deemed') in the first 3–5 years of entry. As Martin (2002) argues, one of the advantages of this is that it had a limited time period attached to the deeming process so that "it did not construct a permanent second class of resident who paid taxes but would not realise any benefits" (Martin 2002: 217). After the passing of PRWORA in which politicians, policy-makers and even some academics framed immigrants as 'undeserving' and guilty of abusing the welfare system (Fujiwara 2005), a number of changes to social entitlements ensued; chief among them was that legal migrants and especially undocumented migrants should not benefit from the same social rights as citizens. AFDC was abolished overnight and whatever cuts were protested by immigrant-friendly advocacy groups and state and local political parties, the basic premise that there should be a distinction between legal immigrants and citizens remained. It is worth noting however that prior to the actual implementation of the law in August 1997, the protest of immigrant rights organizations

Continued

succeeded in restoring SSI benefits to elderly and disabled immigrants especially who had arrived before 1996 (Fujiwara 2005; Levinson 2002; Marchevsky and Theoharis 2006; Martin 2002; Viladrich 2011).

Consider the impact of the PRWORA by a person's entry category or legal status, and the time of their entry. 'Legal permanent residents (LPRs)' who entered the US after August 1996, would no longer be eligible for food stamps or SSI. Accessibility to SSI and Food Stamps only became possible after 40 quarters (roughly 10 years) and with the added restriction that during those 40 quarters, LPRs would have to demonstrate a record of more or less continuous formal employment, which would prove difficult, given that many worked informally. Exceptions were created for children and the elderly (as noted above), and a few other categories of persons, such as military veterans. Furthermore, and as we elaborate further below, those LPRs who arrived after August 1996 could apply for TANF (or Temporary Assistance for Needy Families – the successor to AFDC) five years after entering the country legally (Fujiwara 2005; Levinson 2002; Martin 2002; Viladrich 2011).

Asylum-seekers and refugees (including those entering under Temporary Protected Status or TPS) would only be eligible for benefits after seven years of their date of entry (although refugees would be entitled to other financial and social benefits through state and private organizations reimbursed by the Federal government) (Levinson 2002; Martin 2002). This lack of social entitlements (including access to regular and preventative medical care) can lead to chronic illness for immigrants too poor to afford medical insurance in the US, as Bailey et al. (2002) find in their study of Salvadoran migrants within the US TPS programme. The effects on people's health are certainly not limited to those with TPS though, and considerable research has documented the negative consequences for the physical and mental health of (elderly) refugees, undocumented children and undocumented women, some of whom have taken their own lives in the absence of benefits for themselves or their families (Fujiwara 2005; Hadley et al. 2008; Nam and Jung 2008; Van Hook and Balistreri 2006; Viladrich 2011).

Before PRWORA, undocumented immigrants never had access to means-tested federal programmes, although they did have access to a cash assistance programme managed by local governments (Viladrich 2011). After PRWORA, it became clear that individual US states were now fully responsible for addressing the medical and social care issues associated with undocumented immigrants. However, they were eligible for some emergency medical care (Medicaid) and for other health benefits for reasons of public health or safety (Levinson 2002; Martin 2002; Viladrich 2011).

Alongside this sudden restriction to access to social benefits at the Federal level, individual US states were mandated to provide coverage from the State Children's Health Insurance Program (SCHIP) to child LPRs who had arrived before August 1996. Furthermore, some US states have tried to address the problems of 'welfare reform' by cobbling together resources from Federal and state funds and some of the states with the largest immigrant populations such as California and New York, have enacted state-only Food Stamp and other aid programmes to fill the void created by the Federal government. For those arriving after 1996, individual US states could offer TANF and were allowed to determine if legal migrants can continue to receive TANF, despite the 5-year threshold as mentioned above. Beyond these openings in social entitlements for immigrants at the state level, the Farm Security and Rural Investment Act of 2002 restored immigrants' access to food stamps, some housing and other emergency support at the local level as well as in-state tuition for certain state universities (Levinson 2002; Martin 2002).

In sum, the actual access to social and medical care that migrants receive seems to be dependent upon the visa through which one enters the US, their legal status, and the length of time spent in the US, among other 'variables' (e.g. Durden 2007). Bloemraad (2006) finds that this context of weak economic, political, and social rights for migrants in the US accounts for the low rate of US naturalization compared to Canada, which has relatively more generous rights. In contrast, Nam and Kim (2012) suggest that one must examine different age groups to make such an assertion.

Box 6.5 MIGRANTS AND SOCIAL RIGHTS IN A ONCE STRONG WELFARE STATE: SWEDEN

As is the case with other Scandinavian countries, Sweden has had a long-standing reputation as a strong welfare state, geared towards universal principles in which everyone is in theory entitled to support. By the mid-2000s, Schierup *et al.* (2006) concluded that Sweden continued to have *relatively* strong welfare policies, but that social protection had weakened considerably since the 1980s, and may be irrelevant for the growing number of undocumented immigrants (see also Sainsbury 2012). In contrast, Andersson and Nilsson (2011) argue that in fact, the rights of asylum-seekers and undocumented immigrants have actually *increased* in a number of domains since the 1990s, including in terms of work, health care, and education, while this may not be the case for housing and economic rights. We should scrutinize this more then, given the apparent disagreement between these authors.

Beginning in 1950s, the economic, political and social rights of immigrants became wide-ranging, and rights seemed to increase over time. Immigrants received the right to vote and to run for office in local and regional elections in 1975, and while citizenship was based on *jus sanguinis* until 2001, *jus domicili* began to replace the former as a criterion for access to rights during the 1980s. In fact, up until the late 1980s, most immigrants and asylum-seekers received more or less the same generous benefits that Swedish citizens received, including a guaranteed basic minimum income for families (social assistance) regardless of employment status, unemployment and occupational insurance, national health insurance, pensions for disability and retirement, illness compensation, and parental allocations for children, to name just some of the benefits.

Despite this egalitarian progress, in the late 1980s, the government replaced the 'social minimum' (a guaranteed basic income) that Swedish citizens were entitled to, by a special allowance for asylum-seekers, which was nonetheless indexed (identical) to the 'social minimum'. However, by the early 1990s, the Swedish government began to reduce social benefits for both citizens and migrants alike, which is usually attributed to an acute recession in

Sweden in the early 1990s, rather than to neoliberalism. At any rate, the government argued that because other benefits were being reduced for citizens, asylum-seekers and migrants should not be spared either. In other words, the revisions to Swedish welfare were not aimed at immigrants specifically, at least not overtly. The 'reform' of welfare policies did however have the most detrimental effect on asylum-seekers, labour/family migrants, and refugees, as the sharp decline in the amount of welfare provision affected people on low incomes in particular, and especially asylum-seekers and refugees whose numbers grew during the 1990s. This coincided with rising unemployment during the 1990s so job prospects worsened, and this only exacerbated their right to benefits as asylum-seekers and migrants lost those benefits that were tied to employment. In 1992, the government reduced asylum benefits (asylum allowances) by 10 per cent, consistent with cuts to citizens. However, in the same year, the government also exempted asylum-seekers from requiring a work permit if the processing of an asylum-seekers' claim was likely to take more than four months, and in 1994, it allowed asylum-seekers to reside with their friends or relatives rather than in reception centres while still accessing housing benefits and even receiving an additional housing grant between 1994 and 2005 for this purpose (Andersson and Nilsson 2011; Bevelander and Pendakur 2012). However, in 1994, the asylum allowance became unmoored from the citizen social minimum, as the government argued that during the asylum process, asylum-seekers should not be entitled to the same amount of protection as Swedish citizens. This allowance was also not raised after the recession in the early 1990s and well into the 2000s to reflect changes in the standard of living (Andersson and Nilsson 2011; Sainsbury 2012). Furthermore, the Swedish government responded to an increase in the number of asylum-seekers by issuing only temporary rather than permanent resident cards. Issuing permanent residency cards meant that asylum-seekers and refugees could not access general social insurance. As Sainsbury (2006) puts it, "In this instance the principle of domicile was converted from a mechanism of inclusion to one of social exclusion" (p. 239).

Continued

Nonetheless, a 1996 bill gave asylum-seekers and refugees the right to work, and bolstered educational opportunities for their children. The requirement that elderly migrants joining their families should be 'self-sustaining' in terms of income and accommodation was dropped on the basis that Swedish citizens were not required to support their own elderly parents, which would lead to discrimination between citizens and non-citizens (Sainsbury 2006; Schierup et al. 2006).

As noted above, automatic citizenship in Sweden had been based on *jus sanguinis*, but the foundation of formal citizenship came to be seen as discriminatory from the 1950s to the 1980s, and gradually over the course of the 1980s and 1990s, social and economic rights became increasingly tied to *jus domicili*. Since the 2001 Nationality Law, naturalization became relatively liberal (unlike in the US and many other European states, there is no citizenship test, and dual citizenship became a possibility through the 2001 Law). The 2001 law made naturalization through *jus domicili* official; the government reduced the length of residence from 7 to 5 years, and eligibility begins at 18 years of age. As a consequence, length of residence and employment status replaced nationality or legal citizenship as the criterion for access to rights. This separates Sweden from other European countries (such as Austria, Belgium, France, and Germany, where asylum-seekers are not allowed to work, or where there is a delay of at least one year (in the UK)). In terms of education, the Swedish government ruled in the early 2000s that asylum-seekers should be entitled to the same level of primary and secondary education as Swedish citizens, Around the same time, debates took place in parliament concerning health care for various categories of immigrants. While health care remained unchanged for legal labour migrants and their families, undocumented immigrants, the government reasoned, should be given emergency medical treatment, but they would be billed for it. For undocumented *children* or those whose claim for asylum was rejected, they would be entitled to full medical care. For asylum-seekers, emergency medical care and preventative health care would be guaranteed without additional payments, or at least an administrative fee that was lower than what citizens paid. In contrast, in 2004, the government would reduce the daily allowance for any individual who could not prove their identity.

In the late 2000s, the difficulty that refugees encountered in trying to find work led to the nationalization of standardized benefits from previously locally variable benefits. To encourage the search for work, additional benefits obtained from working did not affect introductory benefits, and the benefits would be more generous than the social minimum, but their duration would be reduced to two years. As noted in Chapter 5, asylum-seekers and refugees would have to follow a course in Swedish society and labour market training measures, and if they failed to do so, would be subject to sanctions, such as having some of their benefits removed (Bevelander and Pendakur 2012; Sainsbury 2012).

In sum, while many social benefits remain intact for migrants, and some have even increased, there has been an increase in the issuance of temporary permits, thus restricting access to benefits, and even in a putatively strong welfare state such as Sweden, the reduction in social benefits has had a profound effect on immigrant lives. Whether one can attribute all these changes (and some of the resilience of benefits) to 'neoliberalism' is quite another matter; some might argue that the above discussion only confirms the validity of this concept, while others might be less bold, content to focus more modestly on actual policy changes and their contradictions, without recourse to a grand concept such as neoliberalism. After all, in the early 1990s, the Swedish government did consider that increasing asylum-seekers rights might have increased asylum-seeking to Sweden, and thus, 'protecting' the level of benefits seems to have taken priority over market solutions (commodification), despite the renewed emphasis on active labour market policies for asylum-seekers and refugees. Given then all the debates around neoliberalism and given those studies which simply do not engage with the concept of neoliberalism, the academic jury seems, on the whole, to be undecided on the neoliberalization of social rights.

Whatever the value of neoliberalism for understanding the evolution of migrant rights in Europe (such as in Sweden above) or in the United States, migrant rights have arguably increased in countries beyond Europe and North America, including in Japan and South Korea. In Japan, this is especially the case for long-term resident Koreans since the end of the Second World War, and particularly since the 1990 Immigration Control and Refugee Recognition Act. Though there are plenty of exceptions, such as the access to political representation among migrants and the continued lack of racial discrimination legislation in Japan (as of 2015), overall the expansion of rights has progressed to such an extent for legal migrants that permanent residency seems to be even more desirable than citizenship in Japan, given the parity of rights between citizens and permanently resident migrants. Surak (2008) conjectures that much of this expansion of rights is attributable, first, to the compliance on the part of the Japanese government to international human rights conventions since the 1980s, and second, to local officials, whose practical outlook towards ensuring the welfare of their migrant constituencies has stimulated them to lobby for rights to jobs, health insurance, and pensions. National immigration policy-makers have taken their cues from this local landscape of expansive rights, and not necessarily the other way around. Yet, we might add that the Japanese government also considered caring services for the elderly in a rapidly 'greying' Japan to be a significant issue which needed to be addressed through the liberalization of immigration policy (Peng 2016). While this says nothing about the actual content associated with these rights, especially for refugees (Dean and Nagashima 2007), the narrative of neoliberalization in terms of migrant and refugee rights seems somewhat misplaced in the context of Japan. This seems equally true of Korea, where migrants enjoyed few rights for the first two thirds of the twentieth century. While legal migrant workers' rights are now broadly consistent with citizens in Korea, especially through the 2007 Basic Act and the First Basic Plan for Immigration Policy (2008–12), rights have expanded especially for 'ethnic Koreans' through unique work permits designed to attract them to Korea and to facilitate their permanent settlement (Surak 2008; Kim 2008; Lee 2008). Nonetheless, while the Basic Livelihood Security Law (a welfare bill) signed in 1999 to provide for basic social welfare regardless of age and employment status was extended to cover marriage migrants with minority-age children who are Korean citizens, it did not cover labour migrants who constitute the bulk of immigrants in Korea.

Thus, as Song (2016) maintains, the emphasis lay on families rather than individuals and "Korea's immigrant incorporation policies were based on the principle of the patriarchal family system and cultural paternalism, as opposed to emphasizing the equally important value of cultural diversity" (p. 13). Furthermore, despite the emollient rhetoric of the two governments in Japan and South Korea in terms of the expansion of rights, this 'expansion' leaves the rights of undocumented migrants on the margins, who have been an allegedly increasing percentage of the migrant population in countries such as Japan and Korea, at least until the late 2000s. Thus, migrants' rights are stratified in Korea and Japan, as they are elsewhere in the world.

In other, especially poorer countries, it may be similarly problematic to deploy the term neoliberalization in terms of the actual content of rights, since economic, social, and political rights were never institutionalized to the same degree in the first place for migrants. And whatever rights existed for migrants in the past, structural adjustment in the poorest countries over the last 25 years or so has eviscerated these already skeletal rights. Consequently, most low-income migrants, especially those who are undocumented must fend for themselves in the burgeoning and desperate informal economies of the ever mushrooming 'mega-cities' and their sometimes equally impoverished suburban peripheries of the global south (e.g. Davis 2004).

In sum, it is easy to exaggerate the claim of post-national rights or transnational citizenship. The content of these rights seem to have eroded at the very moment migrants have obtained access to them, especially in the EU. As we have seen in the US, even access has been constrained, combined with a more general transformation of welfare. That migrants experience different bundles of rights suggests that their opportunities for participation and belonging in the countries of immigration will also be shaped, but not determined by such rights, and it is to this question of belonging that we turn to in the next section.

CITIZENSHIP AS BELONGING

While citizenship may be defined by legal status and social rights, it is also about 'belonging'. By 'belonging', we mean subjective senses and practices of citizenship that hinge upon migrant identities, and this third part of the chapter explores migrants' senses of belonging in relation to their country or countries of origin and immigration. We start however, by elaborating on the terms 'social exclusion', 'cultural marginalization', and 'differential exclusion', which may seem like an odd way of beginning a chapter section on belonging, but depending on the degree of social/differential exclusion and marginalization, migrants may reject the host society and thus harbour feelings and engage in practices which are antithetical to belonging, sometimes referred to as a 'reactive ethnicity'. After this discussion of social/differential exclusion, we then review some of the distinctions between 'assimilation', 'multiculturalism', 'integration' (sometimes called 'acculturation' or 'adaptation'), 'differential inclusion' (Carmel et al. 2012; Mezzadra and Nielson 2013) 'diversity', and 'transnational belonging'. The final section moves to a discussion of citizenship as political participation. This part of the chapter mixes a 'top-down' understanding of citizenship, in which governments shape belonging through discourses, policies and practices, and a 'bottom-up' understanding, in which migrants carve out their own complex identities in relation to these discourses, policies and practices.

Social exclusion, cultural marginalization, differential exclusion/inclusion

The concept of 'social exclusion' gained its popularity in Europe in the 1980s as an alternative to the Anglo-American idea of poverty or poverty lines, partly because it emphasizes the range of social processes that result in "inadequate social participation, lack of social integration and lack of power" (Room 1995: 5). Elsewhere in the world, 'marginalization' or 'vulnerability' (particularly in the context of asylum-seekers, refugees and internally displaced persons) seem to be more widely used terms, although these were not inscribed in state policies in the more concrete way they were in countries such as France and the UK. While policy discourses in Europe gradually shifted during the 2000s to focus either on 'inequality', social cohesion, or 'social inclusion', the processes described by the concept of social exclusion have not disappeared and it serves as a useful if very general and flawed metaphor for the problems that migrants face in cities, towns, villages, detention centres, refugee camps, and other places in both the global north and south (for some critical assessments of the term, see e.g. Madanipour et al. 1998; Mezzadra and Nielson 2013; Goicoechea 2005; Samers 1998a). Let us focus then on processes of social exclusion that operate through a combination of ethnicism, nationalism, racism, xenophobia and various forms of negative prejudice around the many axes of differentiation that we have spoken about throughout this text (e.g. age, class, disability, ethnicity, gender, nationality, sexuality, and so forth). Again, taking into account societal processes beyond 'western' countries, this involves more than 'white privilege', and does not simply involve the racialized exclusion by 'whites' of 'black/brown' peoples, although that continues to be a pivotal form of exclusion across and within many borders (but compare Grosfoguel et al. (2015), Winant (2015), and Wimmer (2015) for the significance of 'race' within and beyond the 'west').

Social exclusion can involve material, legal, and discursive dimensions (Musterd et al. 2006; Samers 1998a). Material exclusions are often related to legal exclusions (those based on one's legal status or denizenship) and might include exclusion from formal waged employment or entrepreneurial possibilities, from banks and other financial institutions, from basic schooling (especially in poorer countries) or higher or prestigious education, and from 'adequate' housing, health and social services, training, and leisure spaces such as parks. It might also involve the inability of migrants to vote in local or national elections, and more generally to 'sit down at the table' with government officials in various local, national, and international contexts. Discursive exclusions may entail the invisibility of migrants in certain policy reports, and the absence or misrepresentations of their individual or collective voices in nationally and internationally dominant entertainment and media, from the most popular blogs to major (news) programmes on television and the internet. Yet migrant-centred and pro-migrant NGO publications and media programming which may span multiple localities and countries and which target migrant and pro-migrant communities suggest that migrants may be socially excluded from some spheres but included in others, virtual or otherwise. This has at least two implications for the concept of social exclusion: exclusion and inclusion are closely related, and there is a complex geography to exclusion which occurs through particular places and territories, as well as through virtual spaces and communities, which are themselves shaped through territory. A better term than social exclusion might be 'socio-spatial exclusion' then (e.g. Sibley 1995).

The socio-spatial exclusion of migrants may also be related to their 'cultural marginalization' or 'othering' insofar as they are viewed as 'too different', 'culturally distant' or 'incapable of assimilating or integrating'. Such fear or anger (or both) among many citizens against certain cultural practices of 'foreigners', might include anything from speaking a foreign language to practising a religion deemed to be heretical or dangerous. 'Foreigners' may also be associated with 'strange' or unacceptable practices that range from patriarchy to polygamy. Indeed, many migrants do not necessarily enjoy a growing acceptance in the twenty-first century. Notably, the US government has scrutinized and used repressive measures against Latino and Muslim migrants (or those perceived to be Latino or Muslim) after the events of 11 September 2001 (known as 9/11) (e.g. Ashutosh 2008; Howell and Shryock 2003; Iyer 2015; Shoeb et al. 2007; Staeheli and Nagel 2006). European governments too have responded in similar ways to Muslim migrants across the EU, including in terms of the construction of mosques (e.g. Césari 2005). For example, following a referendum in Switzerland in 2009, the Swiss government banned minarets, seeing them as a clear symbol of the Islamization of the public sphere (Antonsich and Jones 2010). Beyond the actions of national governments (remember the discussion of Foucault and governmentality in Chapter 4, and the ways in which governance exceeds the realm of the state), there are also other forms of Islamophobic and anti-immigrant violence in public spaces. This may be directed at veiled Muslim women in particular, such as in the city of Malmo, Sweden (Listerborn 2015); in Athens, Greece at Muslim and other immigrants by the neo-Nazi political party 'Golden Dawn' (Hatziprokopiou and Evergeti 2014), or online among private citizens in Europe and North America alike (Ekman 2015; Gemignani and Hernandez-Albujar 2015). Cultural marginalization is hardly limited to Europe and North America. The Malaysian government and many Malaysians welcomed Indonesian migrants in the 1980s as a neighbouring and fellow Muslim work force, but anti-Indonesian sentiment has become widespread in the twenty-first century and related to, among other issues, a resurgent Malay nationalism in which Indonesian migrant workers are seen as 'foreign Islamic nationals' (Spaan et al. 2002), but other migrants, such as Burmese, have also been subject to police repression in both Malaysia and Thailand.

Migrants feel this cultural, racial, ethnic or xenophobic hostility which eventually has an effect on their sense of belonging. It should be clear then that socio-spatial exclusion and marginalization can be manifested in a variety of ways. The consequences may be that migrants reject the cultural discourses and practices of the dominant culture, it may lead to the adoption of these discourses of practices, it may lead to new negotiations and challenges to dominant representations and celebrations of migrants' cultures of origin. This last process is often expressed in parades for example, which are commonplace displays of migrant cultures and very subtle forms of protest against racism, stigmatization, and other exclusionary or marginalizing processes throughout the world. They involve what Mitchell (1995) calls 'spaces for representation'. To provide one illustration, Veronis (2006) shows the significance of the Canadian Hispanic Day Parade in Toronto for various Latino communities that exist in the 'Jane and Finch' section of north-western Toronto. The parade, which proceeds down a central boulevard in the area and passes through commercial areas, is infused with music, dancing (including a dance group of children) food, and a beauty pageant. The parade is not an act of rebellion against living in Canada or Canadians, but a much more celebratory and subtle form of protest against the stigmatization of Latinos, the impoverished spaces of public housing which they inhabit, and the lack of possibility for social mobility. It displays the diversity of Latin Americans but also their unity and pride in being Latin American, while embracing the value of 'multiculturalism' in Canadian political discourse. It is aimed at challenging dominant representations of migrants as "poor, uneducated, corrupted, lazy, and violent" (p. 1665). As Veronis argues however, the parade is also inflected by 'neoliberal' ideas insofar as it is designed to support local businesses, and to provide opportunities for disadvantaged youth. Whether or not the support of businesses and employment is necessarily 'neoliberal' is another matter, but the parade also fosters the values of 'individual responsibilization', in other words, the neoliberal idea that people should be responsible for their own lives and welfare, rather than rely on the state for support. Again, in the parade, there is a celebration of Latino culture, but that does not mean that Latino migrants in Toronto necessarily reject all facets of Canadian life, and it would be difficult to do so given the weight of assimilation pressures. We will return to these 'pressures' momentarily, but let us now divert our attention to an idea related to social exclusion, namely 'differential inclusion'.

Despite all of this exclusion and marginalization, some immigrants are seen as more 'acceptable' or 'deserving' in the minds of citizens than others, depending on their age, language use, skin colour, national and ethnic background, perceptions of their 'work ethic', their real or imagined economic success, and their willingness to embrace political liberalism (e.g. Chauvin 2014; Chauvin and Garcés-Mascareñas 2012). Mezzadra and Neilson (2013) might refer to this as 'differential inclusion'. In fact, the 'acceptability' or 'deservingness' of migrants changes over time. For example, in Ignatiev's (1995) fascinating book How the Irish became White, he discusses the period before the American Civil War, in which the Irish were viewed as 'unacceptable immigrants' but later 'whitened' (accepted) as part of the dominant racial class after the Civil War. Acceptability then may be a question of 'moments' or periods of history.

In any case, for Mezzadra and Neilson (2013), 'inclusion' may not be as 'positive' as we imagine, since first we have to ask 'inclusion into what exactly'? Second, 'differential inclusion', they suggest, might simply be a means of discipline and controlling immigrants. In this regard, Lee (2015) shows how NGOs in South Korea mirror the state by creating a certain 'disciplinary citizenship' (a form we might say of 'neoliberal governmentality' – see Chapter 4) whereby North Korean refugees are urged into economic self-sufficiency and 'civic contributions' such as volunteering as a means of countering their reputation as welfare recipients, while being expected to passively accept their position in low-paying jobs.

Assimilation

In the immigration studies literature during most of the twentieth century, 'assimilation' seemed to have at least three meanings: immigrants adapt to or adopt the cultural ideas and practices of the dominant culture over time; immigrants achieve the same socio-economic status measured in terms of some 'mean' for the 'native born' (Zhou et al. 2008: 41), and immigrants develop a spatial pattern in terms of residence and employment that is indistinguishable from the dominant or more dominant cultural groups.[9] Although the term 'assimilation' still has some currency within European countries, it is in countries such as the United States, Japan, or South Korea (prior to the 2000s) where 'assimilation' (or some version of it) appears to have had the strongest resonance in academic and public debate. In the United States, the preoccupation with 'assimilation' in part reflected the long-standing idea of the United States as a 'melting pot' in which successive waves of immigrants would 'melt' into an idealized conception of 'Americanness'. During the nineteenth century, assimilation meant adopting the cultural practices and expectations of the nationally dominant culture (white, Anglo-Saxon Protestantism, or WASP culture). Assimilationist ideals shifted over time during the twentieth century in the US as waves of immigrants (especially Catholics and to a much lesser extent Jews) re-shaped the cultural landscapes of the United States, and a nationally dominant culture seemed less visible. In fact, Gordon (1964) refers to the US during the twentieth century as a 'triple melting pot' composed of Protestants, Catholics and Jews. The twenty-first century has brought even more diversity including significant numbers of Bhuddist, Hindu and Muslim immigrants, and by 2005, the Protestant population had fallen below 50 per cent (Portes and Rumbaut 2014). The idea of assimilation as public discourse and practice has not disappeared, and as in other countries such as Japan, candidates for naturalization in the US must take a citizenship exam which attempts to re-assert a certain vision of 'Americanness', from questions on what battle happened when, to particular amendments to the US constitution.

One would have thought that 'assimilation' in academic discourse would have disappeared in seemingly ever diverse 'western' countries, but this is not the case. However, the nature of assimilation arguments has changed. In fact, against the 'straight-line' assimilation arguments of the nineteenth and twentieth centuries in which over time, all immigrants' lives eventually mirror those of citizens, Portes and Zhou (1993) develop a less linear conception of assimilation. They argue in fact that the differences in the achievement of socio-economic status and socio-cultural practices among *second generation* immigrants (understood in terms of national groups) can be called 'segmented assimilation' (Portes and Zhou 1993; Portes and Fernandez-Kelly 2008; Zhou et al. 2008). In Portes and Zhou's (1993) initial statement, they portrayed segmented assimilation as moving beyond a 'uniform mainstream' with particular 'mores and prejudices' that would create a national script for integration. Rather, they found several different forms of 'adaptation'. As they claim, one involves "growing acculturation and parallel integration into the white middle class; a second leads straight in the opposite direction to permanent poverty and assimilation into the underclass; still a third associates rapid economic advancement with deliberate preservation of the immigrant community's values and tight solidarity" (p. 82).

Portes and Zhou do not neglect space in their analysis. In fact for them, it is the paradox of Haitians' apparent *cultural* assimilation into a particular African-American dominated area of Miami on one hand, and on the other, their lack of *socio-economic* assimilation (measured in terms of education, occupation, income, home ownership, etc.), that in many ways motivates their call for the idea of 'segmented assimilation'. In order to explain these segmented outcomes, Portes and Zhou refer to 'modes of incorporation' by which they mean "the complex formed by policies of the host government; the values and prejudices of the receiving society; and the characteristics of the coethnic community" (p. 83). These modes of incorporation depend upon how citizens view the skin colour or what they call 'racial type' of migrants, their location, and the absence of 'mobility ladders' (meaning how deindustrialization in the United States wiped out many highly paying, unionized manufacturing jobs). Not satisfied with this understanding of segmented assimilation nor the 'traditional' literature on assimilation in general, Zhou *et al.* (2008) develop the concept of segmented assimilation further by suggesting that the conventional measures of assimilation are problematic, insofar as what is equally important is a measure of 'inter-generational progress', that is, how far migrant groups move beyond their parents' attainment. Yet they take it even one step further by proposing a 'subject-centred' approach to assimilation that accords priority to the way in which migrants themselves "perceive, define, and measure mobility and success" (p. 42). This 'subject-centred' approach is really an exercise in ethnography and other forms of qualitative research that are familiar to anthropologists, other sociologists, and critical human geographers, and it is only weakly elaborated upon in their analysis. Their results from this approach however, can be stimulating for understanding immigrant aspirations. For example, one of their 'subject-centred' findings in the US concerns differences between Chinese, Vietnamese, and Mexican respondents. For Chinese and Vietnamese respondents, it became clear that whatever their socio-economic attainment measured in conventional ways, they did not perceive themselves to be 'successful', while Mexicans with less 'success' in terms of conventional measures, were more likely to report feeling successful.

In any case, to the original elements of incorporation then, they add such 'variables' as family structure, cultural, economic, human, and social capital, the legal status of parents, the expectations and investment priorities of parents, the legal status of the second generation individual (this is akin to civic stratification, as discussed earlier in the context of the EU), the cultural memory of difficult lives in the country of origin, and access to public resources and services, especially for disadvantaged migrants.

Again, this updated, but generally speaking still standard assimilation literature seems to assume national differences from the start, given that its research design uses a comparison of national groups. We have seen from other literatures on labour markets cited in Chapter 5 that axes of differentiation such as gender, but also other 'variables' such as time of arrival, generation, and the political and social contexts of the countries of emigration, and even return policies, may shape standard measures of socio-economic status. Other studies that focus more on 'cultural assimilation' or 'selective acculturation' (Portes and Rumbaut 2014) expect that citizens as well as immigrants change their own cultural and social practices over time in this process of assimilation (see Levitt and Jaworsky (2007) for a review). Another very different assimilation literature has emerged over the last decade which seeks to place assimilation back on the agenda for critical social scientists in the wake of the obsession with transnationalism (see the special issue edited by Leitner and Ehrkamp (2006)). This literature however, focuses not so much on the measurement of adaptation as on examining assimilation *as a discourse* and the way in which these discourses are received.

What this latter literature on assimilation *makes explicit* is that who and what is not acceptable is also a matter of space and the complex identities of citizens and denizens alike. The same practices that might be viewed as acceptable in the city of Dearborn, Michigan and the surrounding towns, with the largest concentration of Arab (if not necessarily Muslim) immigrants in the US, may seem rather odd, and perhaps even unacceptable in, let us say, the predominantly 'white' state of West Virginia. In certain places then, migrants are viewed as 'out of place' (Cresswell 1996). Conversely, migrants may be viewed as 'in place' in certain cities, towns, and neighbourhoods in the US (as elsewhere) that are dominated by particular national, ethnic or religious groups who may be considered minorities in the country of immigration as a whole, but who form a dominant culture in those cities, towns, neighbourhoods and so forth. In fact, the predominance of the Spanish language may not only be acceptable but actively encouraged in Miami or on the southern Texas border with Mexico. Thus, 'assimilation' (and the discourses of assimilation) become more complex, as the cultural benchmarks and practices of immigrants are complicated by nationally dominant cultural discourses and practices, but with local territories dominated by 'minority' cultures.

This is thoroughly documented in Ehrkamp's (2006) ethnographic study of native German and Turkish identity in the small city of Marxloh (about 20,000 people) in the deindustrialized Ruhr region of Germany. In the late 1990s, Turkish immigrants accounted for about 25 per cent of Marxloh's residents. Ehrkamp is concerned not so much with assessing the degree of assimilation than with the presence of assimilation *discourses*, and she mobilizes the work of the late philosopher Jacques Derrida who likens 'assimilation' or 'integration' to a matter of 'hospitality'. The host decides what the rules are and the guests must obey. In the context of Germany as elsewhere, the 'guest-workers' (this is a term contested by many Turkish migrants) are subject to assimilation discourses at the national and local scales which demand that Turkish and other migrants assimilate. There is however, a perverse logic at work, as Ehrkamp (2006) explains insightfully, "By asking immigrants to become similar, nonimmigrant residents portray them as more different, thereby making adaptation or *Integration* [the common term used in German political discourse] almost impossible" (p.1678).

The consequence is that politicians and the mass media then portray Turkish immigrants as 'unassimilable'. At the same time, the 'resident majority' demands that immigrants adopt German norms, without 'disturbing the identity of the majority' (*ibid*). Put differently, these discourses transform Turkish migrants into 'exotic others'; it 'orientalises' them (Said 1978) as culturally distinct and perhaps even ultimately unassimilable.

The discourses of assimilation that circulate throughout Germany are manifested in local German political leaders' and residents' stances against the demand for an Islamic call to prayer in the city of Marxloh, and practices such as bilingualism or speaking Turkish only in the street, using Turkish on store fronts, and so forth. In exploring these two issues, Ehrkamp argues that the construction of a Turkish identity can only be understood in relation to a sense of German identity, and both are produced through assimilation discourses through the national and local spaces. These discourses construct Turkish migrants in opposition to 'German-ness'. Thus while Turkish migrants are perceived to want to 'stick to themselves' in 'ghettos', one of the migrants that Ehrkamp interviews claims that 'living apart' is not desirable, but forced on Turkish migrants because of housing discrimination.

While Turkish migrants internalize the negative discourses about them, Turkish migrants also contest the social and spatial representations of them by the different levels of government, by the 'mass media' and by German citizens. Yet their contestation moves beyond simply representation. Turkish migrants complain of Germans' unwillingness to 'integrate', that is to take part in Turkish cultural practices, instead of Turks having to always adapt to German cultural practices. Given this discursively imagined sense of difference, the migrants that Ehrkamp studies see themselves as 'not German'. As one Turkish migrant claimed, "We Turks are no Germans".

Another principal argument of Ehrkamp's paper is how certain behaviours are not applauded by fellow migrants in Turkish areas of Marxloh, such as acting 'too European', or being 'too conformed' or 'too Germanized'. In these places, migrant assimilation takes on a very different form. When they visit larger cities such as Essen or Dusseldorf, Turkish immigrants dress and act differently. In these cities, the same local assimilation discourses do not seem to prevail or at least do not shape Turkish cultural practices. Ehrkamp concludes then that "The spatiality of assimilation discourses also generates a spatiality of (immigrant) identities" (p. 1688).

Multiculturalism

Multiculturalism is a set of discourses, ideologies, political philosophies, forms of governance, and policies geared towards recognizing group rights. At the same time, it can refer to the aim of political movements in the name of recognition and representation, in a context of diversity and pluralism defined in group terms through which people feel a sense of belonging (e.g. Hall 2000; Korteweg and Triadafilopoulos 2014; Kymlicka 1995; Parekh 2006; Vertovec 2007). Stuart Hall (2000) also thought it important to distinguish between 'multicultural' (that is a condition of diversity) and multiculturalism which attempts to 'fix' the meaning of diversity (see Penrose 2013). Thus, while it may not be controversial to label 'western' countries such as Australia, Canada, Germany, the Netherlands, the US, or the UK as 'multicultural', Joppke (2007a) insists that 'multiculturalism' as a set of official policies first adopted by Canada in 1971, and later the Netherlands and UK in the 1970s and 1980s, have been officially discredited or abandoned in the Netherlands and the UK after more than two decades. At least three questions emerge from the various definitions of multiculturalism noted above, (including Hall's distinction), and Joppke's claim about the end of multicultural policies: (1) To what extent are societies 'multicultural'? (2) To what extent do multicultural policies still figure in government policies? (3) Can multicultural or multiculturalism be extended conceptually beyond its formulation above?

In the United States, it is a pervasive argument that American society has changed from a 'melting pot' to an extraordinarily diverse 'salad bowl', where most immigrants retain some elements of their cultural practices from the country of origin and mix with others who have retained their own cultural practices. Yet, we should ask whether multiculturalism implies that cultures are internally coherent and homogeneous, and whether this is the most appropriate way of looking at 'difference' and 'belonging'? In earlier understandings of this term – let us say in the 1980s – the discourses and policies of multiculturalism seemed to imply that migrants and their cultural practices were distinctive and sharply delineated from the practices of the cultural majority, and that they were fixed, stable, unchanging, and rooted in national or ethnic attachments. Studies in the twenty-first century on multiculturalism (e.g. Parekh 2006) recognize that while cultural practices are *relatively* enduring, they also change over time and are far from internally coherent. One example might be the concentration of 'South Asian' businesses and residents along West Devon Avenue on the north side of Chicago. While most Indians and Pakistanis had little social contact with each other in this area prior to 9/11, they later perceived each other as mutually 'brown' or 'South Asian' in the context of intense Islamophobia and government scrutiny. Thus, as a brown/South Asian 'culture' was cobbled together in this neighbourhood, it became increasingly difficult to speak of a distinctly Indian or Pakistani identity (Ashutosh 2008).

Similarly, earlier versions of multiculturalism rested on recognizing national cultures, 'minority cultures' (in the Netherlands), or even 'races', as in the UK's former 'race relations paradigm'. Regardless of the problems of using the terms 'minority' (that is, who is a minority, and when and where?) and 'race' (there is no biological basis for race), few would deny that cultural practices can be identifiable with particular nationalities and that nationality and ethnic background contribute to identities and senses of belonging. Yet once again, there are other axes of differentiation, including gender, language skills, generational differences, religious beliefs and practices, individualized experiences of racism or historical relations and affective ties to others from the same village, region, or country of origin that may be as central to one's identity as nationality or ethnicity are. For example, Robins and Aksoy (2001) insist that for Turkish Cypriots in the UK, they have never expressed their identity in national terms, fragmented as they are by the presence of Greek Cypriots in the UK, the age of the immigrants with respect to those who arrived before and after the division of Cyprus into its Greek and Turkish halves, and the more contemporary contexts of the UK and Turkey. In a study with Pakistani girls in Hong Kong, Gu (2015) tells us that though they pride themselves in shaking off the patriarchal relations that they left in Pakistan, they do not necessarily assert their 'Pakistaniness', but rather, that they are Muslim and the importance of Islam. In sum, it is difficult to deny the existence of plural cultural identities. What is at stake though is how to reconcile cultural differences in mixed societies from Austria to South Africa.

In regard to question (2) above then; have governments truly abandoned multiculturalism as a form of reconciling differences? Korteweg and Triadafilopoulos (2014) suggest that multiculturalism is alive and well in *practice* in the Netherlands and even Germany, which never had an official multicultural policy in the first place. Other countries, such as South Korea, which have arguably been 'ethnically homogeneous' prior to the 1980s have actually turned to 'multiculturalism' as a form of governance (e.g. Kim 2015; Lim 2010; Watson 2010). For Korteweg and Triadafilopoulos (2014), multiculturalism is 'a form of governance' insofar as minority-based community organizations "engage with other political actors in the articulation and implementation of policy. They contribute knowledge of their communities' specific cultural strengths to help address their communities' challenges" (p. 2). In that way, multiculturalism evolves into a pragmatic and practical approach to integration which is not based in some sort of multicultural philosophy or a formal set of policies or policy orientation adapted by governments. It is therefore a product of horizontal organization informed by a spirit of diversity, instead of being vertically organized through top-down policy implementation. They illustrate this for example, through consultations between the Dutch government and immigrant-oriented NGOs as a means of addressing the relatively rare occurrences of 'honour violence' within both the Moroccan and Turkish communities in the Netherlands.

We would add here that multiculturalism may be seen as a set of *local* discourses and practices that still function in British cities such as Leicester, despite national political and policy changes over three decades (Jones 2015). Similarly, Lobo (2014) tries to extend and localize the concept of multiculturalism in her encounters on a bus with – among others – Fijian migrants in Darwin, Australia, which she refers to as 'everyday multiculturalism'. The case of multicultural practice in Leicester or multicultural encounters and negotiations in Darwin reinforce our over-arching point in this book that there are limits to a national lens on societal issues, such as multicultural policies. Ultimately, multiculturalism and multicultural policies have vied for the intellectual, policy, and public spotlight with the various discourses, meanings, and practices that are called 'integration', and that is precisely the subject of the next section.

Integration

Among European citizens, it is common to hear the words, 'he or she is well integrated in Finland', or 'he or she is poorly integrated in Austria'. Yet the term 'integration' is as contested as any other in both academic and policy circles. For our purposes here, we might consider the term 'integration' to have at least three principal meanings. The first is closer to 'assimilation' and is often used in the context of EU countries, notably in France, Germany, the Netherlands, and the UK since the early 2000s. When governments or citizens lament that there is a problem of 'integration', they are referring to the extent to which migrants fit into an imagined and idealized set of dominant practices and values of the citizen majority, or to their access to such material goods as housing, employment, education and health (e.g. Ager and Strang 2008). The second meaning of integration is closer to that of multiculturalism, whereby immigrants do not somehow 'lose their culture' but rather retain 'their culture' and join the liberal political culture of the western liberal democracy in question. A young woman who is French but who has Algerian parents illustrates this well:

> Me. I find myself totally integrated in France, so I feel at home everywhere. Given that I was born in France, that I speak French, that my culture is French, that I learned French history – France is my country [. . .] My identity is: French of Algerian origin, of Muslim religion. ('Fatima', cited in Keaton 2006: 40)

The third definition is much less common, and that is the 'coming together' of migrants and citizens, whereby each adopts the cultural practices (language, religion, food, music, and so on) from another. While the popular meanings of the term integration in France and Germany may be closer to assimilation, the European Union has in fact publicly called for this 'coming together' in its 'common basic principles' for immigrant integration policies. The November 2004 European Council Agreement proclaimed that "Integration is a dynamic, two-way process of mutual accommodation by all immigrants and residents of the Member States" (cited in Joppke 2007a: 3). As Joppke points out, this statement by the Council of the European Union that European citizens should accommodate and respect the cultures, languages, and religions of migrants was an "unprecedented stance to take" (Ibid.), and it is also a mark of the emphasis on 'integration' rather than 'assimilation'. After all, the practice of culture and religion is guaranteed under European constitutional law. Nevertheless, the Council's common basic principles' also demands a respect for the 'basic values of the European Union' (Ibid.), which include "the principles of liberty, democracy, respect for human rights and fundamental freedoms, and the rule of law".[10] In addition, the Council insists that migrants should respect the 'equality of women', 'the rights and interests of children', and the freedom to practice or not to practice a particular religion' (Ibid.). This stems from concerns about Islamic fundamentalism in European societies, while the Council fails to acknowledge the oppression of women in 'western' cultures (e.g. the demands placed on western women in terms of beauty and sexuality), and the way in which children's rights and wishes are often neglected.

Returning to the first understanding of integration above, it is clear that, like assimilation, it rests on a notion of citizens and others, though who and to what degree these 'others' are seen as different and unacceptable is a matter of space and place, as discussed in the previous section on assimilation. Nonetheless, the 'othering' of migrants tends to have particular class, racial, ethnic and xenophobic currents within it, so that those who are 'darker-skinned' and/or are Muslim are commonly viewed by western governments and publics as threats to the 'cultural homogeneity' or the apparent 'Judeo-Christian' roots of western liberal democracies (e.g. Balibar and Wallerstein 1991; Staeheli and Nagel 2006). Such 'othering' may have different dimensions in Asian, Latin American, or South African societies, but the processes involved rest on a similar 'inferiorization' or 'cultural distancing' of the 'foreigner', which may or may not be overcome through time. The Netherlands is a paradigmatic case of how 'multiculturalism' (or at least its rhetoric) has now been abandoned in the name of what Joppke (2007a) calls 'civic integration'. This again is closer to assimilation and demands that low-income and family visa applicants to the Netherlands be required to take an 'integration test', including proficiency in the Dutch language. But since Dutch language training is generally not available overseas, it effectively blocks the legal migration of most low-income migrants. France, Germany, the Netherlands, and the UK have adopted similar measures (Joppke 2007a; Lister and Pia 2008; Martin 2014). For example, in Germany, migrants who have been in the country for less than six years and who receive social assistance (welfare, etc. – about 20 per cent of foreigners) must follow German language and culture classes, or the renewal of their work permit may be in jeopardy. This is also the case with foreigners joining their families in Germany; they must pass a German language exam (Martin 2014). In the UK, these tests were meant to create a "sense of belonging around 'core British values'" (Lewis and Neal 2005: 431). And suddenly, after years of the Commission for Racial Equality[11] supporting multiculturalism as an extension of the 1970s 'race relations paradigm', the Chairman of the Commission abandoned the discourse of multiculturalism, and instead asserted the value of 'integration'. This was nothing less than remarkable, as Lewis and Neal (2005) noted. In sum, states want to somehow ensure the 'allegiance' of their potential citizens to not just the state as an abstract concept, but to connect it again with some form of nationhood. European states "are foregrounding the notion of core national values as the spine around which to achieve social cohesion and integration" (Lewis and Neal 2005: 433). Thus, citizenship policies in Europe have shifted since 9/11 from one of multiculturalism to 'civic integration' or what might be called a 'neo-assimilationist' stance (Joppke 2007a; Kofman 2005b).

Yet despite this era of civic integration and neo-assimilation, it would be erroneous to assume that nation-states are homogeneous or that civic integration or neo-assimilation eliminates migrant cultures. Far from it, pluralism, heterogeneity and the support of 'diversity' (see Faist 2009) are incontrovertible features of western liberal democracies, even in southern European countries such as Portugal and Greece that may not be thought of as diverse by outsiders. Rather than envisaging then that migrants simply adopt these assimilative or integrative pressures, perhaps it would be better to return to the cultural theorist Homi Bhabha's (1994) notion of 'negotiation'. Migrants navigate or negotiate different social pressures, from those of their co-ethnics in the country of origin or immigration, to those of citizens in a variety of different settings. It is the transnational dimension of this negotiation of identity that will preoccupy us in a subsequent section, but for the moment, we turn our attention not just to 'diversity' as a feature of countries of immigration, but as *policies and practices* actively pursued by governments, companies, private organizations, individuals and groups, whether of migrant origin or not.

Diversity

In this brief section, we rely on Faist (2009)'s conception of 'diversity as a new mode of incorporation'. For Faist, 'diversity' seems to refer to the condition of diverse languages, religions, and ethnic groups (that is, it is similar to cultural pluralism or multiculturalism) but perhaps equally important, it also signifies a tacit agreement (in other words a societal 'understanding') that organizations should not "discriminate against their staff, their members, or their clientele on the grounds of cultural characteristics, but rather should be sensitive and responsive to these characteristics" (p. 3). The third understanding of diversity concerns the importance of 'cultural skills' such as bi or multi-lingualism as well as social networks as 'competencies' and marketable skills that help organizations with multinational business. Thus, the idea of 'diversity' becomes especially associated with organizations. In terms of the criticisms of diversity, Faist argues that it tends to neglect inequality between individuals and between groups; it tends to disregard certain elements of citizenship (such as what citizenship means in terms of moral obligations as well as rights), and in so doing, 'diversity' ends up being just a technique of management in organizations. Thus, in attempting to compare diversity with assimilation and multiculturalism, 'diversity' assumes that the differences between people are a resource/competency to be used by organizations. Thus, organizations have to 'adapt and accommodate for migrant experiences', but it is an asymmetrical relationship, that is, organizations and the dominant ethnic group have power over other ethnicities.

'Diversity' is constituted according to Faist, in two ways: through 'diversity management' in the private sector (that is, viewing diversity as a resource and a set of cultural competencies as noted above) and an 'inter-cultural approach' in the public sector, in which public organizations such as hospitals have 'culturally sensitive catering and nursing', 'transcultural psychiatry', 'multi-lingual labelling', and so forth. 'Diversity' as a mode of incorporation is not likely to decline 'in the near future', he argues, because of two 'epochal developments': 'The transnationalization of migration consequences'. By this he means the characteristics of massive migration and the strong links with the countries of origin that pervade 'western societies' reinforce 'a need' to respond to the plurality of cultures, and second, 'The universalization of legal claims at the national and international level'; in other words, claims of rights by immigrants and their descendants within nation-states but also at the level of international treaties are converted into more and more rights over time that intersect with others such as the right to practise one's religion.

Transnational belonging?

Keaton (2006) recounts the words of 'Aicha', the pseudonym of a young woman of Moroccan origin living in the suburbs of Paris:

> Despite being French on paper, I'll always be an Arab, and it's not a simple paper that could change my culture. I was born in France. I have French culture, but I live with Moroccans. Every year, for 2 *months*, I go to Morocco. I speak Moroccan, I eat Moroccan food. In fact, I have 2 cultures, French and the other, Moroccan. I practically have to be French in order to succeed in life, otherwise you're screwed [. . .] So I'm Muslim of Moroccan origin. ('Aicha', cited in Keaton 2006: 35)

The feeling of belonging that 'Aicha' feels across two countries, and two nationalities, while also self-identifying herself as a Muslim, speaks to what is called 'transnational belonging' (e.g. Basch et al. 1994; Levitt 2001). It is a sense of belonging both to the 'creolized' (mixed) society of the country of immigration and to the equally complex society associated with the country or countries of origin, or indeed other countries involved in particular national or ethnic diasporas. Some have also referred to this as 'bi-focality' (Rouse 1992) or a 'double engagement' (Grillo and Mazzucato 2008) in which migrants constantly think of the 'here' and the 'there', often in contradictory ways (Turner 2008). In some cases, such as when an Italian immigration law (the Bossi-Fini law of 2002) outlawed the repatriation of national insurance contributions of migrants in Italy, many Senegalese migrants with children in Italy increasingly abandoned the hopes of ever returning to Senegal, and as Riccio (2008) puts it "are now starting to feel that they belong neither here nor there" (p. 230). This 'transnational belonging' (or its absence) will be expressed through a variety of cultural, economic, political and social practices (for thorough reviews, see Levitt and Jaworsky (2007), Waldinger (2015) and Vertovec (2004), and it may be useful to distinguish between senses of belonging and their expression or practices (Glick-Schiller 2003)).

'Transnational belonging' among migrants can be understood as the opposite end of the spectrum from assimilation (Ehrkamp 2006), or segmented assimilation. Others, such as Lucassen (2006) do not see these processes as mutually exclusive, arguing that transnational belonging may be certainly aligned with assimilative practices. Still others (e.g. Glick-Schiller et al. 2006) prefer the term 'incorporation' together with transnationalism, which is probably a more sophisticated way of expressing migrants' position within 'transnational social fields'.[12] For Glick-Schiller et al. (2006) incorporation refers to "the processes of building or maintaining networks of social relations through which an individual or an organized group of individuals becomes linked to an institution recognized by one or more nation-states" (p. 614). Yet understanding such incorporation requires some sort of entry point, which they see as "individual migrants, the networks they form, and the social fields created by their networks" (Ibid.).

To return to the discussion of transnational issues discussed in the Introduction and Chapter 3, transnational networks allow for the development and sustenance of transnational belonging, yet a debate unfolded during the 1990s as to whether or not (and to what degree) transnational belonging could actually be considered a new phenomenon. In Glick-Schiller et al.'s path-breaking *Towards a Transnational Perspective on Migration* (1992) and Basch et al.'s (1994) *Nations Unbound*, they argue that the transmigrants of the last decades of the twentieth century between the Caribbean and the US were different from previous cohorts of immigrants of the nineteenth and early twentieth centuries in terms of the extent of their political and social participation in both the Caribbean countries of origin and the United States. This might be tempered however by at least the extent of return migration between 1910 and 1920, in which Foner (2000) notes that for every 100 immigrants entering the US in this 10-year period, more than one third returned home, even Russian Jews who initially fled political persecution (cited in Levitt 2001: 21).

And as we also learned in the introduction and Chapter 2, senses of belonging are unlikely to be only transnational in character, after all, what exactly is meant by 'national' anyway? Instead, national cultural, economic, political and social ties are conditioned by a set of other spatial and social dimensions. From a spatial dimension, a more sensitive understanding would question the 'methodological nationalism' of transnational academic discourse. Consider once again Ashutosh's (2008) study of 'South Asians' on the north side of Chicago. In one restaurant, Ashutosh notes, a sign bearing the name 'Southall' (a reference to an area of west London well-known as a cultural centre for South Asian life in the UK) appears on the wall. Though hardly unusual as a phenomenon, Indian migrants in Chicago not only have attachments to the sub-continent, but also to the wider spaces of the Indian (or South Asian) diaspora, including Hong Kong and Nairobi. Perhaps then we should not speak so much of uniquely transnational belonging but multiple territorial attachments and practices.

Thus to have a greater appreciation of the complex spatial attachments associated with migrant belonging, anthropologists and sociologists have devised a number of widely employed concepts, including 'transnational social field', 'transnational social space', 'trans-localism', 'trans-locality', 'trans-regionalism', 'transnational urbanism' and 'transnational village' (see Vertovec 2001 for a review of some of these terms). Many of these ideas overlap, and we will not work through all of their definitions. Instead, let us concentrate on the last two of these: 'transnational urbanism' and 'transnational village'. For those concerned with cities and migrants, Michael Peter Smith's 'transnational urbanism' has become especially influential. In this concept, transnational social actors are individuals who take advantage of what cosmopolitan cities as nodes in transnational networks can offer in terms of their cultural opportunities (the chance to pursue particular cosmopolitan lifestyles and images and the pursuit of higher education), economic opportunities (finding employment, putting remittances to work for developing economic capital, such as the acquisition or creation of small businesses), and political opportunities (working for pro-migrant and transnational organizations, for example). All of this is accomplished by migrants exploiting their circuits of travel and communication that extend from major city to major city across borders.

A less urban understanding of transnational belonging is encapsulated in Levitt's (2001) idea of a 'transnational village'. For her, a transnational village has four unique dimensions. First, international migrants are not actually required to be a member of 'the village' (or transnational social field); second, such villages emerge and are sustained through social remittances (social capital, behaviour and ideas) that originate in the country of immigration and flow to the country of emigration. Third, transnational villages both emerge from and create a range of religious, civic, and political organizations. A fourth element is the 'social cost'. That is, some migrants return far wealthier, while others have little more than what they left with, which reinforces previous class, gender and generational divisions.

With respect to these social divisions that Levitt underlines, a small selection of research should suffice to convince us how transnational belonging is simply not just transnational in character. To begin with, McGregor (2008) reminds us of the salience of class distinctions in the relationship between 'home' and the UK in her study of Zimbabwean asylum-seekers and undocumented men and women in the 'abject (poverty) spaces' of London. Class distinctions are sometimes reproduced among Zimbabweans, especially those who migrate with little capital or the requisite skills to open up businesses in London. Others with substantial sums of money manage to start at least small businesses. What is remarkable though is also how these class distinctions are often dissolved, as those who were considered 'highly skilled' in Zimbabwe end up cleaning buildings or performing low-paid caring jobs, alongside Zimbabwean asylum-seekers and refugees. They share neighbourhoods, hometown/homeland associations, and the hope for a more prosperous Zimbabwe. Similarly, McAuliffe's (2008) ethnographic work with second generation Muslim and Baha'i Iranians in Sydney, London and Vancouver, reveals that relationships with the 'homeland' are not predicated so much on national origins, as on the dynamics of class aspirations or self-perceived class position ('low', 'middle', and 'higher'), religion (Muslim or Baha'i), and place (their settlement patterns). For example, Baha'i are required to settle in certain areas of the city to establish what is called 'Home-front pioneering', or the creation of local spiritual assemblies composed of nine adult Baha'is in a designated area of the city. This call to form an assembly in what is sometimes an undesirable area of the city often contradicts with their class aspirations, but Baha'is seem to reconcile these contradictions. While they may live near other Muslim Iranians, they do not necessarily share Muslim's stronger attachment to the 'homeland', given the historical pattern of persecution against Baha'is in Iran.

Once again, what we might mistake for distinctly transnational practices and spaces, may actually be ones of *also* locality, kinship, family relations and gender. Concerning gender, consider for example Ehrkamp's (2008) study of young Turkish and Kurdish immigrant *men* in Marxloh, Germany. She argues that the masculinities performed by these men in relation to Turkish women in public spaces, were racialized as 'Turkish' by the German media, politicians and citizens, rather than reading these masculinist and exclusionary spaces as gendered. Again, apparently transnational cultural practices cannot simply be associated with 'cultures' of the country of origin but may intersect with particular kinds of gendered performances. Although with no specific reference to the importance of cities, towns, and villages as scales or territories, Salih's (2001) account of Moroccan women in Italy underlines gendered differences in transnational belonging. She argues that the gendered expectations and family responsibilities of Moroccan women in Italy create different kinds of transnational attachments to Morocco, depending upon a mixture of their employment, marital and residential status in Italy and the need to care for relatives and children in both countries. Finally Glick-Schiller *et al.* (2006) speak of the problem of the 'ethnic lens' in migration research, arguing that migration research has confused religion with ethnicity, and transnational ties may be more about religious social fields than ethnic ones. For example, Somalis in London and Toronto self-identify as Muslims, rather than as part of a broader African diaspora (Grillo and Mazzucato 2008). This brings us to the question of religion in transnational identity.

In fact, we have barely spoken of religion until now. Religion is a pivotal dimension of transnational identities and practices, and while it is Islam, and especially fundamentalist Islam that has preoccupied the 'western' media, every stripe of religious adherence figures in the daily lives of immigrants, including not only Islam in all its diversity, but Baha'ism, Buddhism, Hinduism, Jainism, Sikhism, and various forms of Christianity, including Pentecostalism, the latter especially significant among African immigrants in 'western' countries. These are just some of the religions practised by migrants.[13]

In this brief section, we cannot but offer a very abbreviated discussion of transnationalism and religion, while seeking to avoid simplistic generalizations about this relationship and the complex spaces and places of religious practices among migrants. There are two related issues that we will examine here, essentially in the context of 'western' societies. The first concerns the religious flows of people and ideas between places of origin and destination, and here we will only provide a few examples to illustrate our point. The second revolves around the question of whether religious practices among 'the second generation' have intensified since the 1970s.[14] With respect to the first, not only do migrants bring certain religious practices with them, but Portes and Rumbaut (2014) recount a number of examples where migrants in the richer countries continue to rely on religious figures in their home countries. For instance, the wealthy southern Indian congregants of the Jyothi Hindu temple in suburban Houston trekked back to India to search for swamis to staff it and to perform ancient rites in Sanskrit at the initial phase of its construction. Similarly, Laotian immigrants in Louisiana had to recruit Monks from Laos to staff their new temples. From another perspective, a river flooded in the town of Jimani in the Dominican Republic in 2004, in which 700 people either disappeared or perished. In response, the Dominican immigrant community in New England and New York rallied around two organizations: the *Alianza Dominicana*, a social service organization in New York and a hometown association in Boston called ASOJIMA to provide funds for the town. Rather than channel money through what was perceived to be a corrupt provincial government, the organizations sent the money to a parish priest who organized the logistics of aid.

One of the most significant findings in the study of this relationship is that religiosity may not simply be an effect of transnational identities, but also constitutive of them. In that sense, considerable research on religion and migration suggests that religious identities and practices have in fact become more intense and even more conservative after migration and settlement, but they are related to other axes of differentiation, such as class and gender (e.g. Cadge and Ecklund 2007; Ehrkamp and Nagel 2014; Hatziprokopiou and Evergeti 2014; Hondagneu-Sotelo 2007; Kepel 1997; Kong 2010; Laurence and Vaisse 2006).

Such renewed religious fervour is visible, for instance, among many Koreans in the practice of evangelical Christianity, as well as among West Africans and Pentecostalism (e.g. Riccio 2008). Likewise, Predelli (2008) notes Muslim women's increasing participation in the social life and spaces of Mosques in Norway, although the Mosques remain the locus of 'patriarchial gender regimes'. Pedersen (2012) recounts how one Iraqi refugee woman that she interviewed had become increasingly preoccupied with being a 'proper Muslim woman' in light of her loss of class status in Denmark (she considered herself to be 'middle class' in Iraq).[15] Religion does not somehow travel 'intact' from country of origin to destination however. It changes and adapts to new circumstances. In fact, we need to think beyond some national context in shaping the transformation of religious practices, to consider how localized policies, events, and spaces, from evangelical churches to Hindu temples to religious processions, and public worship also re-shape such practices. In this sense, Saint-Blancat and Cancellieri (2014) discuss how the Filipino procession of the Santacruzan in Padua, Italy is shaped by the local appropriation of public space in that city. Here, there is an ambiguous interaction between the secular and the post-secular. Similarly, Ahmed et al. (2016) show how faith practices in East London, between Christians, Hindus, Jews, and Muslims are related to ethnicity and class and reconfigure the ambiguities of identity formation.

In order to explain the more religious fervour of the second generation and its relationship to transnational identities, we might rely initially on what Hirschman (2004) calls the three 'R's, that is, religion functions as refuge, as a form of respect, and as a resource. As refuge and a form of respect, renewed religiosity may relate to the perception and bewilderment among migrants that they have settled in more individualistic, secular, choice-ridden, or 'immoral' societies (Pedersen 2012; Shoeb et al. 2007), or that they are the objects of discrimination and hatred, as Maliepaard et al. (2015) show for long-established immigrant groups of Muslim origin in the Netherlands. From a different perspective, Shoeb et al. (2007) demonstrate in a study of Iraqi refugees in Dearborn, Michigan, how Iraqis' war trauma, living in refugee camps in Saudi Arabia, the feeling of exile in the US, the complicated freedom and influences in American society, and the myth of return push Iraqi refugees to be more devout Muslims in order to 'hold their families together'. Riccio (2008) finds that religion seems to serve a similar moral purpose among Senegalese Muslims in northern Italy. More specifically, Riccio (2008) and Kaag (2008) discuss how the Mouride Sufi brotherhood (an Islamic Sufi order) helps to provide an ideological and spiritual compass to Senegalese for the complex transnational lives which they lead, including the racism and discrimination they face. As Riccio (2008) describes, transnational practices within the brotherhood are sustained through conversation and the sale of cassettes. Through these two mediums, prayers and *kasaids* (sacred poems) can be found as well as "information about *ndiguel* (orders, decrees) from the Khalifa (the head of the order)". These transnational practices "are shaped and strengthened mainly by the activities of the numerous *dahiras* (religious circles) widespread in the receiving countries, and by the frequent visits of marabouts (spiritual leaders) from Senegal" (2008: 229).

The intensification of religious practice may also be in part related to the neoliberalization of societies, as states retreat from social welfare provision, and immigrant churches, mosques, and temples of many faiths fill the void – what Ley (2008) in his study of Vancouver calls the 'urban service hub'. Though the role of religion in providing social protection is hardly new, 'neoliberalization' is considered to be part of the picture. Anything from 'political opportunity structures' (we discuss these in the last section of the chapter) within overlapping territories to the diverse social axes of differentiation discussed many times in this book might have an effect on the complex identities forged out of the connections between transnationalism, religion, and identity (e.g. Turner 2008). It might seem the case that religiosity would somehow impede naturalization, but many religious organizations alongside hometown associations actually assist immigrants in obtaining US citizenship. And most research suggests that increased religiosity encourages transnational social and political involvement, including in the form of Catholic-based political protest against the militarized character of the US/Mexico border (Cadge and Ecklund 2007). At the same time, the possibility to practise one's religion can also lead to a form of 'integration'. Fridolfsson and Elander (2013) find this to be the case among Muslims in Sweden, in which the presence (rather than the absence) of Mosques in certain neighbourhoods in Swedish cities and towns, serve to build both 'bonding' and 'bridging capital' in local communities (see Chapter 5).

To complete our discussion of transnational identities, we would underline that the expression of these identities (for example, in pursuing naturalization, as one among many) are often seen as a matter of individual preferences, but they are constrained or enabled by states, organizations, and group practices. We might say then that transnational or 'diasporic belonging' (see Box 6.6), and religious identities and practices *are produced* by states in the both the countries of origin and destination. This is clearly evident in the relationship between France's immigration and foreign policies, and the discourses of the Algerian and Moroccan embassies and consulates which figure centrally in the production of a 'French Islam'. In this complex governmental relationship, the French, Algerian, and Moroccan governments shape the actual practice of Islam in France, and the ethnic and religious identities among Algerian and Moroccan immigrants (e.g. Maussen 2009; Samers 2003c).

Box 6.6 DIASPORIC IDENTITIES?

The term diaspora and the study of diasporas have a long history. It has become common to speak of certain 'classic' diasporas, such as those of Africans, Chinese, Jews, and Palestinians but this term could be extended to a range of 'groups' from Albanians to Tamils (e.g. Brubaker 2005; Cohen 1997; Safran 1991; Sheffer 1986, 2003). Although the meaning of the term 'diaspora' is contested, Brubaker (2005) argues that a diaspora has the following elements: dispersion, homeland orientation, and boundary maintenance. *Dispersion* refers to the spreading out of a group of people (defined either in ethno-cultural or national terms) and usually forced from an initial homeland to another part of the world. This 'another part of the world' does not have to be far; it may simply be over an international boundary, and may occur even within the same country. Some scholars of diaspora see division rather than dispersion, arguing that state borders that divide a group of people can create diasporas. *Homeland orientation* involves a longing for a real or imagined homeland. This implies a certain memory, usually a 'myth' about the homeland, and a 'myth of return'. But as Brubaker (2005) points out, not everyone agrees that diaspora involves a longing for return to the homeland. For example, the work of the anthropologist James Clifford on South Asians suggests that their desire to re-create a sense of culture and community in a new home seems to be considerable. *Boundary maintenance* refers to diasporas maintaining strong identities by distinguishing themselves from other groups and especially the dominant or dominant groups.

There may however be a tension between boundary maintenance and boundary erosion. Boundaries must be maintained over an 'extended time' (p. 7), perhaps over generations. For Brubaker, the "interesting question, and the question relevant to the existence of diaspora, is to what extent and in what forms boundaries are maintained by second, third and subsequent generations" (2005: 7). Is this "the 'dawning of an age of diaspora' or are we seeing simply a proliferation of diaspora talk, a change in idiom rather than in the world"? (Ibid.).

This question is not easily answered, since for one thing, the concept of diaspora itself has been the subject of critical scrutiny. Let me mention briefly four criticisms. First, it may not be an appropriate term to describe labour migrations, which are often temporary and voluntaristic (Faist 2000). Second, the idea of 'diaspora' may in fact be an undesirable Weberian 'ideal type', that is, an abstract concept that cannot capture the contradictory nature of social change among certain groups of people (Wahlbeck 2002). Third, the idea of diaspora hinges upon the importance of 'ethnicity' in the country of origin as central to the construction of identity and solidarity, rather than other axes of differentiation. Fourth and similarly, diaspora has been 'fetishized' (turned into an object over which scholars obsess) which then obscures the importance of other social dimensions or processes such as class, capital accumulation, and intra-national trans-ethnic alliances. Fifth, diasporas are not truly diasporas because they are always mediated by states that follow them (Basch et al. 1994). Sixth, diasporas are not only shaped by national economies, cultures, and politics; there are localized attachments that shape diasporic belonging (for a review, see Blunt 2007). The point to be made here is that it is hard to distinguish between transnational and diasporic identities. It depends on whose definition of diaspora and transnational is used by a researcher, and the actual subjects of the research.

To summarize our discussion in this section, it is difficult to imagine a theory of transnational belonging. Rather, we should expect migrants to have diverse transnational and other attachments (trans-local, trans-regional, etc.)

within and across multiple territories. These attachments are likely to produce different forms of political participation, and we will devote some time to this very issue in the next section.

CITIZENSHIP AS CIVIC AND POLITICAL PARTICIPATION

The final dimension of citizenship to be discussed in this chapter is civic and political participation, including resistance to state practices, what some now refer to as 'insurgent citizenship' (Holston and Appadurai 1999; Leitner and Strunk 2014). In this section then, we consider citizenship to be 'active', rather than 'passively' provided by states to individuals or groups. In this sense of citizenship, political participation may entail either more formal political participation (such as involvement in political parties and electoral politics, voting and formal lobbying, including in the country/ies of origin) or more informal political participation (boycotts, hunger strikes, occupations and sit-ins, parades, street protests, work stoppages, and other forms of public mobilization). The lines between formal and informal participation are often ambiguous and blurred, and involve mobilizing or 'claims-making' for access to services or the quality of those services, electing officials in the country of origin or destination, fighting discrimination, demanding the right to construct places of worship or the right to worship in public spaces, regularization, residence and work permits, supporting hometown/homeland associations, or simply to increase a group's visibility We can also think of immigrant political participation in terms of an interconnected 'host-country politics' and a 'homeland politics' (Levitt and Jaworsky 2007) which also range from the more formal to the informal.

We will begin with a discussion of the significance of 'homeland politics' (or what might be considered a 'transnationalization of politics'). Our interest here is in the relationship between homeland politics and host-country politics, or their absence, and whether they are more informal or formal. As one illustration of this relationship, Koopmans (2004) finds that immigrants are more oriented towards a homeland politics in those towns in Germany where migrants have the slimmest political channels for decision-making and are de-legitimized from the public sphere. Yet, as Portes and Rumbaut (2014) point out, participation in transnational politics may increase, rather than decrease participation in host-country politics. Likewise, if migrants do choose to engage in host-country politics, whatever form they take, they might also rely on *homeland* resources. For instance, the Moroccan, Turkish, and Mexican governments have routinely intervened in the affairs of the French, German, and US governments respectively to ensure particular outcomes for its emigrants. Likewise, politicians in the home country visit the countries of immigration for political support. After all, the President of the Dominican Republic in 2004 was actually raised in New York City (Portes and Rumbaut 2006).

Yet, this homeland politics may work in the opposite direction so to speak, and have a number of different dimensions. We will mention only two. First, immigrants participate in expatriate campaigns and elections abroad, and even run for an elected office, what Collyer (2014c) calls 'extra-territorial citizenship'. This may have a number of different aims, but influencing foreign policies in the country of origin is especially once such aim. Second, it may take the form of an engagement with hometown/homeland associations in order to support economic projects in the village or town of origin, especially through the use of remittances. Thus, homeland politics may be as 'trans-local' as they are 'transnational'. Nevertheless, if such projects seem initially localized and economic in character, they quickly become overtly politicized as the government in the country of origin attempts to regulate or support them (e.g. Levitt and Jaworsky 2007; Mohan 2008; Samers 2003c).

In terms of host-country politics and formal political participation, for more than the first half of the twentieth century in European countries for example, immigrants lacked official political rights, from the ability to vote in local elections, to engaging in trade unions. In fact, European governments and publics largely expected immigrants to be politically silent, ironically labelling them as politically docile when it would be difficult to be otherwise. They were considered 'here to work' – guest-workers to simply serve the economic needs of European countries. We should note at this point that the voting rights of immigrants in national elections are only allowed in four countries in the world (Chile, Malawi, New Zealand, and Uruguay) (Bauböck 2006), which may explain why political mobilization is often highly localized (Hampshire 2013). Since the 1970s in Europe however, the lack of political enfranchisement has changed dramatically in terms of the acquisition of political rights (e.g. local voting rights, the liberalization of nationality laws, and so on).[16] To begin with, a whole range of migrant NGOs and immigrant associations have emerged in Europe (as well as in Asia and North America) to represent migrant interests. Pro-migrant NGOs in the EU have concentrated in Brussels to take advantage of lobbying opportunities with European institutions that are located there, but they exist in all major cities and towns with comparatively large immigrant populations. 'Consultative committees' were also established in the late 1960s in Belgium in order to more thoroughly engage migrants in the political process and spread throughout Europe. Mayors, city and town councils regularly consult these committees as alternatives to direct representation, especially for the purposes of delivering public services. Often so-called immigrant or ethnic organizations, such as those centred on the Moluccan,[17] Moroccan, Surinamese[18] and Turkish 'communities', are largely involved in clamouring for specifically religious rights. Some claim that this has been facilitated by the political structure of the Netherlands (so-called 'pillarization')[19] which provides a certain room for religious diversity. This, along with a long-time commitment to multiculturalism in Dutch policy, explains why a range of Islamic and Hindu schools have been established in the Netherlands. This is not an unusual process by any means, and similar outcomes, from the establishment of religious schools to the construction of churches and mosques, are replicated throughout Belgium, France, Germany and the UK for example (e.g. Dumont 2008; Martiniello 2006; Maussen 2009).

In terms of the consultative committees discussed above however, they seem to exercise very little power. Instead, they largely appear to act as 'lip service' to democracy, a means of legitimating the participatory process for local governments, or they serve as a means of contracting out public service delivery through local immigrant elites, without the effective political inclusion of the bulk of migrants themselves (Dumont 2008; Martiniello 2006). For example, Martiniello discusses some of the reluctance of local officials in southern European countries such as Italy and Spain, where the local political participation of migrants seems more difficult than in northern and western European countries. One representative of the National Government of Catalan (the Catalan-speaking region that includes Barcelona) claimed that "they [immigrants] cannot really understand the history of oppression of the Catalan people". The above suggests that there is still a rather pronounced national geography to these formal political rights, with northern and western European countries (such as the Netherlands and the UK) seemingly providing more opportunities for formal political participation than in most eastern and southern European countries. Table 6.1 provides an indication of formal political rights as indicated by the right to vote in local elections in various EU countries (see Table 6.1).

Table 6.1 The right to vote in local elections in 25 EU countries for 'third country nationals'

Country	Right to vote	Eligibility
Austria	No	No
Belgium	After 5 years	No
Cyprus	No (under discussion in 2006)	
Czech Republic	For those migrants for which the Czech Republic has signed an international agreement (in 2006, no non-EU country)	No
Denmark	After 3 years, but no residence for Nordic citizens	
Estonia	Permanent residents only (minimum residence for PR permit: 3 years) with 5 years' residence in municipality	No
Finland	After two years, minimum residence for Nordic citizens	-
France	No	No
Germany	No	No
Greece	No	-
Hungary	Yes, no minimum residence	No
Ireland	Yes, no minimum residence	-
Italy	No	No
Latvia	No	-
Lithuania	Permanent residents (minimum residence for PR permit: 5 years)	-
Luxembourg	After 5 years	-
Malta	After 6 months of residence, but subject to Council of Europe reciprocity agreement which in 2006 did not include any non-EU state	-
Netherlands	After 5 years	-

Continued

Country	Right to vote	Eligibility
Norway	After 3 years	
Poland	No (under discussion in 2006)	No
Portugal	Nationals of reciprocity agreements after 2–3 years	Nationals of reciprocity agreements after 4–5 years
Slovakia	Permanent residence (minimum residence for PR permit: 3 years)	
Slovenia	Permanent residents (minimum residence for PR permit: 8 years) but cannot be elected mayor	-
Spain	Norwegian nationals after 3 years	No
Sweden	After 3 years; no minimum residence for Nordic citizens	
Switzerland	After 5–10 years' residence in 4 cantons; right can be granted in two more cantons	After 5–10 years' residence in 3 cantons; right can be granted in one more canton
UK	Commonwealth and Irish citizens only; no minimum residency requirements	

Source: Based on the research by Harald Waldrauch (2004), adapted from Martiniello (2006)

As ever, there are also local and regional differences (not least in terms of voting rights) whereby an apparently 'liberal city' such as Berlin is alleged to have more in common with Amsterdam, Paris, and London and other 'global cities' in terms of providing opportunities for migrant political participation, than it does with the conservative cities of Munich or Stuttgart in southern Germany. Counter-intuitively however, such 'global cities' may not offer more enabling environments than smaller 'more provincial' cities, and national states still provide crucial sites for fostering political participation. A paramount question then is to what degree local variation "override[s] the impact of national differences in [political] incorporation regimes" (Koopmans 2004: 451). Koopmans argues that political participation takes place at the 'local level' only insofar as the 'local level' is part of "national repertoires of citizenship and integration policies" (p. 467). Indeed, if the political participation of migrants appears to have increased over the last 40 years, Koopmans (2004) and Martiniello (2006) believe that the chief explanation for this growing political mobilization has been the 'political opportunity structure' (see also Cinalli and El Hariri 2012). For Martiniello (2006: 88), a political opportunity structure refers to the means by which states provide opportunities for political participation "in the management of collective affairs". This is achieved "By granting or denying voting rights to foreigners, by facilitating or impeding access to citizenship and nationality, by granting or constraining freedom of association, by ensuring or blocking the representation of migrants' interests, by establishing or not establishing arenas and institutions for consultative politics".

Yet the political opportunity structure (or 'context of reception' as it is often called in the North American research) is only one among many reasons, and certainly cannot explain completely the varied level and types of civic and political participation – whether formal or informal – among different individuals or groups of people in 'western' societies', if not elsewhere. For Portes and Rumbaut (2006), such 'political integration' (as they call it) by immigrants in the US, is achieved because first, experience with transnational politics (or 'homeland politics') provides expertise for engaging in political activity in the US; second, because dual nationality allows migrants to feel they can become involved in American politics without losing their involvement in home country politics; and third, because immigrants can use American institutions and political culture for export back to their home countries. Yet, for Portes and Rumbaut (2006, 2014), the form of political participation oriented towards both the countries of origin and destination is defined at least in part by one's generation (first, second, and so forth).

Equally pertinent for Portes and Rumbaut are the characteristics of the countries of origin themselves. They distinguish between stateless countries, hostile states, consolidated but indifferent states, or states that promoted emigration, and how they shape the qualities and degrees of political participation among immigrant in the countries of origin and destination. This is illustrated for instance by returning to the 'consultative committees' discussed above. In Bologna (a city in northern central Italy), a militant left democrat pronounced that "they are not accustomed to democracy" (p. 95). Although clearly a stereotype-informed generalization, this claim may not be totally unfounded. Echoing Portes and Rumbaut, Bilodeau

(2008) notes that migrants who emigrate from the most repressive regimes (their 'pre-migration experience') are less likely to participate in protest politics. Similarly, Preston *et al.* (2006) report that migrants from Hong Kong in Canada have low levels of formal political participation, which owes to the only very recent history of representative democracy in Hong Kong. However, they also noted that men were slightly more likely to participate in formal Canadian politics, but women are also involved especially when it concerns local education policy and their children. Most Hong Kong migrants in their study, regardless of gender, mentioned political rights (and especially voting) as the most important aspect of Canadian citizenship. In any case, these findings do however neglect the significance of political opportunity structures in the state and localities of immigration, not to mention countless other possible explanations.

Given Portes and Rumbaut's claim that the characteristics of political regimes in the country or countries of origin shape political participation, perhaps this should encourage us to scrutinize the spatial or territorial assumptions of Koopmans and Martiniello (above). That is, they suggest that local politics are predicated on the political opportunity structure which is in turn dependent on national forms of citizenship, but Leitner and Strunk (2014) emphasize how local territorial politics may be nodes within other networks of local territorialized politics, as they show in their study of immigrant political mobilization in various counties around Washington DC and in the city itself. In this case, immigrant NGOs across suburban counties share political tactics, but also mobilize in prominent national sites in the centre of the city (the park-like 'Mall' that surrounds the Washington Monument, in front of the White House, and so on).

We noted above that immigrants' political participation does not necessarily have to involve formalized politics expressed in electoral or party venues. Leitner and Strunk (2014) and Ehrkamp and Jacobsen (2015) impel us to extend the concept of citizenship beyond 'formal politics' and citizenship associated with national states in order to emphasize more localized and informal forms of political mobilization. Papadopoulous and Tsianos (2013) see this mobilization as part of a 'mobile commons' (a mass of mobile but collectively oriented citizens and denizens) which involves an always changing, fluid, and emerging coalescence of caring groups 'below the radar' that are fighting against immigration control by national states. This is manifested in the widespread political participation of undocumented immigrants ("unauthorized yet recognized", as Sassen

(1999) calls them) who have participated across the richer countries through a host of informal acts. Recall from Chapter 4, the protest for regularization among Malian *sans papiers* (undocumented migrants) in St. Bernard's Church in Paris during the mid-1990s. This gave birth to the *sans papiers* movement in France and to a trans-European movement of undocumented immigrants supported by such Brussels-based NGOs as PICUM, and which either employ the *sans papiers* moniker or have been associated with the 'No borders' movement (e.g. Barron *et al*. 2011; McNevin 2011; Rigby and Schlembach 2013). Another prominent example – also discussed in Chapter 4 – is the hundreds of thousands of undocumented immigrants who marched and protested throughout the United States in 2005 and 2006 against the restrictive immigration policies of the US government. In contrast to these grandiose and heroic acts however, simply participating in neighbourhood associations and similar activities constitutes a more mundane, yet equally significant form of political participation for migrants (e.g. Staeheli *et al*. 2012).

CONCLUSIONS

This chapter has explored the various dimensions of citizenship, including legal status, social rights, belonging, and political participation. We began the chapter by arguing that the migration literature, including by geographers, has sometimes hastily jettisoned the importance of local and national territories in favour of transnational linkages and identities. But the reverse may also be a problem, over-emphasizing local forms of citizenship at the expense of national, transnational, trans-local or other cross-border relationships. Our argument in this chapter is that all these territories matter and they matter in interrelated ways. In short, we are not yet ready to pronounce the death of any of them, although some may be more significant than others for particular issues, in certain places, or for different groups of people. We then examined so-called 'citizenship models'. It became evident that these models are increasingly irrelevant (especially in the EU) as jus domicili becomes ever more significant to both naturalization and to social rights. Two caveats are necessary though. This does not mean the end of national states, since they mediate the provision of jus domicili. Second, those who are undocumented may find themselves indefinitely marginalized, no matter how long they have resided in the country of destination, and no matter how vociferous their protests. States are increasingly allowing dual nationality for political and economic reasons, and migrants are availing themselves of these new possibilities, in part out of fear, in part out of practical utility.

In the second part of the chapter, we argued that despite 'post-national' membership and citizenship (Soysal 1994), or 'rights across borders' (Jacobson 1996), nation-states as actors remains significant. Post-nationalization *may* be occurring more prominently in the EU than anywhere else in the world in terms of formal citizenship, but different categories of migrants access different levels of rights ('denizenship' or 'civic stratification'). In any case, the content of social welfare entitlements in the EU, as elsewhere in the world seems to have been eviscerated in the context of what can arguably be called neoliberalism, although social protection certainly varies across countries and sub-national territories, as we saw between the US and Sweden, as well as within Japan and South Korea.

The third part of this chapter addressed the complex issues of social exclusion and (differential) inclusion, assimilation, multiculturalism, integration, 'diversity', and transnational belonging. We argued that place, territory, and length of residence are vital if we want to consider the strength of these ideas. In any case, all of these represent powerful discourses, policies and practices, and many 'western' countries seemed to have moved from operating with *official* notions of multiculturalism to what is now called 'civic integration' and 'neo-assimilation' after the events of 9/11. The opposite seems to be the case with Korea. Civic integration and neo-assimilation, along with other processes of exclusion (in employment, housing, and so forth) has had profound and often deleterious effects on particular immigrant communities. Diversity has also become a 'watchword' of private and public organizations worldwide, but particularly in the US. Migrants engage these multi-territorial discourses, policies, and practices, sometimes by rejecting the norms and practices of the country of immigration, sometimes by adopting them, sometimes by modifying them, and sometimes by constructing, navigating or negotiating complex transnational (or diasporic) identities, which involve attachments to real or imagined 'homelands'. These attachments may be as much urban, trans-local, trans-regional, gendered, sexualized, classed, kin-based, pan-ethnic, or pan-religious, as they are decidedly transnational in character. Religion is a significant feature of transnational migrant belonging, and what we might take as an expression of transnationalism may instead be co-religious attachments across borders.

Dual nationality, jus domicili, and multiple belongings do not overall seem to dissuade political participation whether formal or informal in the countries of immigration. Quite the contrary, it may stimulate political involvement as countries of origin and destination become tied by large overseas communities and intertwined cultural, political, social and economic networks. It could be tempting to explain political participation simply by national citizenship policies and national, regional, and local 'political opportunity structures' which provide for both formal and less formal opportunities for migrants to become politically enfranchised in the process of citizenship. Yet, political participation may also be explained by a nascent 'mobile commons' out there, involving a collaboration of migrants themselves, empathetic and sympathetic citizens, and supportive NGOs.

FURTHER READING

The literature on citizenship and migration is enormous. Useful overviews include Bauböck 's *Migration and Citizenship* (focused on the EU) and a more recent survey by Ehrkamp and Jacobsen (2015) (emphasizing US literature). Alba and Foner (2014) provide a wide-ranging assessment of so-called 'grand narratives' of immigrant integration. Sigona (2016) provides an unusual discussion of 'stateless peoples' who are largely left out of the citizenship literature. The special issue of the *Journal of Ethnic and Migration Studies* (2001) edited by Steven Vertovec, as well as Levitt and Jaworsky (2007) provide wide-ranging summaries of different aspects of transnationalism, and two special issues offer reviews and case studies of the relationship between diaspora and transnationalism: *Journal of Ethnic and Migration Studies* (2008, Volume 34, 7), and the special issue on 'Africa–Europe: A double engagement' in the *Journal of Ethnic and Migration Studies* (Grillo and Mazzucato 2008). Waldinger (2015) gives us an impressive and comprehensive critique of transnationalism. Cadge and Ecklund (2007) offer a review of literature published since 1990 on migration and religion in the US, and this can be supplemented by Portes and Rumbaut (2014), and the more international and geographically focused reviews in Kong (2010) and Ehrkamp and Nagel (2014). Without a discussion of transnationalism, Brubaker (2005) reviews some of the extensive literature on diaspora, as does Blunt (2007), but with more focus on the work of geographers. The special issue on *Ethnic and Racial Studies* (Volume 39, 2, 2016) updates some of the conceptual questions with innovative case studies. Koopmans (2004) offers a rigorous assessment of the importance of local, regional and national contexts for political claims-making, but Ehrkamp and Jacobsen (2015), Isin (2012) and Leitner and Strunk (2014) assemble a more innovative literature beyond national citizenship, as do McNevin (2011) and Nicholls (2013) for undocumented migrants. Finally, the journals *Asia and Pacific Migration Review*, *Citizenship Studies*, *Ethnic and Racial Studies*, *Gender, Place, and Culture*, *International Migration Review*, *Journal of Ethnic and Migration Studies*, and *Social and Cultural Geography* also offer innumerable case studies and discussions of different conceptual approaches to citizenship, migration and belonging.

SUMMARY QUESTIONS

1. Why are national models of citizenship problematic?
2. What are some of the limitations of Soysal's 'post-nationalization' argument?
3. In what sense has there been a 'neoliberalization' of rights?
4. How does geography matter to the processes of assimilation or integration?
5. Why is transnational belonging not simply transnational?
6. What is a political opportunity structure, and what are its limits for explaining political participation?

NOTES

1 See "A family divided by 2 words, legal and illegal", *New York Times*, 26 April 2009.
2 It is legal for undocumented migrants to graduate from New York City universities, and it is estimated that some 65,000 undocumented immigrants in the US graduate from American high schools (Data from the Urban Institute, in the *New York Times* article cited above).
3 The original Latin spelling appears to be *ius* (with an 'i'), but it seems more common now to be spelled with a 'j'.
4 Though, as Silverman (1992) points out, it is a mistake to believe that the French model of *jus soli* does not have ethno-cultural foundations.
5 In what they call 'citizenship configurations', Vink and Baubock (2013) seek to re-think these national categories. We do not address their paper here however.
6 Elsewhere in Europe, the issue of dual nationality seems to be driven by political and economic elites, rather than by migrants themselves (Kraler 2006). This is not the case in the US however, where migrants from a number of Latin American countries in the United States began a campaign for dual citizenship in the 1990s (Escobar 2007).

7 The ECHR is not associated with EU law per se, but nonetheless has power over a national court if a given national government is signatory to the European Convention on Human Rights.

8 Maas (2016) provides a discussion of some of the emergent ambiguities around this relationship in terms of legal status.

9 For reviews of this literature, one might look at Alba and Nee (2003); Hiebert and Ley (2003); Kivisto (2005); Levitt and Jaworsky (2007); and Waters and Jiminez (2005).

10 Council of the European Union (2004), in Joppke (2007a: 3).

11 This organization was dissolved and has now become the 'Equality and Human Rights Commission'.

12 For Glick-Schiller et al. (2006) "Social fields are networks of networks that may be locally situated or extend nationally or transnationally" (p. 614).

13 For some important overviews and studies of migrants, transnationalism and religion, one might refer to Cadge and Ecklund (2007); Ehrkamp and Nagel (2014); Foner and Alba (2008); Glick-Schiller et al. (2006); Kaag (2008); Hirschman (2004); Hondagneu-Sotelo (2007); Kong (2010); Levitt (2008); Menjívar (1999); Portes and Rumbaut (2014); Saraiva (2008); Volume 15, Issue 6 of Social and Cultural Geography (2014); and Van Tubergen (2006).

14 Some refer to this as the debate around 'post-secularism' in western societies, particularly in the urban public sphere (e.g. Beaumont and Baker 2011), but we do not engage this debate.

15 These are just a few examples of how 'piousness' might increase after immigration, see again Ehrkamp and Nagel (2014) for a more recent and wide-ranging review of such studies

16 However, in the 1960s, 1970s and early 1980s, trade unions became avenues for quite substantial industrial mobilization in manufacturing sites (e.g. Castells 1975), a form of political mobilization which has largely disappeared now in the EU, but not necessarily in the US, for example.

17 Molucca is an island in Indonesia (Indonesia was a Dutch colony) and Moluccans constitute one of the largest 'ethnic groups' in the Netherlands.

18 Surinam is in north-east South America and was also a Dutch colony, and the Surinamese also constitute one of the largest ethnic groups in the Netherlands

19 This idea of 'pillarization' is considered to be a stereotype rather than a real reflection of integration politics in the Netherlands (Vink 2007). Furthermore, Vink questions whether integration politics in the Netherlands were as 'multicultural' as is depicted in the literature.

7

CONCLUSIONS

International migration has been a fundamental feature of nation-states since their emergence in the sixteenth century, and it is likely to continue to shape the economic, political, and social life of societies across the world in the twenty-first, regardless or because of the gyrations of world economic activity, the restrictionist stance of countless national, *as well as regional and local* governments, the hospitality of citizens, or the energy, determination, and wishes of migrants themselves. As we pointed out in the Introduction, there were an unprecedented 244 million migrants in the world in 2015, including 19.5 million refugees, which amounts to approximately 50 million more than in 2005. Yet, even if this is a gross under-estimation of migration and immigration, the percentage of the world's population that lives outside their country of origin probably does not exceed 3.5 per cent. In other words, the majority of the world is not involved in international migration. They suffer from what Carling (2002) calls an 'involuntary immobility'. This seems to contradict the value of the so-called 'new mobilities' paradigm of the mid-2000s that has evolved into an extremely wide-ranging discussion of mobilities across the social sciences and perhaps a preference for the term 'mobility' over 'migration'. The language of mobility is hardly without merit. For one, it seems to capture much of the relationship between spatially and temporally dynamic, non-linear, and sometimes circular patterns of international migration, internal migration, and the ambivalences of settlement (e.g. Skeldon 2015). Second, it is pregnant with a fresh social imagination and a rich corpus of ethnographic studies that reflect the voices of migrants as 'actors' rather than simply as victims of environmental stress or disasters, poverty or war, but international migration should not be conflated uncritically with all forms of mobility. The bulk of governments and publics continue to create obstacles for asylum-seekers, refugees, and low-skilled/low income migrants especially, and international migration is hardly a whimsical project taken lightly by most individuals. Indeed, mobility runs into the problem of territory. International migration is therefore segmented or stratified, with some having the capacity to be more *legally* mobile than others.

So despite the obsession in the social sciences in the twenty-first century with dissolving social categories in academic or policy discourse or at least creating 'dynamic categories' (Collyer and de Haas 2012), we should not be reluctant to draw certain distinctions between let us say 'gold collar' (Casey 2010), high-income migrants moving fairly effortlessly between highly paid jobs in the affluent business districts of so-called 'global cities', and the thousands of asylum-seekers and refugees confined to detention centres, refugee camps, or the millions of undocumented/illegalized immigrants around the world, many of whom are unable or unwilling to return home for fear of never being allowed back into the country of immigration. However, it would be erroneous to draw an equation of the kind: higher income = higher mobility; lower income = lower mobility. Rather, low-income migrants – whether or not they may be considered 'highly skilled' – may suffer from *forced* mobility. Mobility is then a paradoxical strategy for coping in a world where mobility is impeded by the very governments responsible for the foreign policies that may have uprooted migrants in the first place. Impeded mobility in the context of what geographers call 'uneven development' (Smith 1984) is then a generator of further mobility as migrants seek out a 'better life'. This is the paradox of mobility/migration for asylum-seekers, refugees, and other low-income international migrants.

We argued that understanding this relationship between differential migrations and space is best understood through greater attention to spatial metaphors such as networks, place, and territory. In the first edition of *Migration*, Samers (2010a) lamented the 'spatial immaturity' of much of the migration studies literature. This entailed at least four prominent problems; first, even when a place (a city, town, neighbourhood, site, etc.) had been carefully documented, their precise effect was not always articulated clearly; second, migration studies suffered from an uncritical 'methodological nationalism'; third, use of terms such as transnationalism were not interrogated sufficiently for their 'spatiality', or more specifically how they related to other metaphors such as 'trans-local', and so on; lastly, loose references to 'scale' or 'multi-scalar' were omnipresent. In terms of the latter, we argued for the use of the term 'scale' only insofar as it is used either as a synonym for territory or carefully defined otherwise. Often, 'scale' is not explained at all, and vaguely employed as if we knew exactly what it meant and how it shapes migration. Some of these problems still persist, especially outside the discipline of geography, but in some ways, the critique of 'methodological nationalism' among geographers and other social scientists may have proceeded too far by over-emphasizing 'the local' or the 'urban' or the 'public square'. Addressing these difficulties, and precisely how 'geography matters' has been one of the aims and challenges of this book.

Nevertheless, over the last decade, the research on border externalization and internalization in its various forms, and the ways in which local states, cities and public spaces have been incorporated into debates about immigrant policing, repression, immigrant mobilization, and sanctuary cities have, we think, abated to a substantial degree, this spatial inchoateness.[1] In any case, we are inspired by Leitner et al.'s (2008) wariness to privilege a particular spatial metaphor over another, and so we chose to frame migration through a number of different metaphors in this book. We have however, simultaneously focused on the significance of interrelated territorialities for migration and immigration, while acknowledging the tension between defining spatial metaphors in the abstract, rather than in relation to how migrants, NGOs, and so forth, construct their own understandings of space. Yet, the choice of spatial metaphor and thus, how one frames an issue such as migration or immigration, is like the choice of all metaphors, political. We ought to think carefully about the spatial metaphors we choose, and how and why we use them (Dikeç 2012).

As we showed in Chapters 2 and 3, explaining such migrations is far from a simple task, given their enormously variegated character, from family migrants to refugees. We did however argue that using 'push-pull' theories or 'demographic disparity' arguments are not where we should begin to look for answers, or at least ways of framing such movements.

Similarly, facile accusations of how climate change or other environmental 'factors' 'cause' migration need to be subject to more sophisticated studies in a political ecology vein, as we tried to show in Chapter 3. Alternatively, explaining migration by reference to global capitalism and global inequality is a giant step in the right direction, but this can only amount to an incomplete and/or vague picture of the how, why, where, and when of migration. Indeed, explanation would have to extend to all the axes of differentiation or 'difference' discussed in this book, as well as the physical environment, migrant networks, employer networks, and state, non-governmental/voluntary, or private institutions across multiple territories (such as customs and border departments, hometown/homeland associations, labour recruitment and remittance offices, migration departments of every manner, pro-migrant and anti-migrant NGOs of all stripes, security organizations, and so on). In short, migration cannot simply be 'read off' from global capitalism; rather migration shapes the very contours of capitalism and the territories through which it is reproduced.

Chapter 4 outlined the diversity, but also the commonalities between entry policies for low-income migrants, asylum-seekers, refugees and highly skilled migrants in richer and poorer countries. Most countries' migration policies are driven either by an estimation or perception of fluctuating labour market demand for certain skills, security fears, cultural racism or ethnicism, xenophobia, a fear of the poor, foreign policy orientations and dictates, the implications of immigration for public budgets, but also nation-building (in Canada, for example), notions of cultural affinity and humanitarian concerns. However, in most theoretical understandings of migration policies, what is often missing are how complex sub-national territorialities shape national policies and in the same vein, how entry policies are shaped by migrants themselves. Entry and settlement policies are usually viewed as creating migration, but migration creates entry and settlement policies through different territories of regulation.

For richer countries, the mantra of 'migration management' is reflected in the implementation of Canadian or Australian-inspired points systems, or other tiered migration systems, in order to attract the 'right kind' of immigrants, that is those who are 'highly skilled' as determined by the needs of employers, or those who have substantial capital to invest. At the same time, the migration of asylum-seekers, low-income migrants and refugees are restricted to the bare minimum in the name of protecting jobs, welfare systems, 'security', 'cultural homogeneity', or 'social cohesion'.

Yet such policies are hardly new. Since at least the nineteenth century, states have sought highly skilled workers from overseas, while nativist anti-immigration panics, including in countries from East Asia to South Africa have managed to exclude the 'ethnically different', the 'poor', or the 'unhealthy'. At the same time, it would be incorrect to assume that the governments of wealthier countries have completely barred entry to low-skilled/low-income migrants in the second decade of the twenty-first century (the 'Fortress Europe' or 'Fortress USA' idea). While certainly regulations against low-skilled migration are tighter today in European countries than they were in the 1950s and 1960s during the heyday of manual labour demand, many governments including the American, British, Canadian, French, Malaysian, Saudi Arabian, Singaporean, Spanish, and UAE governments have encouraged certain forms of low-skilled labour migration through bi-lateral or multilateral agreements, particularly in agriculture (in the UK and the US) and construction and domestic work (in Malaysia, Singapore, and Saudi Arabia). Many countries in Asia, such as South Korea, have courted both 'marriage migration' and the migration of ethnic Koreans, while in Europe, Germany encouraged the migration of 'ethnic Germans' well into the 1990s, and the US in particular has also accepted certain forms of family reunification as a codified social right, or actively encouraged it since the 1960s. And for all the panic about asylum-seeking and refugees, especially in the EU, millions of asylum-seekers and refugees have found at least a modicum of protection in Asia, Europe and North America over the last 40 years, including North Koreans in China, though they have also faced exclusion, poverty and violence in the places they settle. In this process of 'differential inclusion' then, migration and immigration policies cannot be reduced to governmental preoccupations with labour demand. Even if governments succeed in creating completely class-based restrictive policies, hundreds of thousands of migrants have overstayed their visas or flouted militarized borders and security-obsessed governments to settle semi-permanently or eventually obtain regularization or naturalization over a long period of time. Furthermore, liberal courts will intervene, in some cases at least, to prevent the deportation of undocumented/illegalized immigrants or ensure the reunification of internationally divided families.

The reaction of governments in poorer countries to migration and especially emigration is ambivalent. On the one hand, losing highly skilled labour means losing valuable skills for both economic development and for social services such as nursing, the latter being of special concern to the South African government, for example. On the other hand, many governments see the opportunity for the export of unemployment, the potential for vast remittances, and the possibility for their citizens to learn new skills, with the hope that migrants will some day return. Sometimes they do; sometimes they do not. In some cases, computer engineers trained in Silicon Valley have returned to southern India, and particularly Bangalore, contributing to the development of India's burgeoning software industry. This is now referred to as 'brain circulation' rather than brain drain, but it also might be labelled as a clear illustration of 'brain gain' for poorer countries. But who precisely gains from this circulation – everyone in India, everyone in the urban region of Bangalore, or just those fortunate enough to circulate between India and Silicon Valley? That is a question that cannot easily be answered, although in light of uneven development, critical attention to matters of class and space seem vital here.

In Chapter 5, it became apparent that the fortunes of migrants and immigrants in different countries are as diverse as immigrants themselves. It is probably impossible to generalize about their socio-economic achievement across the world, since conditions in the countries of emigration, time of arrival in the countries of immigration, issues of immigration status, the context of cultural and social reception, the related social networks of migrants themselves, their 'social capital', gender, national and ethnic background, and other axes of differentiation combine with racism, sexism, xenophobia and the embodied expectations and practices of employers and migrants alike. All of these processes have an effect on their prospects for entrepreneurship and waged employment in particular places. Analysing these outcomes by starting with pre-defined ethnic or national groups as countless studies do is problematic however, since they seem to assume from the outset that individuals labelled under these categories are homogeneous or that it is something about their nationality or ethnicity which should make a difference to their socio-economic mobility. We are not arguing however that nationality does not matter for the working lives of migrants. It does, if only because of the way in which migration and immigration discourses, practices, and policies create the social importance of nationality. Nevertheless, the use of surveys and ethnographic work that focus on the processes of 'international labour market segmentation' (ILMS) and what others refer to as the embodied character of labour markets seem to provide an antidote to the limitations of more 'conventional' approaches (let us say, HCT and standard discussions in sociology that rely too heavily on 'social capital' without sufficient attention to the multiple territories involved in the governance of labour markets, and the nature of work).

Use of the concept of ILMS and related approaches suggest that none of these labour market outcomes can be divorced from the social rights, immigration policies, housing policies and the character of urban and rural economies (including employment prospects, wages, and so forth) that are associated with specific, but interrelated territories. The important story here, however, is far more than how different labour market outcomes relate to complex geographies. Rather, as we saw in Chapter 5, it is crucial to underline that for the least fortunate migrants, life is marked by monotonous, low paid and sometimes downright dangerous work, from cleaning homes in Italy and Malaysia, to cutting meat in Nebraskan slaughterhouses; from arduous street peddling in Brazil to mining precious metals in South Africa. For those subject to especially callous employers, bereft of the most meagre social rights and subject to exclusion and marginalization, migration can prove to be a terrible experience, in many ways not much better (and sometimes worse) than the conditions which they chose or were forced to leave. In the wake of low-wage employment or chronic unemployment, their health suffers and their children suffer in turn.

Chapter 6 outlined four different dimensions of citizenship, including legal status, social rights, issues of belonging or 'identity', and political participation. It is clear that dual and even multiple nationalities are proliferating in innumerable countries, although some countries of immigration and emigration continue to be reticent about accepting dual nationality, let alone multiple nationalities. At the same time, national models of citizenship seem to be eroding in favour of differential rights based on the length of residence (jus domicili). As national models fade (if they ever existed in some pure form in the first place?), some scholars have argued for the significance of a 'post-national' citizenship (in the European Union), or 'transnational citizenship', 'rights across borders' or 'global or international human rights' (as they are referred to elsewhere in the world). In the EU, where post-national citizenship has allegedly progressed the furthest, there is actually only limited evidence that this is the case for 'Third Country Nationals' (TCNs) and especially their dependants, never mind for asylum-seekers, or those who are undocumented. As Benhabib (2004 [2007]) laments, "Not having one's papers in order in our societies is a form of civil death" (p. 215). This may be an exaggeration, given important political mobilizations by undocumented/illegalized immigrants that we highlighted in Chapters 4 and 6. At any rate, there also seems to be a general movement towards the increasing incorporation of post-national elements in national citizenship policies, including for long-term resident TCNs or full citizens of another EU country. Needless to say, 'denizenship' and 'civic stratification' are complex in the EU, as they are in Asia and North America.

Despite the emergence of jus domicili, and all the rhetoric about 'post-national rights', 'rights across borders' or a 'global human rights regime', access to social entitlements for legal and undocumented migrants, as well as for asylum-seekers and refugees is generally now more difficult in the EU, and their actual welfare content reduced in most of the richer countries. This is consistent with arguments concerning 'neoliberalization', although Canada, the US and Sweden stand out as exceptions here in terms of refugee protection. At the same time, migrants over the last decade have suffered from deportations based on groundless accusations or the most minimal crimes. But this has impacted on some groups more than others: Muslims (in Europe and North America), those of Latino origin (in the US in particular), Afghanis in Iran, Haitians in the Dominican Republic, Indonesians in Malaysia or Singapore, Zimbabweans in South Africa, Rohingya in Burma, or Bangladeshis, Indians, and Pakistanis in Saudi Arabia, to name just some of the nationalities and religious groups affected by state, public, or private repression. In most of the poorer countries, years of structural adjustment have left citizens with bare-bones social entitlements and, not surprisingly, state-based entitlements for asylum-seekers, refugees, and migrants are generally non-existent. Low-income migrants are by and large reduced to settling in favelas, shantytowns, and other informal settlements devoid of even the skeletal social support that some migrants receive in richer countries. Asylum-seekers, refugees, and millions of IDPs may be supported by international agencies and NGOs, but they may languish for years in refugee camps, suffering from minimal resources, periodic violence, and blocked lives. The acquisition of rights is not simply a top-down process however, in which migrants are passive receptors of citizenship. Rather, through a form of active citizenship, migrants mobilize to acquire rights or to achieve certain ends, from constructing a mosque or a temple, or staving off police repression. This might be accomplished in banal ways by talking to one's neighbuor, but it also may occur through quite vociferous protest for rights and recognition. This coming together of all manner of migrants, immigrants, and NGOs, and their collective achievements is described innovatively by Papadopoulous and Tsianos (2013) as 'the mobile commons'. This commons however, is enabled or constrained by particular networks, nodes, places, and territories. Once again then, 'space' should figure centrally in debates around citizenship and belonging.

A sense of belonging among migrants in the wealthier countries of the twenty-first century is partly tied to their multiple legal statuses and their access to social rights, and partly dependent upon the discourses, practices, and policies of assimilation, integration, multiculturalism, including the contradictory post-multicultural re-emergence of civic integration policies and the emphasis on 'diversity' in North America and many European countries. Yet, Germany, the Netherlands, and the UK provide examples of how the *practice* of multiculturalism is persistently enmeshed with 'neo-assimilation' or 'civic integration' discourses and practices. In that sense, we should be wary however, of the idea that migrants follow a linear path from 'foreign' to 'assimilated'. But there are limits to even the convincing concepts of 'segmented assimilation' or 'selective acculturation'. After all, since the societies of immigration are constantly changing, into what precisely do migrants assimilate or integrate? With the poly-ethnic, polyglot character of cities, assimilation or integration in certain neighbourhoods may mean as much pressure to maintain the practices of one's own co-nationals or co-ethnics, or indeed of another migrant group, as it does to adopt nationally dominant cultural or economic practices. In this respect, migrant identities involve complex localized, trans-nationalized, trans-localized, pan-religious, pan-ethnic, or even kinship practices that may equally be influenced by matters of skin colour and the experience of racism, gender, class, sexual practices, labour market skills, use of the internet, the strength of hometown/homeland associations, and the varied cultural practices of other migrants that they encounter in the country of immigration. In short, the discourses and practices of 'assimilation', 'integration' or 'multiculturalism' are complex and operate across different territories with sometimes unique outcomes. By the same token, through choice or necessity, citizens also learn from, and adapt to, the cultural practices of migrants in particular places (think of food and music, for example) that may be called 'social transformation'. Thus, national, regional, and local 'creolized' practices develop around particular constellations of migrants and citizens through 'bonding' and 'bridging capital'. As these interactions develop over time into permanent settlement, whether legal or illegalized', migrants come to shape the cultural, economic, social, and political character of places and their wider societies.

WHAT CAN BE UN-DONE?

There are innumerable pro-migrant organizations which have offered policy proposals for modifying entry and settlement policies across the world. Apart from 'bottom up' activism and advocacy work with migrants, which we discuss further below, we want to suggest three existing 'top-down' policy ideas that might literally create the space for new, more radical possibilities. The first revolves around the idea of national or global mobility regimes (see e.g. Koslowski 2008; Papademetriou 2007). For proponents of such regimes, governments should not fight, but harness mobility, including circular migration. International mobility is seen as an ineluctable reality, and those governments most capable of facilitating it are those that will benefit from it. Though we object to the neoliberalism or national economic utilitarianism that are often at the core of such arguments, facilitating international mobility and explaining the importance of mobility to reticent publics is likely to offer new opportunities for migrants. However, in order for international mobility to have beneficial outcomes for *migrants* rather than just for citizens, employers or states, these regimes must also address issues of xenophobia, exclusion, and marginalization. For this to happen, international mobility must take place in accordance with what Benhabib (2004 [2007] refers to as the decriminalization of global migration and the granting of the 'dignity of moral personhood' to all individuals (p. 179)).[2] Drawing on Hanna Arendt's idea of the 'right to have rights', Benhabib and other, more radical critics insist that mobility regimes should not consist of the usual 'guest-worker' or temporary migration programmes in which migrants do not share the same rights as citizens. On the contrary, international mobility must be accompanied by international human rights norms that involve universal political and social rights for temporary, semi-permanent, and permanent residents, including voting rights in national, regional, and local territories, and full access to health care and other social entitlements. Voting rights for even the most temporary 'guests' will ensure that international mobility increasingly involves the consent and political participation of migrants themselves. In this respect,

Benhabib calls for a 'cosmopolitan federalism' in which states operate within a cosmopolitanism defined by universal human rights norms. She is therefore calling for porous but not open borders (p. 220). This is so because Benhabib, like so many other political theorists, wonders how democracies can function without 'closure', that is without boundaries (but compare with Cole 2000: 180–8, and Carens 2013: 225–54). Nonetheless, Benhabib recognizes that new non-territorially based models of democratic representation are possible and might come to supplant, or at least complicate existing connections between territory and democracy, and she believes these new 'models' should be encouraged.

Given how migration appears to exceed the territorial limitations of national democracies, why not advocate for a global state? Many social thinkers remain wary of a world or global state. This cold response has its roots in the writings of the eighteenth-century philosopher Immanuel Kant, who believed that a world state would lead to a 'universal monarchy' and a 'soulless despotism' (cited in Benhabib 2004 [2007]: 220). The problem for many, including Benhabib herself, is that a global state would face problems of democratic representation and legitimacy, given the sheer size of the potential state involved. This issue cannot be adequately addressed here, but the real problem is that Benhabib's universal rights norms need more than just a sense of cosmopolitanism. Certainly, cosmopolitan norms do have an effect on the behaviour of states with respect to immigration (for example, Japan) but enforcement as well as norms are necessary. As we noted at the close of Chapter 4, there is a great deal of 'talk' about human rights, and a growing cadre of international conventions, but these have few legal teeth. Is it possible to enforce such conventions without a global state? Perhaps not, but other possibilities have been suggested. For instance, the noted economist Jagdash Bagwati has proposed the formation of a World Migration Organisation to manage migration as a complement to the World Trade Organisation. Some antecedents to this type of organization have already appeared in the form of multilateral commissions, including the creation in 2003 of the Global Commission on International Migration. Similarly, the United Nations High Commissioner for Refugees' 'dream for the future' hopes that rich countries will accept the equivalent of 1 per cent of their population for refugee settlement (see Taylor 2005). With the exception of this UN proposal, one cannot help thinking however that these various multilateral ideas and institutions for migration management have in mind the welfare of citizens and states, rather than of the migrants themselves.

The second policy idea previously mentioned in Chapter 4, and worth highlighting here, is the idea of 'sanctuary cities' or 'cities of refuge' in Canada, the UK and the United States. Sanctuary cities are actually far from entirely 'top down' and have been motivated by endless protests by migrants and sympathetic individuals, groups and NGOs. 'Sanctuary' and refuge are useful metaphors of care, generosity and hospitality, and sanctuary in the largest cities of Canada, the UK or the US should also be extended to small cities and towns and rural areas, as well as other countries. In some respects, this is already happening, especially in smaller university-dominated cities and towns in the US. For example, officials and activists in Eugene, Oregon are seeking to incorporate international human rights principles into the city's operations. Similarly, academics and students at the University of North Carolina at Chapel Hill have successfully pushed for the designation of Chapel Hill and neighbouring Carrboro as Human Rights Cities (Bauder 2014b; Darling 2010; Naples 2009; Nyers and Rygiel 2012).

A third policy vehicle might be the expansion of what Casey (2010) calls 'comfort zones', that is, the often regionally oriented bilateral or multilateral (labour) agreements or procedures which slowly encourage the liberalization of migration. After the so-called 'Syrian refugee crisis', the EU may no longer provide a promising example of this, but we might also look at CARICOM, or the Trans-Tasman Travel Arrangement between Australia and New Zealand, as openings in the closures against migration.

The related ideas of mobility regimes tied to a global international human rights regime (hopefully with some mechanism of enforcement), the expansion of sanctuary cities, and the establishment of 'comfort zones' borne of bilateral or multilateral agreements are but a few of the policies that have the potential to slowly dissolve the severity of international borders, further bolster social rights, foster international cooperation, or modify spaces in such a way that they harbour migrants from national, regional, and local state repression. At this point, rather than continue with a selection of policy ideas however, we will conclude with a more academic route by suggesting a few means of imagining a new world of migration.

Un-doing our imaginations

For those who yearn for a world in which people, and not just capital, have unimpeded mobility, there are intellectual resources from which we can borrow. Sure, any calls for the dissolution of national borders are up against staunch opposition and the still powerful belief that only nation-states and their apparent 'collective sense of identity' can deliver security, economic well-being, and social justice (e.g. Walzer 1983). Yet other liberal theorists remind us that liberal ideas do not consider an 'outside' and so-called liberal states and their policies and practices are 'illiberal' to 'non-members' of a society (e.g. Carens 1987, 2013; Cole 2000). Cole concludes that those who see themselves as 'liberal egalitarians' and who are against the international freedom of movement must ask themselves what is meant by liberal egalitarianism. Echoing a long-time demand for 'open borders', Hayter (2004) wishes to incorporate the free movement of people within any discussion or concrete manifestation of, international human rights. Indeed, Carens (1987) argues that "like feudal barriers to mobility, they [borders] protect unjust privilege" (p. 229), and Benhabib (2008) argues that state borders 'require moral justification' (p. 19).

Those critical of such militarized borders (or borders at all), restrictive immigration policies, and the absence of a truly enforceable international human rights regime also have intellectual spokespersons to which they can turn, including the myriad pro-migrant NGOs such as 'No one is illegal' and 'No borders', other immigrants themselves, and the writings of political theorists and philosophers such as Seyla Benhabib (as above), Joseph Carens, or Philip Cole, but also radical geographers and activists such as Harald Bauder, David Harvey, and Theresa Hayter. Such writings are infused with intellectual fodder for developing new geographical imaginations in light of migration. We do not have the space here to develop their arguments at any length,[3] but we will devote a little bit of attention to some of the arguments of the geographer Harald Bauder. To begin with, in a special issue of the critical journal ACME (2003), he questions the justice of borders on the grounds that they reproduce social injustice through what Van Parijs (1992) calls 'citizenship exploitation', the exploitation of human beings based on their lack of citizenship. Borders allow capital (employers) to exploit migrants, by dividing them and preventing the global political organization of migrants. As a consequence, Bauder calls for the end of borders for the sake of international socialism. Most of the papers in response to Bauder's proclamations considered his vision of a borderless world 'unrealistic', and many critical scholars concede that some borders – even national ones – might be desirable (e.g. Naples 2009). In contrast, Samers (2003c) defended and applauded his willingness to pry open our imaginations and to not let cultural arrogance or fear rule such a promising vision. As Carens (1987) writes, "Free migration may not be immediately achievable, but it is a goal towards which we should strive" (p. 270), and Casey (2010) believes we should actually set up a time-table for doing so. In a later paper, Bauder (2016) draws on the strategies of the 'No one is illegal' organization which brings together various protest groups, Ernst Bloch's understanding of the 'contingent possible' (what can be imagined from actually existing social relationships) and what the urban theorist Henri Lefebvre calls the 'possible impossible' (this is confusingly referred to by Bloch as the 'real possible'). The 'possible impossible' and the 'real possible' are synonymous, and refer to that which does not rely on existing relations, but which can nevertheless be imagined. In this way, Bauder's 'contingent possible' imagines an *urban* citizenship based on the already existing and widespread practices of *jus domicili*, which may defy national citizenship policies. Yet he also conceives of the 'possible impossible' by resurrecting his paper in 2003. That is, he calls for something more revolutionary; something that is unimaginable in 'concrete terms'; something that does not yet exist. That is, he asks us to imagine beyond global capitalism, a world, not just with open or porous borders, but no borders at all. In this world, cities would become battlegrounds for as yet new and unknown human 'subjectivities' (that is, 'identities'). In such a world, the term 'migrant' and 'illegalized' might ultimately disappear.

Another way of undoing our imagination is to question the way in which citizens view certain migrants. In the imagination of citizens, migrants and asylum-seekers and refugees are often viewed as culturally and racially 'out of place' (Cresswell 1996; Mitchell 1996; Sibley 1995), whereas citizens of the rich world might not see the same individuals as dangerous, strange or 'out of place' if they visited the migrants' countries of origin. Thus, by not seeing the cultural landscapes of the rich world as fixed, unchanging, and naturally 'Judeo-Christian', citizens of the global north open up the possibilities that newcomers can contribute to these ever changing landscapes. This vision of 'out-of-placeness' is not confined to the rich 'west', but to all countries at different times and with different registers. In the wealthy countries, the fear among many citizens of the 'the other' may stem from not just racism, culturalism, and xenophobia but all of these territorially mediated expressions together with what is sometimes called 'post-colonial guilt'. There may be ways, however, of transforming such post-colonial guilt into post-colonial engagement. As Doreen Massey has argued, "We are responsible to areas beyond the bounds of place not because of what we have done, but because of what we are" (Massey 2004: 16). In a sense, she is calling for a 'relational' way of thinking that sees the 'other' as part of 'the rich world'. By deploying this wisdom in the context of migration and immigration, we might say that the rich world not only borrows and learns so much from the poor world, but the rich world is created through the impoverishment of the poor world through policies that favour the former. Inequality begets wealth, and migration is in part a manifestation of this inequality. By un-doing or de-colonizing citizens' imaginations of migrants as (post-colonial) 'burdens' or 'out of place', then we may begin to imagine a more socially just world where migration is more of an opportunity than an unfortunate necessity.

FURTHER READING

The literature on the ethics of 'free movement' or 'open borders' is extensive. For this reason, we will simply point to some of the references mentioned above, including Benhabib's *The Rights of Others* (2004 [2007]), Cole's *Philosophies of Exclusion* (2000), Joseph Carens' (2013) award-winning book *The Ethics of Immigration*, the Special Issue entitled "Engagements: borders and immigration" in ACME (Bauder 2003), Theresa Hayter's eye-opening plea for open borders from the perspective of an activist in the UK in *Open Borders: The case against immigration controls* (2004), two essays by Harald Bauder (2015, 2016), and Massey (2004). Popke (2007) provides critical discussions of some of the problems of cosmopolitan ethics, justice, and responsibility.

NOTES

1   We do not mean that (critical) social scientists who engage with migration/immigration have not thought about space at all, nor about its effects, and ironically the literature in the 1980s and 1990s on ethnic enclaves for example, seemed to be more thoughtful about space than some subsequent work by many social scientists involved in 'transnationalism' debates.

2   But see Popke (2007) for one discussion of how universal or cosmopolitan norms and ethics can be deeply problematic.

3   See 'Further reading' at the end of the Conclusions.

GLOSSARY

Assimilation – This seems to have at least three meanings: immigrants adapt to or adopt the cultural ideas and practices of the dominant culture over time; immigrants achieve the same socio-economic status measured in terms of some 'average' for the 'native-born' and immigrants develop a spatial pattern in terms of residence and employment that is indistinguishable from the dominant or more dominant cultural groups.

Asylum-seeker – Refers to a migrant who enters a country clandestinely or by legal means and then requests asylum. An individual may also request asylum from outside the country, and thus enters a country *as* an 'asylum-seeker'. An asylum-seeker may or may not in turn be granted asylum or refugee status by a particular national government.

Border externalization – The process of extending 'border-like practices' (such as migration control and customs) beyond the cartographic or 'physical' border of (usually) a country.

Brain circulation – Used to describe the movement back and forth among 'highly skilled' migrants between the country (or countries) of emigration and immigration. Skills are acquired in both (or several) countries and transferred ('circulated') between these countries.

Brain drain – This refers to the loss of 'skilled' or 'highly skilled' migrants from a particular country of origin. It is generally used to describe the effects of emigration from poorer countries on the economies of these poorer countries.

Brain gain – The opposite of 'brain drain'. It generally refers to a country of immigration which benefits economically from an in-migration of 'skilled' or 'highly skilled' labour.

Capitalism – A now arguably global system of social relations combining at least the wide-spread use of wage labour, private property, and the extraction of surplus value (exploitation).

Circular migration – Refers to the process by which migrants move back and forth between a country of emigration and a country of immigration. This typically involves seasonal stays in either country, often related to temporary or seasonal patterns of work. The term is also commonly used for constant internal migration from rural areas to urban centres, and back again.

Denizenship – This describes the various shades of legality and access to cultural, economic, political and social rights among migrants.

Diaspora – A highly contested term in meaning and scope, it generally refers to migrants from some sort of 'homeland' who in turn spread out across the world and re-settle in various countries of immigration and re-establish inter-linked communities.

Diversity – Seems to refer to the condition of diverse languages, religions, and ethnic groups; a societal 'understanding' that organizations should not discriminate on the grounds of cultural characteristics, but should engage those characteristics; and that cultural skills should be seen as marketable 'competencies'.

Forced migrant – This is a general term to describe an individual who is forced from their country or countries of origin. 'Forced' is an imprecise term that may have economic, environmental, political, or social origins, or a combination of all or any of these.

Friction of distance – The time and cost of overcoming distance.

Geneva Convention – A convention signed in 1951 that offers protection to individuals who are persecuted because of their religion, ethnic background, political affiliation, colour of skin, tribal affiliation, and so forth. If a migrant fears that they risk harm or persecution upon returning to their country of origin, that individual is entitled to refugee status by those countries of immigration who are signatories to the Convention.

Governance – This refers to the process by which migration and other social processes are regulated, controlled, enforced, encouraged, or mitigated by various 'levels' of government. These 'levels' can have a geographic basis (i.e. international, national, regional, local, and so forth) or they can be functionally based. That is, a whole range of different organizations with various functions can be involved in regulating migration, for example.

Highly skilled migrants – There is no agreed definition of 'highly skilled'. However, it seems to consist of at least two groups of people. The first is those who are immediately recruited into what are considered highly skilled positions in the country of immigration because these individuals have the required educational backgrounds, qualifications, or skills. A second and less common usage of the term is to describe those who are considered highly skilled in their countries or origin but who may end up performing menial jobs in the country of immigration.

'High-income' migrants – Individuals who are granted admission to a country based on their net wealth; individuals who might be granted admission to a country based on both their net wealth and their willingness to invest or start a business in the country of immigration; or less commonly individuals who might literally buy their citizenship.

'Illegal' (or clandestine/irregular/illegalized) migrant/immigrant – See 'Undocumented migrant'.

Immigrants – (see also 'migrants') There is no precise definition of an 'immigrant'. In this book the term 'immigrant' is used interchangeably with 'migrant', although the term 'immigrant' implies more permanent residence. Some may also refer to those citizens who have recently naturalized as immigrants, given their origins.

Integration – A contested term that has at least three principal meanings. The first is closer to 'assimilation' and refers to the extent to which migrants fit into an imagined and idealized set of dominant practices and values of the citizen majority, or to their access to such material goods as housing, employment, education and health. The second meaning of integration is closer to that of multi-culturalism, whereby immigrants do not somehow 'lose their culture' but rather retain 'their culture' and join the liberal political culture of the western liberal democracy in question. The third less common definition is the 'coming together' of migrants and citizens, whereby each adopts the cultural practices (language, religion, food, music, and so on) from another.

International labour market segmentation (ILMS) – A term used to describe the 'segmenting' of labour in three ways; first, through supra-national and national immigration policies that sort migrant labour based on international grounds (their nationality); second, by 'segmenting' migrant labour within national economies through their confinement into specific sectors; and third, by 'segmenting' migrants within firms and organizations.

Jus domicili – The acquisition of citizenship or access to social rights based on the length of time a migrant has resided in the country of immigration. This is usually subject to other stipulations (such as not being convicted of crime, uninterrupted residence in the country of immigration, and so on).

Jus sanguinis (or 'law of blood') – The acquisition of citizenship based on descent, or the ethnic ties one has to a country of immigration. This is usually based on the origins of the parents, though it may be combined with elements of *jus soli*.

Jus soli (or 'law of soil') – The acquisition of citizenship based on the birthplace of a migrant, or of the migrant's parents. It may be combined with elements of *jus sanguinis*.

'Low-skilled' migrants – Like other migration categories, the definition of 'low-skilled' is imprecise. It is relative to the skill demands of the countries of immigration, but also may be associated with low wages.

'Low-income' migrants – Like other migration categories, the definition of a 'low-income' migrant is imprecise. Low-income migrants may be defined by what they are not, that is, not high-income migrants as perceived by the policies of the countries of immigration. Their low income may or may not be related to their skills.

'Low-wage' migrants – Like other migration categories, there is no precise definition of a 'low-wage' migrant. It may refer to a migrant who is either recruited into a 'low-wage' job in the country of immigration; someone who emigrates and expects or searches only for 'low-wage work', or someone who finds only low-wage work. 'Low-wage work' is itself a relative term, although it generally refers to work with low barriers to entry (requiring less capital or skills etc.).

Migrants – (see also 'immigrants') There is no precise definition of a 'migrant', although some international institutions refer to international migrants as individuals who reside in another country for more than three months. In this book, the term 'migrants' is used interchangeably with 'immigrant', although the term 'migrant' also implies a more temporary sense of residence.

Migration-development nexus – A term used to describe the relationship between migration and '(economic) development', usually in the context of poorer countries of emigration.

Migration management – A term that emerged in the 1990s and is used to describe the way in which countries regulate the migration of specific categories of migrants.

Multi-culturalism – Multi-culturalism is a set of discourses, ideologies, political philosophies, policies, the aims of political movements in the name of recognition and representation, and a context of pluralism defined in group terms through which people feel a sense of belonging.

Myth of return – A psychological condition in which a migrant constantly dreams of returning to the country of origin. Use of the term 'myth' refers to the real problems of settling more permanently in the country of origin, and the increasing unlikelihood of returning as residence in the country of immigration lengthens. Returning permanently is possible, however, and the 'myth of return' only refers to a common condition.

Naturalization – The process by which a non-citizen migrant becomes a legal citizen of a country (or even countries) of immigration.

Neoliberalism – This generally refers to a set of policies, programmes, and discourses (sometimes the term 'ideology' is also used) that favour markets over government intervention and social welfare programmes as a 'solution' to societal or economic development problems. It is normally associated with a reduction in social welfare and in its place the promotion of 'capital welfare' (favouring businesses).

Networks (social networks or migrant networks) – Social or migrant networks are webs of personal relations and interactions across space, that involve both individual migrants and institutions.

Node – This is a particular location in a network. In the context of migration, it is sometimes used to describe the 'cultural hearth' or 'cultural centre' of a transnational community.

Place – A geographic term typically used to describe a city, town, village, neighbourhood, and so forth, which has meaning for individual migrants.

Political opportunity structure – A term used to describe the structure of social relations in any given territory (country, city, and so forth) which allows migrants to participate politically.

Refugees – Often ethnically or nationally defined *groups*, granted refugee status by a state or international organization, and recognized and inscribed in international law, prior to their arrival in another country. Yet *individuals* may be also granted refugee status after a certain period of seeking asylum.

Remittances – A term used to describe money that is brought or sent home to country of emigration from a country of immigration.

Scale – This is a difficult term in the geographical literature. It is used more or less synonymously in this book with 'territory' (e.g. the urban scale or the national scale). In other words, scales are seen as porous, flexible 'containers' such as national states, macro-regions such as the EU, but also the human body, and so forth.

Scalar (or scalar spatiality) – An adjective that describes when a process is subject to, but also transcends particular scales.

Smuggling – A term used to describe the clandestine movement of migrants from one country or another using a variety of agents and institutions.

Social exclusion – Used to describe a set of processes which exclude or marginalize people from different facets of social life (e.g. work, housing, education, etc.).

Social remittances – Refer to the ideas, practices, but also finances that migrants bring home or send home, and which contribute specifically to the construction of schools, roads, religious institutions, hometown institutions, and other social institutions (see also 'remittances').

Social reproduction – Social reproduction is a Marxist-Feminist inspired term which has come to mean the process by which people are housed, fed, clothed, educated and generally raised to become workers and/or citizens under capitalism. In short, people need to be reproduced in particular ways to make them 'ready' for capitalism.

Sojourner migration – See 'temporary migration'.

Substantive citizenship – Refers to the issues that concern the daily lives of immigrants, including family matters, finding a place to live and work, schooling, participating in organizations and events, the challenges of finding appropriate legal advice, and access to health care.

Temporary migrants – As defined by the OECD, international migrants are those whose duration of stay in a given country does not exceed three months.

Territory – An area of geographic space that is occupied and controlled by an individual, a group, or (an) institution(s)/state(s) for the purposes of control and influence.

Territoriality – The ability, practice, and strategy of exercising control over a particular geographic space.

Trafficking – A term used to describe the clandestine movement of migrants from one country or another involving trafficking agents, and which also involves some form of 'forced' employment in order to pay back the costs of being smuggled.

Transnationalism – In the context of migration, this usually refers to the multiple cultural, economic, political and social ties that bind migrants across one or more countries.

Undocumented migrant (also 'irregular'/'illegalized') – The term probably preferred by many migrants and most critical scholars, which refers to a range of individuals, but most commonly those migrants who have entered a country clandestinely without the required papers; or those who have entered legally, but overstayed their visas or other residence regulations.

Voluntary migrant – This is a general term to describe an individual who is *not* forced to migrate from their country or countries of origin. 'Voluntary' is an imprecise and relative term, and there are various degrees of voluntarism. Voluntary migrants may migrate for marriage purposes, to be closer to family, friends, or for other relationships. They also may migrate for a particular job, or simply to experience other cultures. In short, the reasons are very diverse.

BIBLIOGRAPHY

ACME: An International E-journal for Critical Geographies (2003) Engagements, borders, and immigration: A symposium. Special Issue. 2, 2.

Adamson, F.B. (2006) Crossing borders – international migration and national security, *International Security*. 31, 1: 165–99.

Adey, P. (2010) *Mobility (Key Ideas in Geography)*. London: Routledge.

Ager, A., and Strang, A. (2008) Understanding integration: A conceptual framework, *Journal of Refugee Studies*. 21, 2: 166–91.

Agnew, J. (1994) The territorial trap: The geographical assumptions of international relations theory, *Review of International Political Economy*. 1, 1: 53–80.

Agustín, L. (2006) The disappearing of a migration category: Migrants who sell sex, *Journal of Ethnic and Migration Studies*. 32, 1: 29–47.

Ahmad, A.N. (2008a) Dead men working: Time and space in London's ('illegal') migrant economy, *Work, Employment and Society*. 22: 301–18.

Ahmad, A.N. (2008b) The labour market consequences of human smuggling: 'Illegal' employment in London's migrant economy, *Journal of Ethnic and Migration Studies*. 34, 6: 853–74.

Ahmed, N., Garnett, J., Gidley, B., Harris, A., and Keith, M. (2016) Shifting markers of identity in East London's diasporic religious identities, *Ethnic and Racial Studies*. 39, 2: 223–42.

Alba, R.D., and Foner, N. (2014) Comparing immigrant integration in North America and Western Europe: How much do the grand narratives tell us? *International Migration Review*. 48: S263–S291.

Alba, R.D., and Nee, V. (2003) *Remaking the American Mainstream: Assimilation and Contemporary Immigration*. Cambridge: Harvard University Press.

Alienikoff, A., and Klusmeyer, D. (2001) Plural nationality: Facing the future in a migratory world, in *Citizenship Today: Global perspectives and practices*. Washington DC: Carnegie Endowment for International Peace/Migration Policy Institute.

Ambrosini, M. (2014) Migration and transnational commitment: Some evidence from the Italian case, *Journal of Ethnic and Migration Studies*. 40, 4: 619–37.

Amin, A. (2002) Spatialities of globalisation, *Environment and Planning A*. 34, 3: 385–99.

Amoore, L. (2006) Biometric borders: Governing mobilities in the war on terror, *Political Geography*. 25: 336–51.

Anderson, B. (2001a) Why Madam has so many bathrobes? Demand for migrant workers in the EU, *Tijdschrift voor economische en social geografie*. 92, 1: 18–26.

Anderson, B. (2001b) *Doing the Dirty Work? The Global Politics of Domestic Labour*. London and New York: Zed Books.

Anderson, B. (2007) A very private business: Exploring the demand for migrant domestic workers, *European Journal of Women's Studies*. 14, 3: 247–64.

Anderson, B. (2010) Migration, immigration controls and the fashioning of precarious workers, *Work Employment and Society*. 24, 2: 300–17.

Anderson, B. (2013) *Us and Them? The Dangerous Politics of Immigration Control*. Oxford: Oxford University Press.

Anderson, B., Gibney, M., and Paoletti, E. (2011) Citizenship, deportation and the boundaries of belonging, *Citizenship Studies*. 15, 5: 547–63.

Anderson, B., and Hancilová, B. (2011) Migrant labour in Kazakhstan: A cause for concern? *Journal of Ethnic and Migration Studies*. 37, 3: 467–83.

Anderson, B., and Rogaly, B. (2005) Forced labour and migration to the UK. Study prepared by COMPAS in collaboration with the Trades Union Congress.

Anderson, B., Ruhs, M., Rogaly, B., and Spencer, S. (2006) Fair enough? Central and Eastern European migrants in low-wage employment in the UK, Study Prepared by COMPAS in collaboration with the Trades Union Congress.

Andersson, H.E., and Nilsson, S. (2011) Asylum seekers and undocumented migrants' increased social rights in Sweden, *International Migration*. 49, 4: 167–88.

Andersson, R. (2014) Time and the migrant other: European border controls and the temporal economics of illegality, *American Anthropologist*. 116, 4: 1–15.

Andreas, P. (2000) *Border Games: Policing the U.S.–Mexico Divide*. Ithaca: Cornell University Press.

Anthias, F. (1998) Evaluating 'diaspora': Beyond ethnicity? *Sociology*. 32, 3: 557–80.

Antonsich, M. and Jones, P.I. (2010) Mapping the Swiss referendum on the minaret ban, *Political Geography*. 29, 2: 57–62.

Anzaldúa, G. (1987) *Borderlands/La Frontera*. San Francisco: Aunt Lute Books.

Arango, J. (2000) Explaining migration: A critical view, *International Social Science Journal*. 52, 165: 283–96.

Ashutosh, I. (2008) Re-creating the community: South Asian transnationalism on Chicago's Devon Avenue, *Urban Geography*. 29, 3: 224–45.

Ashutosh, I., and Mountz, A. (2011) Migration management for the benefit of whom? Interrogating the work of the International Organization of Migration, *Citizenship Studies*. 15, 6/7: 21–38.

Askola, H. (2007) Violence against women, trafficking, and migration in the European Union, *European Law Journal*. 13, 2: 204–17.

Athwal, H. (2015) 'I don't have a life to live': Deaths and UK detention, *Race and Class*. 56, 3: 50–68.

Audit Commission (2000) *Another Country: Implementing Dispersal Under the Immigration and Asylum Act 1999*. London: The Audit Commission.

Aziz, A. (2014) Urban refugees in a graduated sovereignty: The experiences of the stateless Rohingya in the Klang Valley, *Citizenship Studies*. 18, 8: 839–54.

Bailey, A. (2001) Turning transnational: Notes on the theorisation of international migration, *International Journal of Population Geography*. 7: 413–28.

Bailey, A. (2005) *Making Population Geography*. Oxford: Oxford University Press.

Bailey, A. (2013) Migration, recession and an emerging transnational biopolitics across Europe, *Geoforum*. 44: 202–10.

Bailey, A., Wright, R., Mountz, A., and Miyares, I. (2002) (Re)producing Salvadoran transnational geographies, *Annals of the Association of American Geographers*. 92, 1: 125–44.

Bailey, T., and Waldinger, R. (1991) Primary, secondary, and enclave labor markets: A training systems approach, *American Sociological Review*. 56: 432–45.

Bakewell, O. (2008) 'Keeping them in their place': The ambivalent relationship between development and migration in Africa, *Third World Quarterly*. 29, 7: 1341–58.

Bakewell, O. (2010) Some reflections on structure and agency in migration theory, *Journal of Ethnic and Migration Studies*. 36, 10: 1689–708.

Bakewell, O. (2015) Migration makes the Sustainable Development Goals agenda – time to celebrate? International Migration Institute blog 11.12.2015. Available at: http://www.imi.ox.ac.uk/news/oliver-bakewell-migration-makes-the-sustainable-development-goals-agenda-2013-time-to-celebrate

Bal, E. (2014) Yearning for faraway places: The construction of migration desires among young and educated Bangladeshis in Dhaka, *Identities*. 21, 3: 275–89.

Balch, A. (2015) Assessing the international regime against human trafficking, in Talani, L.S., and McMahon, S. (eds) *Handbook of the International Political Economy of Migration*. Cheltenham and Northampton: Edward Elgar.

Balibar, E., and Wallerstein, I. (1991) *Race, Nation, Class: Ambiguous Identities*. London: Verso.

Banerjee, R., and Phan, M.B. (2015) Do tied movers get tied down? The occupational displacement of dependent applicant immigrants in Canada, *Journal of International Migration and Integration*. 16, 2: 333–53.

Barkan, E. (2004) America in the hand, homeland in the heart: Transnational and translocal immigrant experiences in the American West, *Western Historical Quarterly*. 35, 3: 331–56.

Barnett, C. (2006) The consolations of neo-liberalism, *Geoforum*. 36, 1: 7–12.

Barron, P., Bory, A., Tourette, L., Chauvin, S., and Jounin, N. (2011) *On bosse ici, on reste ici! La grève des sans-papiers : Une aventure inedite*. Paris: La Découverte.

Basch, L., Glick Schiller, N., and Szanton Blanc, C. (eds) (1994) *Nations Unbound: Transnational Projects, Postcolonial Predicaments and Deterritorialized Nation-States*. Amsterdam: Gordon Breach Publishers.

Bastia, T. (2013) The Migration-Development Nexus: Current challenges and future research agenda, *Geography Compass*. 7, 7: 464–77.

Batnitzky, A., and McDowell, L. (2011) Migration, nursing, institutional discrimination and emotional/affective labour: Ethnicity and labour stratification in the UK National Health Service. *Social and Cultural Geography.* 12: 181–201.

Bauböck, R. (ed.) (2006) *Migration and Citizenship: Legal Status, Rights and Political Participation.* Amsterdam: Amsterdam University Press.

Bauder, H. (2003) Equality, justice and the problem of international borders: The case of Canadian immigration regulation, *ACME: An International E-journal of Critical Geographies.* 2, 2: 167–82.

Bauder, H. (2005) *Labor Movement: How Migration Shapes Labor Markets.* Oxford: Oxford University Press.

Bauder, H. (2008) Neoliberalism and the economic utility of immigration: Media perspectives of Germany's immigration law, *Antipode.* 40, 1: 55–78.

Bauder, H. (2014a) Why we should use the term 'illegalized' refugee or immigrant: A commentary, *International Journal of Refugee Law.* 26, 3: 327–32.

Bauder, H. (2014b) Domicile citizenship, human mobility and territoriality, *Progress in Human Geography.* 38, 1: 91–106.

Bauder, H. (2015) Perspectives of open borders and no border, *Geography Compass.* 9, 7: 395–405.

Bauder, H. (2016) Possibilities of urban belonging, *Antipode.* 48, 2: 252–71.

Bauer, T.K., and Kunze, A. (2004) The demand for high-skilled workers and immigration policy, Forschungsinstitut zur Zukunft der Arbeit (Institute for the Study of Labor), IZA Discussion Paper No. 999, January.

Baumol, W. (2012) *The Cost Disease Problem.* New Haven and London: Yale University Press.

Beaumont, J., and Baker, C. (2011) Introduction: The rise of the post-secular city, in Beaumont, J., and Baker, C. (eds) *Postsecular Cities: Space, Theory, and Practice.* London: Bloomsbury Academic.

Beaverstock, J.V. (2002) Transnational elites in global cities: British expatriates in Singapore's financial district, *Geoforum.* 33, 4: 525–38.

Beaverstock, J.V. (2005) Transnational elites in the city: British highly-skilled inter-company transferees in New York City's financial district, *Journal of Ethnic and Migration Studies.* 31, 2: 245–68.

Beaverstock, J.V., and Boardwell, J.T. (2000) Negotiating globalization, transnational corporations and global city financial centres in transient migration studies, *Applied Geography.* 20, 3: 277–304.

Beck, U. (2000a) *The Brave New World of Work.* Cambridge: Polity Press.

Beck, U. (2000b) Cosmopolitan legacy: Sociology of the second age of modernity, *British Journal of Sociology.* 51: 79–105.

Becker, G. (1964) *Human Capital.* New York: National Bureau of Economic Research/Columbia University Press.

Beech, S. (2014) Why place matters: Imaginative geography and international student mobility, *Area.* 46, 2: 170–77.

Bendel, P. (2005) Immigration policy in the European Union: Still bring up the walls for fortress Europe? *Migration Letters.* 2, 1: 20–31.

Beneria, L. (2001) Shifting the risk: New employment patterns, informalization, and women's work, unpublished paper. Available at: http://www.arts.cornell.edu/poverty/Papers/Beneria_InformalizationUrbana.pdf

Benhabib, S. (2004 [2007]) The Rights of Others: Aliens, Residents and Citizens. Cambridge: Cambridge University Press.

Benhabib, S. (2008) Another Cosmopolitanism. Oxford: Oxford University Press.

Benton Short, L., Price, M.D., and Friedman, S. (2005) Globalization from below: The ranking of global immigrant cities, International Journal of Urban and Regional Research. 29, 4: 945–59.

Berg, L., and Millbank, J. (2009) Constructing the personal narratives of lesbian, gay and bisexual asylum claimants, Journal of Refugee Studies. 22, 2: 195–223.

Bermudez, A. (2013) A gendered perspective on the arrival and settlement of Columbian refugees in the United Kingdom, Journal of Ethnic and Migration Studies. 39, 7: 1159–75.

Bernard, J., Kostelecky, T., and Patočková, V. (2014) The innovative regions in the Czech Republic and their position in the international labour market of highly skilled workers, Regional Studies. 37, 5: 1691–705.

Betts, A. (2009) Forced Migration and Global Politics. Oxford: Blackwell.

Betts, A. (ed.) (2012) Global Migration Governance. Oxford: Oxford University Press.

Bevelander, P., and Pendakur, R. (2012) The labour market integration of refugee and family reunion immigrants: A comparison of outcomes in Canada and Sweden, IZA DP No. 6924.

Bhabha, H. (1994) The Location of Culture. London: Routledge.

Bialasiewicz, L. (2011) Borders, above all? Political Geography. 30, 6: 299–300.

Bialasiewicz, L. (2012) Off-shoring and out-sourcing the borders of Europe: Libya and EU border work in the Mediterranean, Geopolitics. 17, 4: 843–66.

Bigo, D. (1998). Europe passoire et Europe forteresse. La sécuritisation/humani-tarisation de l'immigration, in Rea, A. (ed.) Immigration et Racisme en Europe. Bruxelles, Editions complexe, 203–41.

Bigo, D. (2002) Security and immigration: Toward a critique of the governmental-ity of unease, Alternatives. 27, 1 (suppl): 63–92.

Bigo, D. (2005) Frontier controls in the European Union: Who is in control?, in Bigo, D., and Guild, E. (eds) Controlling Frontiers: Free Movement Into and Within Europe. Aldershot: Ashgate.

Bigo, D., and Guild, E. (eds) (2005) Controlling Frontiers: Free Movement Into and Within Europe. Aldershot: Ashgate.

Bilodeau, A. (2008) Immigrants' voice through protest politics in Canada and Australia: Assessing the impact of pre-migration political repression, Journal of Ethnic and Migration Studies. 34, 6: 975–1002.

Binational Migration Institute (2013) A continued humanitarian crisis at the border: Undocumented border crosser deaths recorded by the Pima County Office of the Medical Examiner, 1990–2012, unpublished report.

Black, R. (2001) 50 years of refugee studies: From theory to policy, International Migration Review. Special Issue: UNHCR at 50: Past, Present and Future of Refugee Assistance. 35, 1: 57–78.

Black, R. (2003) Breaking the convention: Researching the 'illegal' migration of refugees to Europe, *Antipode*. 35, 1: 34–54.

Black, R., Adger, W.N., Arnell, N.W., Dercon, S., Geddes, A., and Thomas, D. (2011) The effect of environmental change on human migration, *Global Environmental Change*. 21, Supp. 1 (Migration and global environmental change – Review of drivers of migration): S3–S11.

Black, R., and Castaldo, A. (2009) Return migration and entrepreneurship in Ghana and Cote d'Ivoire: The role of capital transfers, *Tijdschrift Voor Economische en Social Geografie*. 100, 1: 44–58.

Blank, N. (2011) Making migration policy: Reflections on the Philippines' bilateral labor agreements, *Asian Politics and Policy*. 3, 2: 185–205.

Blinder, D. (2015) Briefings: Deportations, removals, and voluntary departures from the UK, Migration Observatory.

Bloemraad, I. (2006) *Becoming a Citizen: Incorporating Immigrants and Refugees in the United States and Canada*. Berkeley: University of California Press.

Bloemraad, I., Korteweg, A., and Yurdakul, G. (2008) Citizenship and immigration: Multiculturalism, assimilation, and challenges to the nation-state, *Annual Review of Sociology*. 34: 153–79.

Bloch, A. (2007) Methodological challenges for national and multi-sited comparative survey research, *Journal of Refugee Studies*. 20, 2: 230–47.

Bloch, A. (2010) The right to rights? Undocumented migrants from Zimbabwe living in South Africa, *Sociology*. 44, 2: 233–50.

Blue, S.A. (2004) State policy, economic crisis, gender, and family ties: Determinants of family remittances to Cuba, *Economic Geography*. 80, 1: 63–82.

Blumenberg, E. (2008) Immigrants and transport barriers to employment: The case of Southeast Asian welfare recipients in California, *Transport Policy*. 15, 1: 33–42.

Blunt, A. (2007) Cultural geographies of migration: Mobility, transnationality and diaspora, *Progress in Human Geography*. 31, 5: 684–94.

Boehm, D.A. (2008) 'Now I Am a Man and a Woman!' – Gendered moves and migrations in a transnational Mexican community, *Latin American Perspectives*. 35, 1: 16–30.

Böhning, W.R., and Oishi, N. (1995) Is international migration spreading? *International Migration Review*. 29, 3: 794–9.

Bommes, M., and Geddes, A. (eds) (2000) *Immigration and Welfare: Challenging the Borders of the Welfare State*. London: Routledge.

Bonacich, E., and Modell, J. (1980) *The Economic Basis of Ethnic Solidarity: Small Business in the Japanese American Community*. Berkeley: University of California Press.

Borjas, G. (1989) Economic theory and international migration, *International Migration Review*. 23, 3: 457–85.

Borjas, G., Grogger, J., and Hanson, G.H. (2006) Immigration and African-American employment opportunities: The response of wages, employment, and incarceration to labor supply shocks, NBER Working Paper, No. 12518.

Bosniak, L. (2000) Citizenship denationalized. Symposium: The state of citizenship, *Indiana Journal of Global Legal Studies*. 7: 447–90.

Boswell, C. (2003) The 'external dimension' of EU immigration and asylum policy, *International Affairs*. 79, 3: 619–38.

Boswell, C. (2007a) Theorizing migration policy: Is there a third way? *International Migration Review*. 41, 1: 75–100.

Boswell, C. (2007b) Migration control in Europe after 9/11: Explaining the absence of securitization, *Journal of Common Market Studies*. 45, 3: 589–610.

Boswell, C. (2008a) Combining economics and sociology in migration theory, *Journal of Ethnic and Migration Studies*. 34, 4: 549–66.

Boswell, C. (2008b) Evasion, reinterpretation and decoupling: European Commission responses to the 'external dimension' of immigration and asylum, *West European Politics*. 31, 3: 491–512.

Boswell, C. (2009) Migration, security, and legitimacy: Some reflections, in Givens, T.E., Freeman, G.P., and Leal, D.L. (2009) (eds) *Immigration Policy and Security*. New York and London: Routledge.

Boswell, C., and Geddes, A. (2011) *Migration and Mobility in the European Union*. London: Palgrave Macmillan.

Bourbeau, P. (2013) *The Securitization of Migration: A Study of Movement and Order*. London: Routledge.

Boyd, M. (1989) Family and personal networks in international migration: Recent developments and new agendas, *International Migration Review*. 23, 3: 638–70.

Boyle, M., Halfacree, K., and Robinson, V. (eds) (1998) *Exploring Contemporary Migration*. Harlow: Addison Wesley Longman.

Bradatan, C.E., and Sandu, D. (2012). Before crisis: Gender and economic outcomes of the two largest immigrant communities in Spain, *International Migration Review*. 46, 1: 221–43.

Bradley, H., Erikson, M., Stephenson, C., and Williams, S. (2002) *Myths at Work*. London: Polity Press.

Brenner, N. (2001) The limits to scale? Methodological reflections on scalar structuration, *Progress in Human Geography*. 25, 4: 591–614.

Brenner, N., and Theodore, N. (2002) Cities and the geographies of 'actually existing neoliberalism', *Antipode*. 34, 3: 349–79.

Brenner, N., Peck, J., and Theodore, N. (2010) *Global Networks*. 10, 2: 182–222.

Brettell, C.B. (2008) Theorizing migration in anthropology, in Brettell, C.B., and Hollifield, J.F. (eds) *Migration Theory: Talking Across Disciplines*, 2nd edition. New York and London: Routledge.

Brettell, C.B., and Hollifield, J.F. (2008a) Migration Theory: Talking across disciplines, in Brettell, C.B., and Hollifield, J.F. (eds) *Migration Theory: Talking Across Disciplines*. New York and London: Routledge.

Brettell, C.B., and Hollifield, J.F. (eds) (2008b) *Migration Theory: Talking Across Disciplines*, 2nd edition. New York and London: Routledge.

Brettell, C.B., and Hollifield, J.F. (eds) (2015) *Migration Theory: Talking Across Disciplines*, 3rd edition. New York and London: Routledge.

Brickell, K., and Yeoh, B.S.A. (2014) Geographies of domestic Life: 'Householding' in transition in East and Southeast Asia, *Geoforum*. 51: 259–61.

Broadway, M. (2007) Meatpacking and the transformation of rural communities: A comparison of Brooks, Alberta and Garden City, Kansas, *Rural Sociology*. 72, 4: 560–82.

Brubaker, R. (1992) *Citizenship and Nationhood in France and Germany*. Cambridge: Harvard University Press.

Brubaker, R. (2005) The 'diaspora' diaspora, *Ethnic and Racial Studies*. 28, 1: 1–19.

Brubaker, R., Loveman, M., and Stamatov, P. (2004) Ethnicity as cognition, *Theory and Society*. 33, 1: 31–64.

Bryant, C. (1997) Citizenship, national identity and the accommodation of difference: Reflections on the German, French, Dutch, and British cases, *New Community*. 23: 157–72.

Büchel, F., and Frick, J. (2005) Immigrants' economic performance across Europe – does immigration policy matter? *Population Research and Policy Review*. 24: 175–212.

Buckley, M. (2013) Locating neoliberalism in Dubai: Migrant workers and class struggle in the autocratic city, *Antipode*. 45, 2: 256–74.

Burawoy, M. (1976) The functions and reproduction of migrant labor: Comparative material from South Africa and the United States, *American Journal of Sociology*. 81, 5: 1050–87.

Burchell, G., Gordon, C., and Miller, P. (1991) *The Foucault effect: Studies in governmentality*. Chicago: University of Chicago Press.

Burgers, J., and Engbersen, G. (1996) Globalisation, migration, and undocumented immigrants, *New Community*. 22, 4: 619–35.

Cadge, W., and Ecklund, E. (2007) Immigration and religion, *Annual Review of Sociology*. 33: 359–79.

Caglar, A. (2006) Hometown associations, the rescaling of state spatiality and migrant grassroots transnationalism, *Global Networks*. 6, 1: 1–22.

Canales, A.I. (2003) Mexican labour migration to the United States in the age of globalization, *Journal of Ethnic and Migration Studies*. 29, 4: 741–61.

Carens, J. (1987) Aliens and citizens: The case for open borders, *Review of Politics*. 49, 2: 251–73.

Carens, J. (2013) *The Ethics of Immigration*. Oxford: Oxford University Press.

Carle, R. (2007) Citizenship debates in the new Germany, *Society*. 44: 147–54.

Carling, J. (2002) Migration in the age of involuntary immobility: Theoretical reflections and Cape Verdean experiences, *Journal of Ethnic and Migration Studies*. 28, 1: 5–42.

Carmel, E., Cerami, A., and Papadopoulos, T. (eds) (2012) *Migration and Welfare in the New Europe*. Bristol: The Policy Press.

Carr, E. (2005) Placing the environment in migration: Environment, economy, and power in Ghana's Central Region, *Environment and Planning A*. 37: 925–46.

Carrera, S. (2014) How much does EU citizenship cost? The Maltese citizenship-for-sale affair: A breakthrough for sincere cooperation in citizenship of the union? CEPS Papers in Liberty and Security, No. 64/April.

Casas-Cortes, M., Cobarrubias, S., and Pickles, J. (2013) Re-bordering the neigh-bourhood: Europe's emerging geographies of non-accession integration, *European Urban and Regional Studies*. 20, 1: 37–58.

Casey, J.P. (2010) Open borders: Absurd chimera or inevitable future policy? *International Migration*. 48, 5: 14–62.

Castells, M. (1975) Immigrant workers and class struggles in advanced capitalism – Western European experience, *Politics and Society*. 5, 1: 33–66.

Castells, M. (1996) *The Rise of the Network Society*, Volume I. Oxford: Basil Blackwell.

Castells, M., and Portes, A. (1989) Worlds underneath: The origins, dynamics, and effects of the informal economy, in Portes, A., Castells, M., and Benton, L. (eds) *The Informal Economy: Studies in Advanced and Less Developed Countries*. Baltimore: Johns Hopkins University Press.

Castles, S. (1984) *Here for Good: Western Europe's New Ethnic Minorities*. London: Pluto Press.

Castles, S. (2003) Towards a sociology of forced migration and social transforma-tion, *Sociology*. 37: 1: 13–34.

Castles, S. (2004) The factors that make and unmake migration policies, *International Migration Review*. 38, 3: 852–84.

Castles, S. (2006) Guestworkers in Europe: A resurrection? *International Migration Review*. 40, 4: 741–66.

Castles, S. (2010) Understanding global migration: A social transformation per-spective, *Journal of Ethnic and Migration Studies*. 36, 10: 1565–86.

Castles, S., De Haas, H., and Miller, M. (2014) *The Age of Migration: International Population Movements in the Modern World*, 5th edition. New York: Guilford Press.

Castles, S., and Kosack, G. (1973) *Immigrant Workers and Class Structure in Western Europe*. London: Oxford University Press.

Castles, S., and Miller, M. (1993) *The Age of Migration*. London: Macmillan

Castles, S., and Miller, M. (2003) *The Age of Migration: Population Movements in the Modern World*, 3rd edition. New York: Guilford Press.

Castles, S., and Miller, M. (2009) *The Age of Migration: International Population Movements in the Modern World*, 4th edition. New York: Guilford Press.

Castree, N. (2004) Differential geographies: Place, indigenous rights and 'local' resources, *Political Geography*. 23: 133–67.

Caviedes, A. (2010) The sectoral turn in labour migration policy, in Menz, G., and Caviedes, A., *The Changing Face of Labour Migration in Europe*. Basingstoke: Palgrave Macmillan, 54–75.

Caviedes, A. (2016) European integration and the governance of migration, *Journal of Contemporary European Research*. 12, 1: 552–65.

CBP (2014) Border Patrol Agent Staffing by Fiscal Year (as of 20 September 2014). Available at: http://www.cbp.gov/sites/default/files/documents/BP%20Staffing %20FY1992-FY2014_0.pdf

CEC (2005) Green Paper on an EU approach to managing economic migration, COM (2004) 811 Final.

Cerna, L. (2010) The EU Blue Card: A bridge too far? Available at: http://www. jhubc.it/ecprporto/virtualpaperroom/041.pdf

Cerna, L., and Chou, M.-H. (2014) The regional dimension in the global competition for talent: Lessons from framing the European scientific visa and blue card, *Journal of European Public Policy*. 21, 1: 76–95.

Césari, J. (2005) Mosque conflicts in European cities: Introduction, *Journal of Ethnic and Migration Studies*. 31, 6: 1015–24.

Chatty, D. (2010) *Displacement and Dispossession in the Modern Middle East*. Cambridge: Cambridge University Press.

Chauvin, S. (2014) Becoming less illegal: Deservingness frames and undocumented migrant incorporation, *Sociology Compass*. 8, 4: 422–32.

Chauvin, S., and Garcés-Mascareñas, B. (2012) Beyond informal citizenship: The new moral economy of migrant illegality, *International Political Sociology*. 6: 241–59.

Chee, H.L., Yeoh, B.S.A, and Shuib, R. (2012) Circuitous pathways: Marriage as a route toward (il)legality for Indonesian migrant workers in Malaysia, *Asian and Pacific Migration Journal*. 21, 3: 317–44.

Chee, H.L., Yeoh, B.S.A. and Vu, T.K.D. (2012) From client to matchmaker: Social capital in the making of commercial matchmaking agents in Malaysia, *Pacific Affairs*. 85, 1: 89–114.

Chiang, L.H.N. (2004) The dynamics of self-employment and ethnic business ownership among Taiwanese in Australia, *International Migration*. 42, 2: 153–73.

Chiswick, B.R., and Miller, P.W. (2010) The effects of educational-occupational mismatch on immigrant earnings in Australia, with international comparisons, *International Migration Review*. 44, 4: 869–98.

Chung, E.A. (2014) Japan and South Korea, in Hollifield, J., Martin, P., and Orrenius P.M. (eds) *Controlling Immigration: A Global Perspective*, 3rd edition. Stanford: Stanford University Press.

Cinalli, M., and El Hariri, A.E. (2012) Contentious opportunities: Comparing metropolitan policymaking for immigrants in France and Italy, in Carmel *et al.* (eds) *Migration and Welfare in the New Europe*. Bristol: The Policy Press.

Clark, C.L. (2009) Environment, land, and rural out-migration in the southern Ecuadorian Andes, *World Development*. 37, 2: 457–68.

Clark, W.A.V. (1986) *Human Migration*. Beverly Hills, CA: Sage Publications.

Clark, W.A.V. (1998) *California Cauldron: Immigration and the Fortunes of Local Communities*. New York: Guilford Press.

Clark, W.A.V. (2003) *Immigrants and the American Dream*. New York and London: Guilford Press.

Cloke, P., Crang, P., and Goodwin, M. (2014) *Introducing Human Geographies*, 3rd edition. London and New York: Routledge.

Cloke, P., Philo, C., and Sadler, D. (1991) *Approaching Human Geography*. London: Paul Chapman Publishing.

Cohen, R. (1987) *The New Helots: Migrants in the International Division of Labour*. Aldershot: Gower Publishers.

Cohen, R. (1997) *Global Diasporas: An Introduction*. Seattle: University of Washington Press.

Cohen, R. (2006) *Migration and its Enemies: Global Capital, Migrant Labour and the Nation-State*. Aldershot: Ashgate.

Cohen, S. (2003) *No One is Illegal*. Stoke on Trent: Trentham Books.

Cole, P. (2000) *Philosophies of Exclusion: Liberal Political Theory and Immigration*. Edinburgh: Edinburgh University Press.

Coleman, M. (2005) U.S. statecraft and the U.S.–Mexico border as security/economy nexus, *Political Geography*. 24: 185–209.

Coleman, M., and Stuesse, A. (2016) The disappearing state and the quasi-event of immigration control, *Antipode*. 48, 3: 524–43.

Collier, P. (2013) *Exodus: How Migration is Changing our World*. Oxford: Oxford University Press.

Collier, S. (2012) Neoliberalism as big Leviathan, or . . .? A response to Wacquant and Hilgers, *Social Anthropology*. 20, 2: 186–95.

Collyer, M. (2005) When do social networks fail to explain migration? Accounting for the movement of Algerian asylum-seekers to the UK, *Journal of Ethnic and Migration Studies*. 31, 4: 699–718.

Collyer, M. (2007) In-between places: Undocumented Sub-Saharan African migrants in Morocco and the fragmented journey to Europe, *Antipode*. 39, 4: 620–35.

Collyer, M. (2010) Stranded migrants and the fragmented journey, *Journal of Refugee Studies*. 23, 3: 273–93.

Collyer, M. (2014a) Introduction: Locating and narrating emigration nations, in Collyer, M. (ed.) in *Emigration Nations: Policies and Ideologies of Emigrant Engagement*. Basingstoke: Palgrave.

Collyer, M. (2014b) Geography and Forced Migration, in Fiddian-Qasmiyeh, E., Loescher, G., Long, K. and Sigona, N. (eds) *The Oxford Handbook of Forced Migration*. Oxford: Oxford University Press.

Collyer, M. (2014c) Inside out? Directly elected 'special representation' of emigrants in national legislatures and the role of popular sovereignty, *Political Geography*. 41: 64–73.

Collyer, M., and de Haas, H. (2012) Developing dynamic categorisations of migration, *Population, Space and Place*. 18, 4: 468–81.

Collyer, M., Düvell, F., and de Haas (2012) Editorial Introduction: Critical approaches to transit migration, *Population, Space and Place*. 18, 4: 407–14.

Collyer, M., and King, R. (2015) Producing transnational space: International migration and the extra-territorial reach of state power, *Progress in Human Geography*. 39, 2: 185–204.

Columbo Process (2016) Columbo Process website. Available at: http://www.colomboprocess.org/media-centre

Conlon, D. (2010) Ties that bind: Governmentality, the state, and asylum in contemporary Ireland, *Environment and Planning D*. 28, 1: 95–111.

Connell, J. (2008) Niue: Embracing a culture of migration, *Journal of Ethnic and Migration Studies*. 34, 6: 1021–40.

Connor, P., and Koenig, M. (2013) Bridges and barriers: Religion and immigrant occupational attainment across integration contexts, *International Migration Review*. 47, 1: 3–38.

Conway, D. (2007) Caribbean transnational migration behavior: Reconceptualising its 'strategic flexibility', *Population, Space and Place*. 13: 415–31.

Cornelius, W. (2004) Spain: The uneasy transition from labor exporter to labor importer, in Cornelius, W., Tsuda, T., Martin, P.L., and Hollifield, J. (eds) *Controlling Immigration: A Global Perspective*. Stanford: Stanford University Press.

Cornelius, W. (2005) Controlling 'unwanted' immigration: Lessons from the United States, 1993–2004, *Journal of Ethnic and Migration Studies*. 31, 4: 775–94.

Cornelius, W., and Tsuda, T. (1994) Controlling immigration: The limits of government intervention, in Cornelius, W., Tsuda, T., Martin, P.L., and Hollifield, J.F. (eds) *Controlling Immigration: A Global Perspective*, 2nd edition. Stanford: Stanford University Press.

Correia, A., do Valle, P.O., and Moco, C. (2007) Modeling motivations and perceptions of Portuguese tourists, *Journal of Business Research*. 60, 1: 76–80.

Costello, T. (2001) Summary of paper delivered, 'The underground economy in North America', Conference of 21–22 May 2001, Harvard University.

Council of Europe/Parliamentary Assembly (2003) Migrants in irregular employment in the agricultural sector of southern European countries, Doc. 9883, July 18.

Coutin, S. (2015) Deportation studies: Origins, themes and directions, *Journal of Ethnic and Migration Studies*. 41, 4: 671–81.

Cox, K. (1997) *Spaces of Globalization: Reasserting the Power of the Local*. New York: Guilford Press.

Cravey, A. (2003) Toque una ranchera, por favor, *Antipode*. 35, 3: 603–21.

Creese, G. and Wiebe, B. (2012) 'Survival employment': Gender and deskilling among African Immigrants in Canada, *International Migration*. 50, 5: 56–76.

Cresswell, T. (1996) *In Place/Out of Place: Geography, Ideology and Transgression*. Minneapolis: University of Minnesota Press.

Cresswell, T. (2004) *Place: A Short Introduction*. Oxford: Basil Blackwell.

Cresswell, T. (2006) *On the Move: Mobility in the Modern Western World*. London: Routledge.

Cresswell, T. (2015) *Place: A Short Introduction*, 2nd edition. Chichester: John Wiley and Sons.

Crinis, V. (2010) Sweat or no sweat: Foreign workers in the garment industry in Malaysia, *Journal of Contemporary Asia*. 40, 4: 589–611.

Crush, J. (2011) Complex movements, confused responses: Labour migration in South Africa, SAMP Policy Brief No 25, AUGUST. Available at: http://www.queensu.ca/samp/sampresources/samppublications/policybriefs/brief25.pdf

Cwerner, S. (2001) The times of migration, *Journal of Ethnic and Migration Studies*. 27, 1: 7–36.

Czaika, M., and de Haas, H. (2014) The globalization of migration: Has the world become more migratory? *International Migration Review*. 48, 2: 283–323.

Dahi, O.S. (2014) Syria in fragments: The politics of the refugee crisis, *Dissent*. 61, 1: 45–8.

Dannecker, P. (2005) Transnational migration and the transformation of gender relations: The case of Bangladeshi labour migrants, *Current Sociology*. 53: 655–74.

D'Aoust, A-M. (2013) In the name of love: Marriage migration, governmentality, and technologies of love, *International Political Sociology*. 7, 3: 655–74.

Darling, J. (2009) Becoming bare life: Asylum, hospitality, and the politics of encampment, *Environment and Planning D: Society and Space*. 27: 183–9.

Darling, J. (2010) A city of sanctuary: The relational re-imagining of Sheffield's asylum politics, *Transactions of the Institute of British Geographers*. 35: 125–40.

Darling, J. (2011) Domopolitics, governmentality and the regulation of asylum accommodation, *Political Geography*. 30, 5: 263–71.

Darling, J., and Squire, V. (2012) Everyday enactments of sanctuary: The UK City of Sanctuary movement, in Lippert, R., and Rehaag, S. (eds) *Sanctuary Practices in International Perspective: Migration, Citizenship and Social Movements*. London: Routledge, 2012.

Datta, K., McIlwaine, C., Wills, J., Evans, Y., Herbert, J., and May, J. (2007) The new development finance or exploiting migrant labour? Remittance sending among low-paid migrant workers in London, *International Development Planning Review*, 29, 1: 43–68.

Davis, M. (1999) Magical urbanism: Latinos reinvent the US big city, *New Left Review*. 234: 3–43.

Davis, M. (2004) Planet of slums. Urban involution and the informal proletariat, *New Left Review*. 26: 5–34.

Davis, M. (2006) Fear and money in Dubai, *New Left Review*. 41: 47–68.

Dawson, C. (2014) The discourse of hospitality and the rise of immigration detention in Canada, *University of Toronto Quarterly*. 83, 4: 826–46.

Dean, M., and Nagashima, M. (2007) Sharing the burden: The role of government and NGOs in protecting and providing for asylum seekers and refugees in Japan, *Journal of Refugee Studies*. 20, 3: 481–508.

DeChaine, D.R. (2009) Bordering the civic imaginary: Alienization, fence logic, and the Minuteman Civil Defense Corps, *Quarterly Journal of Speech*, 95, 1: 43–65.

De Genova, N. (2002) Migrant 'Illegality' and deportability in everyday life, *Annual Review of Anthropology*. 31: 419–47.

De Genova, N., and Peutz, N. (eds) (2010) *The Deportation Regime: Sovereignty, Space and the Freedom of Movement*. Durham: Duke University Press.

de Haas, H. (2006) Migration, remittances and regional development in southern Morocco, *Geoforum*. 37: 565–80.

de Haas, H. (2007) Turning the tide? Why development will not stop migration, *Development and Change*. 38, 5: 819–41.

de Haas H. (2010) Migration and development: A theoretical perspective, *International Migration Review*. 44: 227–64.

de Haas, H. (2011) The internal dynamics of migration processes: A theoretical inquiry, *Journal of Ethnic and Migration Studies*. 36, 10: 1587–617.

Delaney, D. (2005) *Territory: A Short Introduction*. Oxford: Blackwell.

de Lange, A. (2007) Child labour migration and trafficking in rural Burkina Faso, *International Migration*. 45, 2: 147–67.

Delano, A. (2009) From 'shared responsibility' to a migration agreement? The limits for cooperation in the Mexico–United States case (2000–2008), *International Migration*. 50: e42–e59.

Dell'Olio, F. (2004) Immigration and immigrant policy in Italy and the UK: Is housing policy a barrier to a common approach towards immigration in the EU? *Journal of Ethnic and Migration Studies*. 30, 1: 107–28.

Demireva, N. (2011) New migrants in the UK: Employment patterns and occupational attainment, *Journal of Ethnic and Migration Studies*. 37, 4: 637–55.

Department of Homeland Security (2013) *Yearbook of Immigration Statistics: 2013. Enforcement Actions*.

Der Spiegel On-line (2015a) 'Germany shows signs of strain from mass of refugees'. October 17. http://www.spiegel.de/international/germany/germany-shows-signs-of-strain-from-mass-of-refugees-a-1058237.html

Der Spiegel On-line (2015b), 'Merkel slowly changes tune on refugee issue'. November 20. http://www.spiegel.de/international/germany/angela-merkel-changes-her-stance-on-refugee-limits-a-1063773.html

Der Spiegel On-Line (2016) 'Stretched to the limit: Has the German state lost control?' January 21. http://www.spiegel.de/international/germany/germans-ask-if-country-is-still-safe-after-cologne-attacks-a-1073165.html

Devitt, C. (2011) Varieties of capitalism, variation in labour immigration, *Journal of Ethnic and Migration Studies*. 37, 4: 579–96.

Dickenson, J., and Bailey, A.J (2007) (Re)membering diaspora: Uneven geographies of Indian dual citizenship, *Political Geography*. 26, 7: 757–74.

Dikeç, M. (2012) Space as a mode of political thinking, *Geoforum*. 43: 669–76.

Dines, N., Montagna, N., and Ruggiero, V (2015) Thinking Lampedusa: Border construction, the spectacle of bare life and the productivity of migrants, *Ethnic and Racial Studies*. 38, 3: 430–445.

Donato, K., and Sisk, B (2015) Children's migration + to the United States from Mexico and Central America: Evidence from the Mexican and Latin American migration projects, *Journal of Human Migration and Security*. 3, 1: 58–79.

Douglas, M. (2006) Global householding in Pacific Asia, *International Development Planning Review*. 28, 4: 421–45.

Douglas, M. (2012) Global householding and social reproduction: Migration research, dynamics and public policy in East and Southeast Asia, Asia Research Institute, Working Paper Series No. 188, paper available at: http://www.ari.nus.edu.sg/docs/wps/wps12_188.pdf

Drever, A., and Blue, S. (2011) Surviving sin papeles in Post-Katrina New Orleans: An exploration of the challenges facing undocumented Latino immigrants in new and re-emerging Latino destinations, *Population Space and Place*. 17, 1: 89–102.

Dumont, A. (2008) Representing voiceless migrants: Moroccan political transnationalism and Moroccan migrants' organizations in France, *Ethnic and Racial Studies*. 31, 4: 792–811.

Dunn, T. (1996) *The Militarization of the US–Mexican Border, 1978–1992: Low-Intensity Conflict Doctrine Comes Home*. Austin: University of Texas Press.

Durden, E. (2007) Nativity, duration of residence, citizenship, and access to health care for Hispanic children, *International Migration Review*. 41, 2: 537–45.

Düvell, F. (2012) Transit migration: A blurred and politicised concept, *Population, Space and Place*. 18, 4: 415–27.

Dwyer, P. (2005) Governance, forced migration and welfare, *Social Policy and Administration*. 39, 6: 622–39.

Eastmond, M. (2007) Stories as lived experience: Narratives in forced migration research, *Journal of Refugee Studies*. 20, 2: 248–64.

Eckstein, S. and Nguyen, T.N. (2011) The making and transnationalization of an ethnic niche: Vietnamese manicurists, *International Migration Review*. 45, 3: 639–74.

Economist (2015a) 'Apartheid on the Andaman Sea', 13 June. http://www.economist.com/news/leaders/21654055-myanmar-treats-rohingyas-badly-old-south-africa-treated-blacks-world-should

Economist (2015b) 'The most persecuted people on Earth', 13 June. http://www.economist.com/news/asia/21654124-myanmars-muslim-minority-have-been-attacked-impunity-stripped-vote-and-driven

Economist (2015c) 'How much room at the inn?', 1 September. http://www.economist.com/blogs/graphicdetail/2015/09/daily-chart

Economist (2015d) 'A city that wants more refugees', 31 October. http://www.economist.com/news/united-states/21677240-hardscrabble-baltimore-finds-kindness-brings-its-own-rewards-city-wants-more

Economist (2015e) 'Looking for a home', 29 August. http://www.economist.com/news/europe/21662597-asylum-seekers-economic-migrants-and-residents-all-stripes-fret-over-their-place-looking

Economist (2016a) 'Forming an orderly queue', 6–12 February. http://www.economist.com/news/briefing/21690066-europe-desperately-needs-control-wave-migrants-breaking-over-its-borders-how

Economist (2016b) '1.2 billion opportunities', 16–22 April. http://www.economist.com/news/special-report/21696792-commodity-boom-may-be-over-and-barriers-doing-business-are-everywhere-africas

Edin, P-A., Fredriksson, P., and Aslund, O. (2004) Settlement policies and the economic success of immigrants, *Journal of Population Economics*. 17, 1: 133–55.

Ehrenreich, B., and Hochschild, A. (2004) (eds) *Global Women: Nannies, Maids, and Sex Workers in the New Economy*. New York: Metropolitan Books.

Ehrkamp, P. (2006) 'We Turks are no Germans': Assimilation discourse and the dialectical construction of identities in Germany, *Environment and Planning A*. 38: 1673–92.

Ehrkamp, P. (2008) Risking publicity: Masculinities and the racialization of public neighborhood space, *Social and Cultural Geography*. 9, 2: 117–33.

Ehrkamp, P., and Jacobsen, M. (2015) Citizenship, in Agnew, J., Mamadouh, V., Secor, A., and Sharp, J. (eds) *The Wiley Blackwell Companion to Political Geography*. Chichester: Wiley Blackwell.

Ehrkamp, P., and Nagel, C. (2014) 'Under the radar': Undocumented immigrants, Christian faith communities, and the precarious spaces of welcome in the U.S. South, *Annals of the Association of American Geographers*. 104, 2: 319–28.

Ekman, M. (2015) Online Islamophobia and the politics of fear: Manufacturing the green scare, *Ethnic and Racial Studies*. 38, 11: 1986–2002.

Elias, J. (2008) Struggles over the rights of foreign domestic workers in Malaysia: The possibilities and limitations of 'rights talk', *Economy and Society*. 37, 2: 282–303.

Ellis, M. (2006) Unsettling immigrant geographies: US immigration and the politics of scale, *Tijdschrift voor Economischeen Sociale Geografie*. 97, 1: 49–58.

Ellis, M., Wright, R., and Parks, V. (2007) Geography and the immigrant division of labor, *Economic Geography*. 83, 3: 255–81.

Ellis, M., Wright, R., and Townley, M. (2014) The great recession and the allure of new immigrant destinations in the United States, *International Migration Review*. 48, 1: 3–33.

El Qadim, N. (2014) Postcolonial challenges to migration control: French-Moroccan cooperation practices on forced returns, *Security Dialogue*. 45, 3: 242–61.

Engelen, E. (2003) Conceptualizing economic incorporation: From 'institutional linkages' to 'institutional hybrids', paper written for the conference on 'Conceptual and Methodological Developments in the Study of International Migration', Princeton University, 23–25 May.

England, K. (2015) Nurses across borders: Global migration of registered nurses to the US, *Gender Place and Culture*. 22, 1: 143–56.

Enke, S. (1962) Economic development with unlimited and limited supplies of labour, *Oxford Economic Papers*. 14, 2: 158–72.

Ersanilli, E., and Koopmans, R. (2010) Rewarding integration? Citizenship regulations and the socio-cultural integration of immigrants in the Netherlands, France and Germany, *Journal of Ethnic and Migration Studies*. 36, 5: 773–91.

Escobar, C. (2007) Extraterritorial political rights and dual citizenship in Latin America, *Latin American Research Review*. 42, 3: 43–75.

European Council (1999) Tampere European Council. Presidency Conclusions, 15–16 October 1999. Available at: http://www.europarl.europa.eu/summits/tam_en.htm

EU (European Union) Commission (2016) Available at: http://ec.europa.eu/dgs/home-affairs/what-we-do/policies/legal-migration/family-reunification/index_en.htm

Faist, T. (1995) Boundaries of welfare states: Immigrants and social rights on the national and supranational level, in Miles, R., and Tranhardt, D. (eds) *Migration and European Integration: The Dynamics of Inclusion and Exclusion*. London: Pinter Publishers.

Faist, T. (2000) Transnationalization in international migration: Implications for the study of citizenship and culture, *Ethnic and Racial Studies*. 23: 189–222.

Faist, T. (2008) Migrants as transnational development agents: An inquiry into the newest round of the migration-development nexus, *Population, Space and Place*. 14: 21–42.

Faist, T. (2009) Diversity – a new mode of incorporation? *Ethnic and Racial Studies*. 32, 1: 171–90.

Faist, T. (2010) Towards transnational studies: World theories, transnationalisation and changing institutions, *Journal of Ethnic and Migration Studies*. 36, 10: 1665–87.

Fan, C.C. (2001) Migration and labor-market returns in urban China: Results from a recent survey in Guangzhou, *Environment and Planning A*. 33: 479–508.

Fassin, D. (2011) Policing borders, producing boundaries. The governmentality of immigration in dark times, *Annual Review of Anthropology*. 40, 1: 213–26.

Favell, A. (1998 [2001, 2nd edn.]) *Philosophies of Integration*. Basingstoke: Palgrave.

Favell, A. (2008) Rebooting migration theory: Interdisciplinarity, globality and postdisciplinarity in migration studies, in Brettell, C.B., and Hollifield, J.F. (eds) *Migration Theory: Talking Across Disciplines*, 2nd edition. New York and London: Routledge.

Favell, A. (2015) Migration theory rebooted? in Brettell, C.B., and Hollifield, J.F. (eds) *Migration Theory: Talking Across Disciplines, 3rd edition*. New York and London: Routledge.

Favell, A., and Hansen, R. (2002) Markets against politics: Migration, EU enlargement and the idea of Europe, *Journal of Ethnic and Migration Studies*. 28, 4: 581–601.

Fawcett, J.T. (1989) Networks, linkages, and migration systems, *International Migration Review*. 23, 3: 671–80.

Feldblum, M. (1993) Paradoxes of ethnic politics: The case of Franco-Maghrebis in France, *Ethnic and Racial Studies*. 16: 52–74.

Feldblum, M. (1998) Reconfiguring citizenship in Western Europe, in Joppke, C. (ed.) *Challenge to the Nation-state*. Oxford: Oxford University Press.

Feldblum, M. (1999) *Reconstructing Citizenship: The Politics of Nationality Reform and Immigration in Contemporary France*. Albany: SUNY Press.

Fetzer, J.S., and Soper, J.C. (2005) *Muslims and the State in Britain, France, and Germany*. Cambridge: Cambridge University Press.

Fiddian-Qasmiyeh, E., Loescher, G., Long, K., and Sigona, N. (eds) (2014) *The Oxford Handbook of Refugee and Forced Migration Studies*. Oxford: Basil Blackwell.

Findley, S.E. (1994) Does drought increase migration? A study of migration from rural Mali during the 1983–1985 drought, *International Migration Review*. 28: 539–53.

Finotelli, C., and Arango, J. (2011) Regularisation of unauthorised immigrants in Italy and Spain: Determinants and effects, *Documents d'Anàlisi Geogràfica*. 57, 3: 495–515.

Fitzgerald, D. (2006) Towards a theoretical ethnography of migration, *Qualitative Sociology*. 29, 1: 1–24.

FitzGerald, S.A. (2016) Vulnerable geographies: Human trafficking, immigration and border control in the UK and beyond, *Gender, Place, and Culture*. 23, 2: 181–97.

Fleischmann, F., and Dronkers, J. (2010) Unemployment among immigrants in European labour markets: An analysis of origin and destination effects, *Work, Employment and Society*. 24, 2: 337–54.

Foner, N. (2000) *From Ellis Island to JFK: New York's Two Great Waves of Immigration*. New Haven: Yale University Press.

Foner, N. (2001) Transnationalism then and now: New York immigrants today and at the turn of the twentieth century, in Cordero-Guzmán, H.R., Smith, R.C., and Grosfoguel, R. (eds) *Migration, Transnationalisation, and Race in a Changing New York*. Philadelphia: Temple University Press.

Foner, N., and Alba, R. (2008) Immigrant religion in the US and Western Europe: Bridge or barrier to inclusion? *International Migration Review*. 42, 2: 360–92.

Forcese, C. (2006) The capacity to protect: Diplomatic protection of dual nationals in the 'war on terror', *European Journal of International Law*. 17, 2: 369–94.

Ford, M. (2006) After Nunukan: The regulation of Indonesian migration to Malaysia, in Kaur, A., and Metcalfe, I. (eds) *Mobility, Labour Migration and Border Controls in Asia*. Springer, online.

Ford, M., and Kawashima, K. (2013) Temporary labour migration and care work: The Japanese experience, *The Journal of Industrial Relations*. 55, 3: 430–44.

Fossland, T. (2013) Crossing borders – getting work: Skilled migrants' gendered labour market participation in Norway, *Norsk Geografisk Tidsskrift–Norwegian Journal of Geography*. 67, 5: 276–83.

Foucault, M. (1977) *Discipline and Punish*. Harmondsworth: Penguin.

Foucault, M. (2007) *Security, Territory, Population: Lectures at the Collège de France, 1977–78*. New York: Palgrave Macmillan.

Foucault, M. (2004 [2008]) *The Birth of Biopolitics Lectures at the Collège de France, 1978–79*. Basingstoke: Palgrave-Macmillan

Freeman, G. (1995) Modes of immigration politics in liberal democracies, *International Migration Review*. XXIX, 4: 881–902.

Fridolfsson, C., and Elander, I. (2013) Faith and place: Constructing Muslim identity in a secular Lutheran society, *Cultural Geographies*. 20, 3: 319–37.

Friedmann, J., and Wolff, G. (1982) World City Formation: An agenda for research and action, *International Journal of Urban and Regional Research*. 15, 1: 269–83.

Frey, W. (1998) *Emerging Demographic Balkanization: Toward One America or Two?* Ann Arbor: Population Studies Center.

Fröbel, F., Heinrichs, J., and Kreye, O. (1977) *The New International Division of Labour*. Cambridge: Cambridge University Press.

FRONTEX (2015a) EUROSUR, available at: http://frontex.europa.eu/intelligence/eurosur/

FRONTEX (2015b) Frontex launches Joint Operation Triton, available at: http://frontex.europa.eu/search-results/?q=Triton

Fudge, J., and Strauss, K. (eds) (2014) *Temporary Work, Agencies and Unfree Labour*. London: Routledge.

Fujiwara, L.H. (2005) Immigrant rights are human rights: The reframing of immigrant entitlement and welfare, *Social Problems*. 52, 1: 79–101.

Fuller, S. (1994) Making agency count: A brief foray into the foundations of social theory, *American Behavioral Scientist.* 37: 741–53.

Gabriel, C., (2010) A 'healthy' trade? NAFTA, labour mobility and Canadian nurses, in Gabriel, C. and Pellerin, H. (eds) *Governing International Labour Migration.* London: Routledge.

Gabriel, C. (2013) NAFTA, skilled migration, and continental nursing markets, *Population, Space and Place.* 19: 389–403.

Gabriel, C., and Pellerin, H. (eds) (2010) *Governing International Labour Migration.* London: Routledge.

Gaetano, A.M., and Yeoh, B.S.A. (2010) Introduction to the Special Issue on Women and Migration in Globalizing Asia: Gendered Experiences, Agency and Activism, *International Migration.* 48, 6: 1–12.

Gallie, D. (1991) Patterns of skill change – upskilling, deskilling or the polarization of skills, *Work Employment and Society.* 5, 3: 319–51.

Galvin, T.M. (2015) 'We deport them but they keep coming back': The normalcy of deportation in the daily life of 'undocumented' Zimbabwean migrant workers in Botswana, *Journal of Ethnic and Migration Studies.* 41, 4: 617–34.

Gamlen, A. (2008) The emigration state and the modern geopolitical imagination, *Political Geography.* 27, 8: 840–56.

Gamlen, A., and Délano, A. (eds) (2014) Special Issue: Comparing and theorizing state-diaspora relations, *Political Geography,* 41: 43–53.

GCIM (Global Commission on International Migration) (2005) *Migration in an Interconnected World: New Directions for Action.* Report of the Global Commission on International Migration.

Geddes, A. (2000a) Denying access: Asylum-seekers and welfare benefits in the UK, in Bommes, M., and Geddes, A. (eds) *Immigration and Welfare: Challenging the Borders of the Welfare State.* London: Routledge.

Geddes, A. (2000b) *Immigration and European Integration: Towards Fortress Europe?* Manchester: Manchester University Press.

Geddes, A. (2003) *The Politics of Migration and Immigration in Europe.* London: Sage.

Geddes, A. (2014) Supranational governance and the remaking of European migration policy and politics, in Hollifield, J., Martin, P.L., and Orrenius, P.M. (eds) *Controlling Immigration: A Global Perspective,* 3rd edition. Stanford: Stanford University Press.

Geddes, A. (2015) The state and the regulation of migration, in Talani, L.S., and McMahon, S. (eds) *Handbook of the International Political Economy of Migration.* Cheltenham and Northampton: Edward Elgar.

Geiger, M., and Pecoud, A. (eds) (2010) *The Politics of International Migration Management.* Basingstoke: Palgrave Macmillan.

Geiger, M., and Pecoud, A. (eds) (2013) *Disciplining the Transnational Mobility of People.* Basingstoke: Palgrave Macmillan.

Gemenne, F. (2011) Why the numbers don't add up: A review of estimates and predictions of people displaced by environmental changes, *Global Environmental Change.* 21: S41–S49.

Gemignani, M., and Hernandez-Albujar, Y. (2015) Hate groups targeting undocumented immigrants: Discourses, narratives and subjectivation practices on
their websites, *Ethnic and Racial Studies*. 38, 15: 2754–70.

Ghosh, B. (1998) *Huddled Masses and Uncertain Shores: Insights into Irregular
Migration*. The Hague: M Nijhoff.

Gibney, M. (2008) Asylum and the expansion of deportation in the United
Kingdom, *Government and Opposition*. 43, 2: 146–67.

Gibney, M., and Hansen, R. (2003) Deportation and the liberal state: The forcible
return of asylum-seekers and unlawful migrants in Canada, Germany, and
the United Kingdom. *New Issues in Refugee Research*, Working Paper. Geneva:
UNHCR EPAU.

Gibson-Graham, J.K. (1996) *The End of Capitalism (as we knew it)*. Oxford: Basil
Blackwell.

Gibson-Graham, J.K. (2002) Beyond global vs. local: Economic politics outside
the binary frame, in Herod, A., and Wright, M.W. (eds) *Geographies of Power:
Placing Scale*. Oxford: Blackwell Publishers.

Giddens, A. (1984) *The Constitution of Society*. Cambridge: Polity.

Gidwani, V., and Sivaramakrishnan, K. (2003) Circular migration and the spaces
of cultural assertion, *Annals of the Association of American Geographers*. 93, 1:
186–213.

Gill, N. (2009) Presentational state power: Temporal and spatial influences over
asylum sector decisionmakers, *Transactions of the Institute of British
Geographers*. 34: 215–33.

Gill, N. (2010) New state theoretic approaches to asylum and refugee geographies, *Progress in Human Geography*. 34, 5: 626–45.

Gill, N., and Bialski, P. (2010b) New friends in new places: Network formation
during the migration process among Poles in the UK, *Geoforum*. 42, 2: 241–9.

Gill, N., and Conlon, D. (2015) Editorial: Migration and activism, *Acme: An
International E-Journal for Critical Geographies*. 14, 2: 442–52.

Giordano, C. (2008) Practices of translation and the making of migrant subjectivities in contemporary Italy, *American Ethnologist*. 35, 4: 588–606.

Givens, T.E., Freeman, G.P., and Leal, D.L. (2009) (eds) *Immigration Policy and
Security*. New York and London: Routledge.

Glick-Schiller, N. (2003) The centrality of ethnography in the study of transnational migration, in Foner, N. (ed.) *American Arrivals: Anthropology Engages
the New Immigration*. Santa Fe: School of American Research Press.

Glick-Schiller, N., Basch, L., and Blanc-Szanton, C. (1992) (eds) *Towards a
Transnational Perspective on Migration: Race, Class, Ethnicity, and Nationalism
Reconsidered*. Volume. 465. New York: New York Academy of Sciences.

Glick-Schiller, N., and Caglar, A. (eds) (2010) *Locating Migration: Re-scaling Cities
and Migrants*. Ithaca: Cornell University Press.

Glick-Schiller, N., Caglar, A., and Guldbransen, T.C. (2006) Beyond the ethnic
lens: Locality, globality, and born-again incorporation, *American Ethnologist*.
33, 4: 612–33.

Global Commission on International Migration (2005) *Migration in an Interconnected World: New Directions for Action*. Report of the Global Commission on International Migration Commission.

Goicoechea, E.R. (2005) Immigrants contesting ethnic exclusion: Structures and practices of identity, *International Journal of Urban and Regional Research*. 29, 3: 654–69.

Goldin, I., Cameron, G., and Balarajan, M. (2011) *Exceptional People: How Migration Shaped Our World and Will Define Our Future*. Princeton and Oxford: Princeton University Press.

Goldring, L. (2001) Dis-aggregating transnational social spaces: Gender, place and citizenship in Mexico–U.S. transnational spaces, in Pries, L. (ed.) *New Transnational Social Spaces: International Migration and Transnational Companies in the Early Twenty-First Century*. London and New York: Routledge.

Gomberg-Muñoz, R. (2011) *Labor and Legality: An Ethnography of a Mexican Immigrant Network*. Oxford: Oxford University Press.

Goodwin-Gill, G. (2014) The International Law of Refugee Protection, in Fiddian-Qasmiyeh, E., Loescher, G., Long, K., and Sigona, N. (eds) *The Oxford Handbook of Refugee and Forced Migration Studies*. Oxford: Basil Blackwell.

Gordon, D., Edwards, R., and Reich, M. (1982) *Segmented Work, Divided Workers*. Cambridge: Cambridge University Press.

Gordon, J. (2005) *Suburban Sweatshops: The Fight for Immigrant Rights*. Cambridge: Belknap Press.

Gordon, M. (1964) *Assimilation in American Life*. New York: Oxford University Press.

Gordon, S. (2015) Xenophobia across the class divide: South African attitudes towards foreigners 2003–2012, *Contemporary African Studies*. 33, 4: 494–509.

Goss, J., and Lindquist, B. (1995) Conceptualising international labor migration, *International Migration Review*. 29, 2: 317–51.

Grabher, G. (2002) Cool projects, boring institutions: Temporary collaboration in social context, *Regional Studies*. 36, 3: 205–14.

Graham, S. (2002) Bridging Urban Digital Divides? Urban Polarisation and Information and Communications Technologies (ICTs), *Urban Studies*. 39: 33–56.

Granotier, B. (1970) *Les Travailleurs Immigrés en France*. Paris: Francois Maspero.

Granovetter, M. (1973) The strength of weak ties, *American Journal of Sociology*. 78, 6: 1360–80.

Grant, R.G., and Thompson, D. (2015) City on edge: Immigrant business and the right to urban space in inner-city Johannesburg, *Urban Geography*. 36, 2: 181–200.

Gray, C.L. (2009) Rural out-migration and small holder agriculture in the southern Ecuadorian Andes, *Population and Environment*. 30, 4–5: 193–217.

Green, S. (2001) Immigration, asylum, and citizenship in Germany: The impact of unification and the Berlin Republic, *West European Politics*. 24, 4: 82–104.

Green, S. (2012) Much ado about not-very-much? Assessing ten years of German citizenship reform, *Citizenship Studies*. 16, 2: 173–88.

Greenhill, K. (2010) *Weapons of Mass Migration: Forced Displacement, Coercion, and Foreign Policy*. Ithaca: Cornell University Press.

Grillo, R., and Mazzucato, V. (2008) Africa-Europe: A double engagement, *Journal of Ethnic and Migration Studies*. 34, 2: 75–198.

Grosfoguel, R., Oso, L., and Christou, A. (2015) 'Racism', intersectionality and migration studies: framing some theoretical reflections, *Identities-Global Studies in Culture and Power*. 22, 6: 635–52.

Grugel, J., and Riggirozzi, P. (2012) Post-neoliberalism in Latin America: Rebuilding and reclaiming the state after crisis, *Development and Change*. 46, 4: 585–1022.

Grzymala-Kazlowska, A. (2005) From ethnic cooperation to in-group competition: Undocumented Polish workers in Brussels, *Journal of Ethnic and Migration Studies*. 31, 4: 675–97.

Gu, M. (2015) A complex interplay between religion, gender and marginalization: Pakistani schoolgirls in Hong Kong, *Ethnic and Racial Studies*. 38, 11: 1934–51.

Guardian (2003) 'Rotterdam plans to ban poor immigrants from moving in', 2 December.

Guardian (2014) 'Lampedusa boat tragedy: a survivor's story', 22 March.

Guardian (2015a) 'Which EU countries had the most asylum seekers?', 11 May.

Guardian (2015b) 'Lifejackets going cheap: people smugglers of Izmir, Turkey, predict drop in business', 24 September.

Guardian (2015c) 'Hungary closes border to refugees as Turkey questions EU deal to stem crisis', 17 October.

Guild, E. (2009) *Security and Migration in the 21st Century*. London: Polity Press.

Guiraudon, V. (2000) *Les Politiques d'Immigration en Europe: Allemagne, France, Pays-Bas*. Paris: L'Harmattan.

Guiraudon, V. (2003) The constitution of a European immigration policy domain: A political sociology approach, *Journal of European Public Policy*. 10, 2: 263–82.

Guiraudon, V., and Lahav, G. (2000) The state sovereignty debate revisited: The case of migration control, *Comparative Political Studies*. 33, 2: 163–95.

Gurak, D.T., and Caces, F. (1992) Migration networks and the shaping of migration system, in Kritz, M., Lim, L., and Zlotnik, H. (eds) *International Migration Systems: A Global Approach*. Oxford: Clarendon Press.

Hadley, C., Galea, S., Nandi, V., Nandi, A., Lopez, G., and Strongarone, S. (2008) Hunger and health among undocumented Mexican migrants in a US urban area, *Public Health Nutrition*. 11, 2: 151–8.

Hagan, J., Eschbach, K., and Rodriguez, N. (2008) US deportation policy, family separation, and circular migration, *International Migration Review*. 42, 1: 64–88.

Halfacree, K. (1995) Household migration and the structuration of patriarchy: Evidence from the U.S.A., *Progress in Human Geography*. 19: 159–82.

Halfacree, K., and Boyle, M. (1993) The challenge facing migration research: The case for a biographical approach, *Progress in Human Geography*. 17: 333–48.

Hall, S. (2000) Conclusion: The multicultural question, in Hesse, B. (ed.) *Unsettled Multiculturalisms: Diasporas, Entanglements, Transruptions*. London: Zed Books

Hamilton, L.C., Colocousis, C.R., and Johansen, S.T. (2004) Migration from resource depletion: The case of the Faroe Islands, *Society and Natural Resources*. 17: 443–53.

Hammar, A. (2014) *Displacement Economies in Africa: Paradoxes of Crisis and Creativity (Africa Now)*. London: Zed Books.

Hammar, T. (1990) *Democracy and the Nation State: Aliens, Denizens, and Citizens in a World of International Migration*. Aldershot: Avebury.

Hammar, T., Brochman, G., Tamas, K., and Faist, T. (1997) *International Migration, Immobility and Development: Multidisciplinary Perspectives*. Oxford: Berg Press.

Hammermesh, D., and Bean, F. (1998) *Help or hindrance? The Economic Implications of Immigration for African-Americans*. New York: The Russell Sage Foundation.

Hampshire, J. (2009) Disembedding Liberalism? Immigration politics and security in Britain since 9/11, in Givens, T.E, Freeman, G.P, and Leal, D.L. (eds) *Immigration Policy and Security*. New York and London: Routledge.

Hampshire, J. (2013) *The Politics of Immigration*. Cambridge: Polity Press.

Hannan, C.-A. (2015) Illegalized migrants, in Bauder, H., and Shields, J. (eds) *Immigrant Experiences in North America: Understanding Settlement and Integration*. Toronto: Canadian Scholars Press.

Hansen, R. (2002) The dog that didn't bark: Dual nationality in the United Kingdom, in Hansen, R., and Weil, P. (eds) *Dual Nationality, Social Rights and Federal Citizenship in the U.S. and Europe*, New York and Oxford: Berghahn Books.

Hansen, R., and Weil, P. (eds) (2002a) *Dual Nationality, Social Rights and Federal Citizenship in the U.S. and Europe*, New York and Oxford: Berghahn Books.

Hansen, R., and Weil, P. (2002b) Dual citizenship in a changed world: Immigration, gender and social rights, in Hansen, R., and Weil, P. (eds) *Dual Nationality, Social Rights and Federal Citizenship in the U.S. and Europe*, New York and Oxford: Berghahn Books.

Hanson, S., and Pratt, G. (1991) Time, space, and the occupational segregation of women – a critique of human-capital theory, *Geoforum*. 22, 2: 149–57.

Hardwick, S. (2008) Place, space, and pattern: Geographical theories in international migration, in Brettell, C.B., and Hollifield, J.F. (eds) *Migration Theory: Talking Across Disciplines*, 2nd edition. New York: Routledge.

Hardwick, S. (2015) Coming of age: Migration theory in geography, in Brettell, C.B., and Hollifield, J.F. (eds) *Migration Theory: Talking Across Disciplines*, 3rd edition. New York: Routledge.

Harney, N.D., and Baldassar, L. (2007) Tracking transnationalism: Migrancy and its futures, *Journal of Ethnic and Migration Studies*. 33, 2: 189–98.

Harris, J.R., and Todaro, M.P. (1970) Migration, unemployment and development: A two-sector analysis, *The American Economic Review*. 60, 1: 126–42.

Harriss-White, B. (2003) Inequality at work in the informal economy: Key issues and illustrations, *International Labour Review*. 142, 4: 459–69.

Hart, K. (1973) Informal income opportunities and urban employment in Ghana, *Journal of Modern African Studies*. 11, 1: 61–89.

Harvey, D. (1982 [1989 reprint edition]) *Limits to Capital*. Chicago: Chicago University Press.

Harvey, D. (1996) *Justice, Nature and the Geography of Difference*. Oxford: Basil Blackwell.

Harvey, D. (2000) *Spaces of Hope*. Edinburgh: Edinburgh University Press.

Harvey, D. (2005) *A Brief History of Neoliberalism*. Oxford: Oxford University Press.

Hathaway, J.C. (2007) Forced migration studies: Could we agree just to 'date'? *Journal of Refugee Studies*. 20, 3: 349–69.

Hatziprokopiou, P., and Evergeti, V. (2014) Negotiating Muslim identity and diversity in Greek urban spaces, *Social and Cultural Geography*. 15, 6: 603–26.

Hayter, T. (2004) *Open Borders: The Case Against Immigration Controls*, 2nd edition. London: Pluto Press.

Hazán, M. (2014) The uneasy transition from labor exporter to labor importer and the new emigration challenge, in Hollifield, J., Martin, P.L., and Orrenius, P.M. (eds) *Controlling Immigration: A Global Perspective*, 3rd edition. Stanford: Stanford University Press.

Hazans, M. (2011) Informal workers across Europe: Evidence from 30 countries, IZA Discussion paper, July.

Hazen, H.D., and Alberts, H.C. (2006) Visitors or immigrants? International students in the United States, *Population, Space and Place*. 12: 201–16.

Hedberg, C., Hermelin, B., and Westermark, K. (2014) Transnational spaces 'from above' – The role of institutions in promoting highly skilled labour migration from India to Sweden, *Tijdschrift voor Economische en Sociale Geografie*. 105, 5: 511–25.

Hedman, E.-L.E. (2008) Refuge, governmentality and citizenship: Capturing 'illegal migrants' in Malaysia and Thailand, *Government and Opposition*. 43, 2: 358–83.

Held, D., McGrew, A., Goldblatt, D., and Perraton, J. (1999) *Global Transformations: Politics, Economics and Culture*. London: Polity Press.

Heller, C., and Pezzani, L. (2016) Ebbing and Flowing: The EU's Shifting Practices of (Non-) Assistance and Bordering in a Time of Crisis. Available at: http://nearfuturesonline.org/ebbing-and-flowing-the-eus-shifting-practices-of-non-assistance-and-bordering-in-a-time-of-crisis/

Hero, R.E., and Preuhs, R.R. (2007) Immigration and the evolving American welfare state: Examining policies in the US states, *American Journal of Political Science*. 51, 3: 498–517.

Herod, A. (2010) *Scale* (Key Ideas in Geography Series). London: Routledge.

Heyman, J. (ed.) (1999) *States and Illegal Practices*. Oxford: Berg Publishers.

Hickey, M. (2015) Modernisation, migration and mobilization: Relinking internal and international migrations in the 'migration and development nexus', *Population, Space and Place*. DOI: 10.1002/psp.1952

Hiebert, D. (2002) The spatial limits to entrepreneurship: Immigrant entrepreneurs in Canada, *Tijdschrift voor Economische en Sociale Geographie*. 93, 2: 173–90.

Hiebert, D., and Ley, D. (2003) Assimilation, cultural pluralism, and social exclusion among ethno-cultural groups in Vancouver, *Urban Geography*. 24, 1: 16–44.

Hiemstra, N. (2012) Geopolitical reverberations of US migrant detention and deportation: The view from Ecuador, *Geopolitics*. 17, 2: 293–311.

Higham (1955) *Strangers in the Land: Patterns of American Nativism, 1860–1920*. New Brunswick: Rutgers University Press.

Hilsdon, A-M. (2006) Migration and human rights: The case of Filipino Muslim women in Sabah, Malaysia, *Women's Studies International Forum*. 29, 4: 405–16.

Hirschman, C. (2004) The role of religion in the origins and adaptation of immigrant groups in the United States, *International Migration Review*. 38, 3: 1206–33.

Hirst, P., and Thompson, G. (1996) *Globalization in Question*. Oxford: Polity Press.

Hjarnø, J. (2003) *Illegal Immigrants and Developments in Employment in the Labour Markets of the EU*. London: Ashgate Publishing.

Ho, C., and Alcorsco, C. (2004) Migrants and employment – challenging the success story, *Journal of Sociology*. 40, 3: 237–59.

Hochschild, A. R. (1983) *The Managed Heart: Commercialization of Human Feeling*. Berkeley, CA: University of California Press.

Hochschild, A.R. (2000) Global care chains and emotional surplus value, in Hutton, W., and Giddens, A. (eds) *On The Edge: Living with Global Capitalism*. London: Jonathan Cape.

Hofmann, E. (2015) Choosing your country: Networks, perceptions and destination selection among Georgian labour migrants, *Journal of Ethnic and Migration Studies*. 41, 5: 813–34.

Holgate, J. (2004) Organizing migrant workers: A case study of working condition and unionization in a London sandwich factory, *Work, Employment and Society*. 19, 3: 463–80.

Hollifield, J. (1992) *Immigrants, Markets, and States: The Political Economy of Postwar Europe*. Cambridge: Harvard University Press.

Hollifield, J. (2000) Immigration and the politics of rights: The French case in comparative perspective, in Bommes, M., and Geddes, A. (eds) *Immigration and Welfare: Challenging the Borders of the Welfare State*. London: Routledge.

Hollifield, J. (2004) The emerging migration state, *International Migration Review*. 38, 3: 885–912.

Hollifield, J., Martin, P.L., and Orrenius, P.M. (2014 (eds) *Controlling Immigration: A Global Perspective*, 3rd edition. Stanford: Stanford University Press.

Holston, J., and Appadurai, A. (1999) Cities and citizenship, in Holston, J. (ed.) *Cities and Citizenship*. Durham: Duke University Press.

Home Office (2005) *Controlling our Borders: Making Migration Work for Britain. Five-year Strategy for Asylum and Immigration.* Norwich: HMSO.

Hondagneu-Sotelo, P. (1994) *Gendered Transitions: Mexican Experiences of Immigration.* Berkeley: University of California Press.

Hondagneu-Sotelo, P. (2002) Families on the frontier: From braceros in the fields to braceras in the home, in Suárez-Orozco, M.M., and Páez, M.M. (eds) *Latinos: Remaking America.* Berkeley: University of California Press.

Hondagneu-Sotelo, P. (2007) *Religion and Social Justice for Immigrants.* New Brunswick: Rutgers University Press.

Howell, S., and Shryock, A. (2003) Cracking down on diaspora: Arab Detroit and America's 'War on Terror', *Anthropological Quarterly.* 76: 443–62.

Hu, R. (2015) Competitiveness, migration, and mobility in the global city: Insights from Sydney, Australia, *Economies.* 3, 1: 37–54.

Hubbard, P. (2005a) Inappropriate and incongruous: Proposition to asylum centres in the English countryside, *Journal of Rural Studies.* 21, 1: 3–17.

Hubbard, P. (2005b) Accommodating otherness: Anti-asylum centre protest and the maintenance of white privilege, *Transactions of the Institute of British Geographers.* 30, 1: 52–65.

Hughes, D.M. (1999) Refugees and squatters: Immigration and the politics of territory on the Zimbabwe-Mozambique border, *Journal of Southern African Studies.* 25, 4: 533–52.

Hugo, G. (1996) Environmental concerns and international migration, *International Migration Review.* 30, 1: 105–31.

Hugo, G. (2006) Immigration responses to global change in Asia: A review, *Geographical Research.* 44, 2: 155–72.

Hugo, G. (2014) Change and continuity in Australian international migration policy, *International Migration Review.* 48, 3: 868–90, Fall 2014.

Hunter, L.M., Luna, J.K., and Norton, R.M. (2015) Environmental dimensions of migration, *Annual Reviews of Sociology.* 41: 377–97.

Huysmans, J. (2000) The European Union and the securitization of migration, *Journal of Common Market Studies.* 38, 5: 751–77.

Hyndman, J. (2000) *Managing Displacement.* Minneapolis: University of Minnesota Press.

Hyndman, J. (2003) Aid, conflict and migration: The Canada-Sri Lanka connection, *The Canadian Geographer.* 47, 3: 251–68.

Hyndman, J. (2012) The geopolitics of migration and mobility, *Geopolitics.* 17: 243–55.

Hyndman, J., and Mountz, A. (2008) Another brick in the wall? Neo-*refoulement* and the externalization of asylum by Australia and Europe, *Government and Opposition.* 43, 2: 249–69.

Ignatiev, N. (1995) *How the Irish Became White.* London: Routledge

IIE (International Institute of Education)/Open Doors (2015a) 'Project Atlas Fast Facts'.

IIE (International Institute of Education)/Open Doors (2015b) 'Fast Facts'.

Indra, D. (1999) *Engendering Forced Migration.* New York and Oxford: Berghahn Books.

IOM (International Organization for Migration) (2003) *World Migration 2003*. Geneva: IOM.

IOM (2005) *The Millennium Development Goals and Migration*, Migration Research Series 20. Geneva: IOM.

IOM (2008a) *World Migration 2008*. Geneva: IOM

IOM (2008b) IOM's Activities on Migration Data: An Overview, Geneva: International Organization for Migration. Report available at: http://www.iom.int.

IOM (2010) *Labour migration from Indonesia*. Geneva: IOM.

IOM (2014) Fatal journeys: Tracking lives lost during migration. Geneva: IOM, Report available at: http://publications.iom.int/bookstore/free/FatalJourneys_CountingtheUncounted.pdf

IOM (2015) World Migration Report 2015. Migrants and Cities. New Partnerships to Manage Mobility Geneva: IOM.

IPPR (Institute for Public Policy Research) (2006) Irregular migration in the UK, an IPPR FactFile. London: IPPR.

Iredale, R.R. (2005) Gender, immigration policies and accreditation: Valuing the skills of professional women migrants, *Geoforum*. 36, 2: 155–66.

Iredale, R.R., Voigt-Graf, C., and Khoo, S.-E. (2015) Trends in international and internal teacher mobility in three Pacific island countries, *International Migration*. 53, 1: 97–114.

Isin, E.F. (ed.) (2000) *Democracy, Citizenship and the Global City*. London and New York: Routledge.

Isin, E.F (2012) Citizens without nations, *Environment and Planning D: Society and Space*. 30, 3: 450–67.

Isin, E.F. (2014) Acts, in Anderson, B., and Keith, M. (eds) *Migration: A COMPAS Anthology*. Oxford: COMPAS

Isin, E.F., and Wood, P.K. (1999) *Citizenship and Identity*. London: Sage.

Iskander, N. (2000) Immigrant workers in an irregular situation: The case of the garment industry in Paris and its suburbs, in OECD (ed.) *Combating the Illegal Employment of Foreign Workers*. Paris: OECD.

Iskander, N., Riordan, C., and Lowe, N. (2013) Learning in Place: Immigrants' spatial and temporal strategies for occupational advancement, *Economic Geography*. 89, 1: 53–75.

Iyer, D. (2015) *We Too Sing America: South Asian, Arab, Muslim, and Sikh Immigrants Shape Our Multiracial America*. New York and London: The New Press.

Jacobsen, K. (2006) Editorial introduction: Refugees and asylum seekers in urban areas: A livelihoods approach, *Journal of Refugee Studies*. 19, 3: 273–86.

Jacobsen, K., and Landau, L.B. (2003) The dual imperative in refugee research: Some methodological and ethical considerations in social science research on forced migration, *Disasters*. 27, 3: 185–206.

Jacobson, D. (1996) *Rights Across Borders: Immigration and the Decline of Citizenship*. Baltimore: The Johns Hopkins Press.

Jennissen, R. (2007) Causality chains in the international migration systems approach, *Population Research Policy Review*. 26: 411–36.

Jessop, B. (1997) Capitalism and its future: Remarks on regulation, government and governance, *Review of International Political Economy*. 4, 3: 561–81.

Johannsson, H., and Shulman, S. (2003) Immigration and the employment of African American workers, *The Review of Black Political Economy*. 31, 1–2: 95–110.

Johnson, C., Reece, J., Paasi, A., Amoore, L., Mountz, A., Salter, M., and Rumford, C. (2011) Interventions on rethinking 'the border' in border studies, *Political Geography*. 30: 61–9.

Johnston, D. (2007) Who needs immigrant farm workers? A South African case study, *Journal of Agrarian Change*. 7, 4: 494–525.

Jones, G.W. (2012a) International marriage migration in Asia: What do we know and what we need to know, Working Paper, Asia Research Institute Working Paper Series No. 174.

Jones, G.W. (2012b) Marriage migration in Asia: An introduction, *Asian and Pacific Migration Journal*. 21, 3: 287–90.

Jones, M. (2009) Phase space, geography, relational thinking, and beyond, *Progress in Human Geography*. 33, 4: 487–506.

Jones, M.A. (1960) *American Immigration*. Chicago: University of Chicago Press.

Jones, S.H. (2015) The 'metropolis of dissent': Muslim participation in Leicester and the 'failure' of multiculturalism in Britain, *Ethnic and Racial Studies*. 38, 11: 1969–85.

Jones-Correa, M. (2012) Contested Ground: Immigration in the United States, Migration Policy Institute Report. Available at: http://www.migrationpolicy.org/research/TCM-US-immigration-national-identity

Jónsson, G. (2008) Migration aspirations and immobility in a Malian Soninké village. IMI Working Paper, 10. Oxford: International Migration Institute, University of Oxford.

Joppke, C. (1998a) Immigration challenges the Nation-State, in Joppke, C. (ed.) *Challenge to the Nation-State*. Oxford: Oxford University Press.

Joppke, C. (ed.) (1998b) *Challenge to the Nation-State*. Oxford: Oxford University Press.

Joppke, C. (2007a) Beyond national models: Civic integration policies for immigrants in Western Europe, *Western European Politics*. 30, 1: 1–22.

Joppke, C. (2007b) Transformation of citizenship: Status, rights, identity, *Citizenship Studies*. 11, 1: 37–48.

Jordan, B., and Duvell, F. (2003) *Irregular Migration: Dilemmas of Transnational Mobility*. Cheltenham: Edward Elgar.

Jouin, N. (2006) Les travailleurs immigrés du bâtiment entre discrimination et précarité. L'exemple d'une activité externalisée: le feraillage, *Revue de l'IRES*. 50, 1: 3–25.

Journal of Ethnic and Migration Studies (2008) Diasporic tensions: The dilemmas and conflicts of transnational engagement. Special Issue. 34, 7.

Kaag, M. (2008) Mouride transnational livelihoods at the margins of a European society: The case of residence Prealpino, Brescia, Italy, *Journal of Ethnic and Migration Studies*. 34, 2: 271–85.

Kalleberg, A.L. (2009) Precarious work, insecure workers: Employment relations in transition, *American Sociological Review*. 74, 1: 1–22.

Kanas, A., Chiswick, B.R, van der Lippe, T., and van Tubergen, F. (2012) Social contacts and the economic performance of immigrants: A panel study of immigrants in Germany, *International Migration Review*. 46, 3: 680–709.

Kanas, A., and van Tubergen, F. (2011) The role of social contacts in the employment status of immigrants: A panel study of immigrants in Germany, *International Sociology*. 26, 1: 95–122.

Kaplan, D.H. (1998) The spatial structure of urban ethnic economies, *Urban Geography*. 19, 6: 489–501.

Kapur, D. (2004) Remittances: The new development mantra? UNCTD Discussion Paper G-24.

Kastoryano, R. (2002) *Negotiating Identities: States and Immigrants in France and Germany*. Princeton: Princeton University Press.

Kearney, M. (1986) From the invisible hand to visible feet: Anthropological studies of migration and development, *Annual Review of Anthropology*. 15: 331–61.

Keaton, T.D. (2006) *Muslim Girls and the Other France: Race, Identity Politics and Social Exclusion*. Bloomington and Indianapolis: Indiana University Press.

Keith, M. (1993) From punishment to discipline? Racism, racialization and the policing of social control, in Keith, M., and Cross, M. (eds) *Racism, the City and the State*. London: Routledge.

Kelly, P. (2010) Filipino migration and labor subordination, in McGrath-Champ, S., Herod, A., and Rainnie, A. (eds) *Handbook of Employment and Society*. Cheltenham and Northampton: Edward Elgar.

Kelson, G., and De Laet, D. (1999) *Gender and Immigration*. New York: New York University Press.

Kempadoo, K. (2007) The war on human trafficking in the Caribbean, *Race and Class*. 49: 79–85.

Kepel, G. (1997) *Allah in the West: Islamic Movements in America and Europe*. Cambridge: Polity Press.

Kim, C.S. (2008) Features of international marriage of Korean men to women from four Asian countries, in Doo, S.K. (ed.) *Cross Border Marriage: Process and Dynamics*. Seoul: Institute of Population and Aging Research.

Kim, D.Y. (1999) Beyond co-ethnic solidarity: Mexican and Ecuadorian employment in Korean-owned businesses in New York City, *Ethnic and Racial Studies*. 22, 3: 581–605.

Kim, N.H.J. (2008) Korean immigration policy changes and the political liberals' dilemma, *International Migration Review*. 42, 3: 576–96.

Kim, N.H.J. (2015) The retreat of multiculturalism? Explaining the South Korean exception, *American Behavioral Scientist*. 59, 6: 727–46.

King, R. (ed.) (1993) *New Geography of European Migrations*. London: John Wiley.

King, R. (2012) Geography and migration studies: Retrospect and prospect, *Population, Space, and Place*. 18: 134–53.

King, R., Connell, J., and White, P. (1995) *Writing Across Worlds: Literature and Migration*. London: Routledge.

King, R., Dalipaj, M., and Mai, N. (2006) Gendering migration and remittances: Evidence from London and Northern Albania, *Population, Space and Place.* 12: 409–34.

King, R., Mata-Codesal, D., and Vullnetari, J. (2013) Migration, development, gender and the 'black box' of remittances: Comparative findings from Albania and Ecuador, *Comparative Migration Studies.* 1, 1: 69–96.

King, R., and Raghuram, P. (2013) International student migration: Mapping the field and new research agendas, *Population Space and Place.* 19, 2: 127–37.

King, R., and Skeldon, R. (2010) 'Mind the gap!' Integrating approaches to internal and international migration, *Journal of Ethnic and Migration Studies.* 36: 1619–46.

Kivisto, P. (2005) *Incorporating Diversity: Re-thinking Assimilation in a Multicultural Age.* Boulder: Paradigm.

Kloosterman, R., and Rath, J. (2003) *Immigrant Entrepreneurs: Venturing Abroad in the Age of Globalization,* Oxford: Berg Press.

Kloosterman, R., and Rath, J. (2015) Immigrant Entrepreneurship, in Martiniello, M., and Rath, J. (eds) *An Introduction to Immigrant Incorporation Studies: European Perspectives.* Amsterdam: University of Amsterdam Press.

Kloosterman, R., Van der Leun, J., and Rath, J. (1999) Mixed embeddedness: (In) formal economic activity and immigrant businesses in the Netherlands, *International Journal of Urban and Regional Research.* 23: 253–67.

Klotz, A. (2000) Migration after Apartheid: Deracializing South African foreign policy, *Third World Quarterly.* 21, 5: 831–47.

Klusmeyer, D. (2001) A 'guiding culture' for immigrants? Integration and diversity in Germany, *Journal of Ethnic and Migration Studies.* 26, 3: 519–32.

Knox, P.L., and Marston, S.A. (2007) *Human Geography: Places and Regions in a Global Context,* 4th edition. Upper Saddle River: Pearson Prentice Hall.

Kobayashi, A., and Ley, D. (2005) Back to Hong Kong: Return migration or transnational sojourn? *Global Networks.* 2: 111–127

Kobayashi, A., and Preston, V. (2007) Transnationalism through the life course: Hong Kong immigrants in Canada, *Asia Pacific Viewpoint.* 48, 2: 151–67.

Kobelinsky, C. (2010) *L'accueil des Demandeurs d'Asile: Une Ethnographie de l'Attente.* Paris: Éditions du Cygne.

Kofman, E. (1999) 'Birds of Passage' a decade later: Gender and immigration in the European Union, *International Migration Review.* 33, 2: 269–99.

Kofman, E. (2002) Contemporary European migrations, civic stratification and citizenship, *Political Geography.* 21: 1035–54.

Kofman, E. (2004) Family-related migration: A critical review of European Studies, *Journal of Ethnic and Migration Studies.* 30, 2: 243–62.

Kofman, E. (2005a) Gender and skilled migrants: Into and beyond the work place, *Geoforum.* 36: 149–54.

Kofman, E. (2005b) Citizenship, migration and the reassertion of national identity, *Citizenship Studies.* 9, 5: 453–67.

Kofman, E. (2010) Managing migration and citizenship in Europe: Towards an overarching framework, in Gabriel, C., and Pellerin, H. (eds) *Governing International Labour Migration*. London: Routledge.

Kofman, E. (2012) Rethinking care through social reproduction, *Social Politics*. 19, 1: 142–62.

Kofman, E. (2013) Gendered labour migrations in Europe and emblematic migratory figures, *Journal of Ethnic and Migration Studies*. 39, 4: 579–600.

Kofman, E. (2014) Towards a gendered evaluation of (highly) skilled immigration policies in Europe, *International Migration*. 52, 3: 116–28.

Kofman, E., and Raghuram, P. (2006) Gender and global labour migrations: Incorporating skilled workers, *Antipode*. 38, 2: 282–303.

Kofman, E., and Raghuram, P. (2012) Women, migration, and care: Explorations of diversity and dynamism in the global south, *Social Politics*. 19, 3: 408–32.

Koh, H.H. (1997) How is international human rights law enforced? *Indiana Law Journal*. 74, 4: 1397–417.

Kogan, I. (2004) Last hired, first fired? The unemployment dynamics of male immigrants in Germany, *European Sociological Review*. 20, 5: 445–61.

Kogan, I. (2007) *Working through Barriers: Host Country Institutions and Immigrant Labour Market Performance in Europe*. Dordrecht: Springer.

Kong, L. (2010) Global shifts, theoretical shifts: Changing geographies of religion, *Progress in Human Geography*. 34, 6: 755–76.

Koopmans, R. (2004) Migrant mobilization and political opportunities: Variation among German cities and a comparison with the United Kingdom and the Netherlands, *Journal of Ethnic and Migration Studies*. 30, 3: 449–70.

Koopmans, R. (2010) Trade-offs between equality and difference: Immigrant integration, multiculturalism and the welfare state in cross-national perspective, *Journal of Ethnic and Migration Studies*. 36, 1: 1–26.

Koopmans, R., Michalowski, I., and Waibel, S. (2012) Citizenship rights for immigrants: National political processes and cross-national convergence in Western Europe, 1980-2008, *American Journal of Sociology*. 117, 4: 1202–45.

Köppe, O. (2003) The leviathan of competitiveness: How and why do liberal states (not) accept unwanted immigration? *Journal of Ethnic and Migration Studies*. 29, 3: 431–48.

Korteweg, A.C., and Triadafilopoulos, T. (2014) Is multiculturalism dead? Groups, governments and the 'real work of integration', *Ethnic and Racial Studies*. 38, 5: 663–80.

Koser, K. (2005) Migration and refugees, in Cloke, P., Crang, M., and Goodwin, M. (eds) *Introducing Human Geographies*, 2nd edition. London: Hodder Arnold.

Koser, K. (2007) *International Migration: A Very Short Introduction*. Oxford: Oxford University Press.

Koslowski, R. (2008) Global mobility and the quest for an international migration regime, in Chamie, J. and Dall'Oglio, L. (eds) *International Migration and Development - Continuing the Dialogue: Legal and Policy Perspectives*. Geneva: International Organization for Migration.

Koslowski, R. (2014) Selective migration policy models and changing realities of implementation, *International Migration*. Special Issue, 52, 3: 26–39.

Kostakopoulou, D. (2002) Long-term resident third country nationals in the European Union: Normative expectations and institutional openings, *Journal of Ethnic and Migration Studies*. 28, 3: 443–62.

Koubi, V., Spilker, G., Schafer, L., and Bernauer, T. (2016) Environmental stressors and migration: Evidence from Vietnam, *World Development*. 79: 197–210.

Koulish, R. (2015) Spiderman's web and the governmentality of electronic immigrant detention, *Law Culture and the Humanities*. 11, 1: 83–108.

Kraler, A. (2006) The legal status of immigrants and their access to nationality, in Bauböck, R. (ed.) *Migration and Citizenship: Legal Status, Rights and Political Participation*. Amsterdam: Amsterdam University Press.

Krissman, F. (2005) Sin Coyote Ni Patron: Why the 'migrant network' fails to explain international migration, *International Migration Review*. 39, 1: 4–44.

Kritz, M.M., Keely, C.B., and Tomasi, S.M. (eds) (1981) *Global Trends in Migration*. Staten Island: Centre for Migration Studies.

Kritz, M.M., Lim, L., and Zlotnik, H. (eds) (1992) *International Migration Systems: A Global Approach*. Oxford: Clarendon Press.

Kuusisto-Arponen, A.K., and Gilmartin, M. (2015) The politics of migration, *Political Geography*. Virtual Issue, 48: 143–45.

Kyle, D., and Dale, J. (2001) Smuggling the state back in: Agents of human smuggling reconsidered, in Kyle, D., and Koslowski, R. (eds) *Global Human Smuggling*, Baltimore: Johns Hopkins University Press.

Kyle, D., and Koslowski, R. (2001) *Global Human Smuggling: Comparative Perspectives*. Baltimore: Johns Hopkins University Press.

Kyle, D., and Koslowski, R. (2011) *Global Human Smuggling: Comparative Perspectives*, 2nd edition. Baltimore and London: Johns Hopkins University Press.

Kymlicka, W. (1995) *Multicultural Citizenship*. Oxford: Oxford University Press.

Lafleur, J.-M. (2013) *Transnational Politics and the State: The External Voting Rights of Diasporas*. New York: London and Routledge.

Lam, T., Yeoh, B., and Huang, S. (2006) Global householding in a city-state: Emerging trends in Singapore, *International Development Planning Review*. 28, 4: 475–97.

Lancee, B. (2010) The economic returns of Immigrants' bonding and bridging social capital. The case of the Netherlands, *International Migration Review*. 44, 1: 202–26.

Lankov, A. (2004) North Korean Refugees in Northeast China, *Asian Survey*. 44, 6: 856–73.

Larner, W. (2003) Neoliberalism? *Environment and Planning D: Society and Space*. 21, 5: 509–12.

Laurence, J., and Vaisse, J. (2006) *Integrating Islam: Political and Religious Challenges in Contemporary France*. Washington: Brookings Institution.

Lavenex, S. (2006a) Shifting up and out: The foreign policy of European immigration control, *West European Politics*. 29, 2: 329–50.

Lavenex, S. (2006b) Towards the constitutionalization of aliens' rights in the European Union? *Journal of European Public Policy*. 13, 8: 1284–301.

Lavenex, S. (2007) The Competition state and highly skilled migration, *Society*. 44, 2: 32–41.

Lavenex, S., and Schimmelfennig, F. (2009) EU rules beyond EU borders: Theorizing external governance in European Politics, *Journal of European Public Policy*. 16, 6: 791–812.

Lawson, V. (1999) Questions of migration and belonging: Understandings of migration under neo-liberalism in Ecuador, *International Journal of Population Geography*. 5: 261–76.

Lawson, V. (2000) Arguments within geographies of movement: The theoretical potential of migrants' stories, *Progress in Human Geography*. 24: 173–89.

Layton-Henry, Z. (2004) Britain: From immigration control to migration management, in Cornelius, W., Tsuda, T., Martin, P.L., and Hollifield, J. (eds) *Controlling Immigration: A Global Perspective*. Stanford: Stanford University Press.

Ledeneva, A.V. (1998) *Russia's Economy of Favors: Blat, Networks, and Informal Exchange*. Cambridge: Cambridge University Press.

Lee, D.J.,S and Turner, B.S. (1996) *Conflicts about Class*. Harlow: Longman Group.

Lee, E. (2008) Citizenship in Korea: From ethnic purity to multicultural identity, *Canadian diversity/Diversité Canadienne*. 6, 4: 82–5.

Lee, E.S. (1969) A theory of migration, in Jackson, J.A. (ed.) *Migration*. London: Cambridge

Lee, H-K. (2006) Migrant domestic workers in Korea, *International Development Planning Review*. 28, 4: 499–514.

Lee, J.E. (2015) Disciplinary citizenship in South Korean NGOs' narratives of resettlement for North Korean refugees, *Ethnic and Racial Studies*. 38, 15: 2688–704.

Lee, S-J. (2007) The governance of foreign workers in Korea and Japan, *Korea Observer*. 38, 4: 609–31.

Lefebvre, H. (1974 [1991]) *The Production of Space*. Oxford: Basil Blackwell.

Leitner, H., and Ehrkamp, P. (2006) Transnationalism and migrants' imaginings of citizenship, *Environment and Planning A*. 38: 1615–32.

Leitner, H., and Miller, B. (2007) Scale and the limitations of ontological debate: A commentary on Marston, Jones and Woodward, *Transactions of the Institute of British Geographers*. 32, 1: 116–25.

Leitner, H., Peck, J., and Sheppard, E.S. (eds) (2007) *Contesting Neoliberalism: Urban Frontiers*. New York: Guilford Press.

Leitner, H., Sheppard, E., and Sziarto, K.M. (2008) The spatialities of contentious politics, *Transactions of the Institute of British Geographers*. 33, 2: 157–72.

Leitner, H., and Strunk, C. (2014) Assembling insurgent citizenship: Immigrant advocacy struggles in the Washington DC metropolitan area, *Urban Geography*. 35, 7: 943–64.

Legoux, L. (1999) La politique d'asile, in Dewitte, P. (ed.) *Immigration et Intégration: l'Etat des saviors*. Paris: La Découverte.

Lessard-Phillips, L., Fibbi, R., and Wanner, P. (2013) Assessing the labour market position and its determinants for the second generation, in Crul, M., Schneider, J., and Lelie, F. (eds) *The European Second Generation Compared Does the Integration Context Matter?* Amsterdam: University of Amsterdam Press.

Levinson, A. (2002) Immigration and welfare, Migration Policy Institute, paper available at: http://www.migrationpolicy.org/article/immigrants-and-welfare-use

Levitt, P. (1998) Social remittances: Migration driven local-level forms of cultural diffusion, *International Migration Review*. 32, 4: 926–48.

Levitt, P. (2001) *The Transnational Villagers*. Berkeley: University of California Press.

Levitt, P. (2002) Variations in transnational belonging: Lessons from Brazil and the Dominican Republic, in Hansen, R., and Weil, P. (eds) *Dual Nationality, Social Rights and Federal Citizenship in the U.S. and Europe*. New York and Oxford: Berghahn Books.

Levitt, P. (2003) You Know, Abraham was really the first immigrant: Religion and transnational migration, *International Migration Review*. 37, 3: 847–73.

Levitt, P. (2008) Religion as a path to civic engagement, *Ethnic and Racial Studies*. 31, 4: 766–91.

Levitt, P., and Jaworksy, N. (2007) Transnational migration studies: Past developments and future trends, *Annual Review of Sociology*. 33: 129–56.

Lewis, A. (1954) Development with unlimited supplies of labour, *The Manchester School*. 22: 139–92.

Lewis, G., and Neal, S. (2005) Introduction: Contemporary political contexts, changing terrains and revisited discourses, *Ethnic and Racial Studies*. 28, 3: 423–44.

Ley, D. (2003) Seeking homo economicus: The Canadian state and the strange story of the business immigration program, *Annals of the Association of American Geographers*. 93, 2: 426–41.

Ley, D. (2004) Transnational spaces and everyday lives, *Transactions of the Institute of British Geographers*. 29: 151–64.

Ley, D. (2006) Explaining variations in business performance among immigrant entrepreneurs in Canada, *Journal of Ethnic and Migration Studies*. 32, 5: 743–64.

Ley, D. (2008) The immigrant church as an urban service hub, *Urban Studies*. 45, 10: 2057–74.

Li, M., and Bray, M. (2007) Cross-border flows of students for higher education: Push-pull factors and motivations of mainland Chinese students in Hong Kong and Macau, *Higher Education*. 53, 6: 791–818.

Li, P.S. (2003) *Destination Canada: Immigration debates and issues*. Toronto: Oxford University Press.

Li, W. (1998a) Anatomy of a new ethnic settlement: The Chinese ethnoburb in Los Angeles, *Urban Studies*. 35, 3: 479–501.

Li, W. (1998b) Los Angeles's Chinese ethnoburb: From ethnic service center to global economy outpost, *Urban Geography*. 19, 6: 502–17.

Light, I. (2004) Immigration and ethnic economies in giant cities, *International Social Science Journal*. 181: 385–98.

Light, I. (2005) The ethnic economy, in Smelser, N. and Swedberg, R. (eds) *Handbook of Economic Sociology*, 2nd edition. Princeton: Princeton University Press.

Light, I., Bernard, R.B., and Kim, R. (1999) Immigrant incorporation in the garment industry of Los Angeles, *International Migration Review*. XXXIII, 1: 5–25.

Light, I., Sabagh, G., Bozorgmehr, M., and Der-Martirosian, C. (1994) Beyond the ethnic enclave economy, *Social Problems*. 41, 1: 65–80.

Lillie, N., and Greer, I. (2007) Industrial relations, migration, and neoliberal politics: The case of the European construction sector, *Politics and Society*. 35: 551–81.

Lim, T. (2010) Rethinking belongingness in Korea: Transnational migration, 'migrant marriages' and the politics of multiculturalism, *Pacific Affairs*. 83, 1: 51–71.

Lin, W., and Yeoh, B.S.A. (2011) Questioning the 'field in motion': Emerging concepts, research practices and the geographical imagination in Asian migration studies. *Cultural Geographies*. 18, 1: 125–31.

Lindley, A. (2010) Leaving Mogadishu: Towards a sociology of conflict-related mobility, *Journal of Refugee Studies*. 23, 1: 2–22.

Lindquist, J., Xiang, B., and Yeoh, B.S.A. (2012) Introduction: Opening the black box of migration: Brokers, the organisation of transnational mobility, and the changing political economy in Asia. *Pacific Affairs*. 85, 1: 7–18.

Lister, M., and Pia, E. (2008) *Citizenship in Contemporary Europe*. Edinburgh: Edinburgh University Press.

Listerborn, C. (2015) Geographies of the veil: Violent encounters in urban public spaces in Malmo, Sweden, *Social and Cultural Geography*. 16, 1: 95–115.

Lobo, M. (2014) Everyday multiculturalism: Catching the bus in Darwin, Australia, *Social and Cultural Geography*. 15, 7: 714–29.

Loescher, G. (2014) UNHCR and forced migration, in Fiddian-Qasmiyeh, E., Loescher, G., Long, K., and Sigona, N. (eds) *The Oxford Handbook of Refugee and Forced Migration Studies*. Oxford: Basil Blackwell.

Loescher, G., and Scanlan, J. (1986) *Calculated Kindness*. New York: The Free Press.

Logan, J.R., Alba, R.D., Dill, M., and Zhou, M. (2000) Ethnic segmentation in the American metropolis: Increasing divergence in economic incorporation, 1980–1990, *International Migration Review*. 34, 1: 98–132.

Logan, J.R, Alba, R.D., and Stults, B.J. (2003) Enclaves and entrepreneurs: Assessing the payoff for immigrants and minorities, *International Migration Review*. 37, 2: 344–88.

Loizos, P. (2007) 'Generations' in forced migration: Towards greater clarity, *Journal of Refugee Studies*. 20, 2: 193–209.

Lu, M., and Yang, W.S. (2010) *Asian Cross-border Marriage Migration: Demographic Patterns and Social Issues*. Chicago: University of Chicago Press.

Lucassen, L.A.C.G. (2006) Is transnationalism compatible with assimilation? Examples from Western Europe since 1850, *IMIS-Beiträge*. 29: 15–35.

Lutz, H. (2002) At your service madam! The globalization of domestic service, *Feminist Review*. 70: 89–104.

Lutz, H. (2010) Gender in the migratory process, *Journal of Ethnic and Migration Studies*. 36, 10: 1647–63.

Lutz, H., and Palenga-Möllenbeck, E. (2010) Care work migration in Germany: Semi-compliance and complicity, *Social Policy and Society*. 9, 3: 419–30.

Maas, W. (2016) European governance of citizenship and nationality, *Journal of Contemporary European Research*. 12, 1: 532–51.

Mabogunje, A. (1970) Systems approach to a theory of rural-urban migration, *Geographical Analysis*. 2, 1: 1–92.

MacDonald, J.S., and MacDonald, L.D. (1964) Chain migration, ethnic neighborhood formation, and social networks, *The Milbank Memorial Fund Quarterly*. 42, 1: 82–97.

Macdonald, L., and Ruckert, A. (eds) (2009) *Post-Neoliberalism in the Americas*. Basingstoke: Palgrave-Macmillan.

Madanipour, A., Cars, G., and Allen, J. (1998) *Social Exclusion in European Cities*. London and Philadelphia: Regional Studies Association.

Mahler, S. (1995) *American Dreaming*. Princeton: Princeton University Press.

Mahler, S. (2001) Transnational relationships: The struggle to communicate across borders, *Identities: Global Studies in Culture and Power*. 7, 4: 583–619.

Malecki, E., and Ewers, M.C. (2007) Labor migration to world cities: With a research agenda for the Arab Gulf, *Progress in Human Geography*. 31, 4: 467–84.

Malkii, L. (1995) Refugees and exile: From 'Refugee Studies' to the national order of things, *Annual Review of Anthropology*. 24: 495–523.

Malipaard, M., Gijsberts, M., and Phalet, K. (2015) Islamic gatherings: Experiences of discrimination and religious affirmation across established and new immigrant communities, *Ethnic and Racial Studies*. 38, 15: 2635–51.

Malos, E. (ed.) (1980) *The Politics of Housework*. London: Allison and Busby.

Mansfield, B. (2005) Beyond re-scaling: Reintegrating the 'national' as a dimension of scalar relations, *Progress in Human Geography*. 29, 4: 458–73.

Maochun, L., and Wen, C. (2014) Transnational undocumented marriages in the Sino-Vietnamese border areas of China, *Asian and Pacific Migration Journal*. 23, 1: 113–23.

Marcelli, E., Williams, C.C., and Joassart, O. (eds) (2010) *Informal Work in Developed Nations*. London: Routledge.

Marchevsky, A., and Theoharis, J. (eds) (2006) *Not Working: Latina Immigrants, Low-wage Jobs, and the Failure of Welfare Reform*. New York and London: New York University Press.

Marie, C.V. (2000) Measures taken to combat the employment of undocumented foreign workers in France, in OECD (ed.) *Combating the Illegal Employment of Foreign Workers*. Paris: OECD.

Maron, N., and Connell, J. (2008) Back to Nukunuku: Employment, identity and return migration in Tonga, *Asia Pacific Viewpoint*. 49, 2: 168–84.

Marshall, T.H. (1950) *Citizenship and Social Class*. London: Cambridge University Press.

Marston S.A., Jones III, J.P., and Woodward, K. (2005) Human geography without scale, *Transactions of the Institute of British Geographers*. 30, 4: 416–32.

Martin, D. (2003) Enacting neighborhood, *Urban Geography*. 24, 5: 361–85.

Martin, L.L. (2012a) Governing through the family: Struggles over US noncitizen family detention policy, *Environment and Planning A*. 44, 4: 866–888

Martin, L.L. (2012b) 'Catch and remove': Detention, deterrence, and discipline in U.S. noncitizen family detention practice, *Geopolitics*. 17, 2: 312–34.

Martin, P. (1997) Economic instruments to affect countries of origin, in Munz, R., and Weiner, M. (eds), *Migrants, Refugees and Foreign Policy: US and German Policies Towards Countries of Origin*. Oxford: Berghahn.

Martin, P. (2014) Germany, in Hollifield, J., Martin, P.L., and Orrenius, P.M. (2014) (eds) *Controlling Immigration: A Global Perspective*, 3rd edition. Stanford: Stanford University Press.

Martin, P. (2015) Economic aspects of migration, Brettell, C.B., and Hollifield, J.F. (eds) *Migration Theory: Talking Across Disciplines*, 3rd edition. New York and London: Routledge.

Martin, S. (2002) The attack on social rights: US citizenship devalued, in Hansen, R., and Weil, P. (eds) *Dual Nationality, Social Rights and Federal Citizenship in the U.S. and Europe*. New York and Oxford: Berghahn Books.

Martin, S.F., Weerasinghe, S., and Taylor, A. (eds) (2014) *Humanitarian Crises and Migration: Causes, Consequences and Responses*. London: Routledge.

Martiniello, M. (2006) Political participation, mobilisation and representation of immigrants and their offspring in Europe, in Bauböck, R. (ed.) (2006) *Migration and Citizenship: Legal Status, Rights and Political Participation*. Amsterdam: Amsterdam University Press.

Massey, D. (Doreen) (1994) *Space, Place, and Gender*. Cambridge: Polity Press.

Massey, D. (Doreen) (2004) Geographies of responsibility, *Geografiska Annaler B*. 86: 5–18.

Massey, D. (Doreen) (2005) *For Space*. London: Sage.

Massey, D.S. (1999) International migration at the dawn of the twenty-first century: The role of the state, *Population and Development Review*. 25, 2: 303–22.

Massey, D.S., Alarcón, R., Durand, J., and Gonzáles, H. (1987) *Return to Aztlan: The Social Process of International Migration from Western Mexico*. Berkeley and Los Angeles: University of California Press.

Massey, D.S., Arango, J., Hugo, G., Kouaouci, A., Pellegrino, A., and Taylor J.E. (1993) Theories of international migration: Review and appraisal, *Population and Development Review*. 19, 3: 431–66.

Massey, D.S., Arango, J., Hugo, G., Kouaouci, A., Pellegrino, A., and Taylor J.E. (1994) An evaluation of international migration theory, *Population and Development Review*. 20, 4: 699–751.

Massey, D.S., Arango, J., Hugo, G., Kouaouci, A., Pellegrino, A., and Taylor J.E. (1998) *Worlds in Motion: Understanding International Migration at the End of the Millennium*. Oxford: Oxford University Press.

Massey, D.S., and Capoferro, C. (2006) *Sálvese Quien Pueda*: Structural adjustment and emigration from Lima, *Annals of the American Academy of Political and Social Science*. 606: 116–27.

Massey, D.S, Durand, J., and Malone, N.J. (2002) *Beyond Smoke and Mirrors: Mexican immigration in an era of economic integration*. New York: Russell Sage Foundation.

Massey, D.S., Durand, J., and Pren, K. (2014) Explaining undocumented migration to the U.S., *International Migration Review*. 48, 4: 1028–61.

Mattingly, D. (1999) Job search, social networks, and local labour market dynamics: The case of paid household work in San Diego, California, *Urban Geography*. 20: 46–74.

Maussen, M. (2009) *Constructing Mosques: The Governance of Islam in France and the Netherlands*. Amsterdam: Amsterdam School for Social Science Research.

Mavroudi, E. (2008) Palestinians and pragmatic citizenship: Negotiating relationships between citizenship and national identity in diaspora, *Geoforum*. 39, 1: 307–18.

May, J., Wills, J., Datta, K., Evans, Y., Herbert, J., and McIlwaine, C. (2007) Keeping London working: Global cities, the British state and London's new migrant division of labour, *Transactions of the Institute of British Geographers*. 32: 151–67.

Mazur, R.E. (1989) The political economy of refugee creation in Southern Africa: Micro and macro issues in sociological perspective, *Journal of Refugee Studies*. 2, 4: 441–67.

Mbembe, A. (2003) Necropolitics, *Public Culture*. 15, 1: 11–40.

McAdam, J. (ed.) (2010) *Climate Change and Displacement*. Oxford: Hart Publishers.

McAdam, J. (2014) Conceptualizing 'conflict migration', in Martin, S.F., Weerasinghe, S., and Taylor, A. (eds) *Humanitarian Crises and Migration: Causes, Consequences and Responses*. London: Routledge.

McAuliffe, C. (2008) Transnationalism within: Internal diversity in the Iranian diaspora, *Australian Geographer*. 39, 1: 63–80.

McCann, E. (2002) The urban as an object of study in global cities literatures: Representational practices and conceptions of place and scale, in Herod, A., and Wright, M. (eds) *Geographies of Power: Placing scale*. Oxford: Basil Blackwell.

McCann, E., and Ward, K. (2010) Relationality/territoriality: Toward a conceptualization of cities in the world, *Geoforum*. 41, 1: 175–84.

McDonald, D.A., Zinyama, L, Gay, J., de Vletter, F., and Mattes, R. (2000) Guess who's coming to dinner: Migration from Lesotho, Mozambique and Zimbabwe to South Africa, *International Migration Review*. 34, 3: 813–41.

McDowell, L. (1991) Life without father and Ford: The new gender order of post-Fordism, *Transactions of the Institute of British Geographers*. 16: 400–419.

McDowell, L., Batnitzky, A., and Dyer, S. (2007) Division, segmentation, and interpellation: The embodied labors of migrant workers in a greater London hotel, *Economic Geography*. 83, 1: 1–25.

McGregor (2008) Abject spaces, transnational calculations: Zimbabweans in Britain navigating work, class and the law, *Transactions of the Institute of British Geographers*. 33: 466–82.

McHugh, K. (2000) Inside, outside, upside down, backward, forward, round and round: A case for ethnographic studies in migration, *Progress in Human Geography*. 24, 1: 71–89.

McNevin, A. (2011) *Contesting Citizenship: Irregular Migrants and New Frontiers of the Political*. New York: Columbia University Press.

Meillassoux, C. (1992) *Femmes, Greniers and Capitaux*. Paris: L'Harmattan.

Meissner, D., and Rosenblum, M.R. (2009) *The Next Generation of E-Verify: Getting Employment Verification Right.* Washington, DC: Migration Policy Institute.

Mendoza, C. (2001) The role of the state in influencing African labor outcomes in Spain and Portugal, *Geoforum.* 32: 167–80.

Menjívar, C. (1999) Religious institutions and transnationalism: A case study of Catholic and evangelical Salvadoran immigrants, *International Journal of Politics, Culture and Society.* 12: 589–612.

Menjívar, C. (2014) Immigration law beyond borders: Externalizing and internalizing border controls in an era of securitization, *Annual Review of Law and Social Science.* 10: 353–69.

Mercer, C., Page, B., and Evans, M. (2009) Unsettling connections: Transnational networks, development and African home associations, *Global Networks.* 9, 2: 141–61.

Messina, A.M. (2014) Securitizing immigration in the Age of Terror, *World Politics.* 66, 3: 530–59.

Messina, A.M., and Lahav, G. (eds) (2005) *The Migration Reader.* Boulder: Lynne Reiner Publishers.

Mezzadra, S., and Neilson, B. (2013) *Border as Method, or the Multiplication of Labor.* Durham and London: Duke University Press.

Meyers, E. (2000) Theories of international immigration policy – a comparative analysis, *International Migration Review.* 34, 4: 1245–82.

Miles, R. (1982) *Racism and Migrant Labour.* London: Routledge and Kegan Paul.

Miles, M., and Crush, J. (1993) Personal narratives as interactive texts: Collecting and interpreting migrant life-histories, *Professional Geographer.* 45: 84–94.

Millner, N. (2011) From 'refugee' to 'migrant' in Calais solidarity activism: Re-staging undocumented migration for a future politics of asylum, *Political Geography.* 30: 320–28.

Mitchell, C. (1959) Labour migration in Africa south of the Sahara: The causes of labour migration, *Bulletin of the Inter-African Labour Institute.* 6, 1: 12–46.

Mitchell, D. (1995) The end of public space? People's park, definitions of the public, and democracy, *Annals of the Association of American Geographers.* 85: 108–33.

Mitchell, D. (1996) *The Lie of the Land: Migrant Workers and the California Landscape.* Minneapolis: University of Minnesota Press.

Mitchell, D. (2000) *Cultural Geography: An Introduction.* Oxford: Basil Blackwell.

Mitchell, D. (2003) *The Right to the City.* New York: Guilford Press.

Mitchell, K. (1997), Different diasporas and the hype of hybridity, *Environment and Planning D: Society and space.* 15, 5: 533–53.

Mitchell, K. (2003) Educating the national citizen in neo-liberal times: From the multicultural self to the strategic cosmopolitan, *Transactions of the Institute of British Geographers.* 28, 4: 387–403.

Mittelman, J. (2000) *The Globalization Syndrome: Transformation and Resistance.* Princeton: Princeton University Press.

Mohan, G. (2008) Making neoliberal states of development: The Ghanaian diaspora and the politics of homelands, *Environment and Planning D: Society and Space.* 26, 3: 464–79.

Molho, I. (1986) Theories of migration: A review, *Scottish Journal of Political Economy*. 33: 396–419.

Money, J. (1999) *Fences and Neighbors: The Political Geography of Immigration Control*. Ithaca: Cornell University Press.

Moore, L. (2006) Biometric borders: Governing mobilities in the war on terror. *Political Geography*. 25: 336–51.

Morakvasic, M. (1984) Birds of passage are also women, *International Migration Review*. 18, 4: 886–907.

Moran, D., Gill, N., and Conlon, D. (2013) Carceral Spaces: Linking imprisonment and migrant detention. In Moran, D., Gill, N., and Conlon, D. (eds) *Carceral Spaces: Mobility and Agency in Imprisonment and Migrant Detention*. Farnham: Ashgate.

Morehouse, C., and Blomfield, M. (2011) *Irregular Migration in Europe*. Washington: Migration Policy Institute.

Morris, L. (2001) The ambiguous terrain of rights: Civic stratification in Italy's emergent immigration regime, *International Journal of Urban and Regional Research*. 25, 3: 497–516.

Morris, L. (2002) *Managing Migration: Civic Stratification and Migrant's Rights*. London and New York: Routledge.

Morrisens, A., and Sainsbury, D. (2005) Migrants' social rights, ethnicity and welfare regimes, *Journal of Social Policy*. 34, 4: 637–60.

Mountz, A. (2004) Embodying the nation-state: Canada's response to human struggling, *Political Geography*. 23, 3: 323–45.

Mountz, A. (2010) *Seeking Asylum: Human Smuggling and Bureaucracy at the Border*. Minneapolis: University of Minnesota Press.

Mountz, A. (2011) The enforcement archipelago: Detention, haunting, and asylum on islands, *Political Geography*. 30: 118–28.

Mountz, A., and Hiemstra, N. (2013) Chaos and crisis: Dissecting the spatiotemporal logics of contemporary migrations and state practices, *Annals of the Association of American Geographers*. 104, 2: 382–90.

Mountz, A., and Loyd, J. (2014) Constructing the Mediterranean region: Obscuring violence in the bordering of Europe's migration 'crises', *ACME, International E-Journal for Critical Geographies*. 13, 2: 173–95.

Mountz, A., and Wright, R. (1996) Daily life in the transnational migrant community of San Agustin, Oaxaca and Poughkeepsie, *Diaspora*. 5, 3: 403–28.

MPI (Migration Policy Institute) (2014a) Hazleton immigration ordinance that began with a bang goes out with a whimper, March 28.

MPI (Migration Policy Institute) (2014b) Dramatic surge in the arrival of unaccompanied children has deep roots and no simple solutions, June 13.

MPI (Migration Policy Institute) (2015a) Refugees and asylees in the United States, Migration Policy Institute, October 28.

MPI (Migration Policy Institute) (2015b) No end in sight: The worsening Syrian refugee crisis, May 4.

Mueller, C.F. (1981) *The Economics of Labor Migration*. New York: Academic Press.

Mumford, K., and Smith, P.N. (2004) Job tenure in Britain: Employee characteristics versus workplace effects, *Economica*. 71: 275–98.

Murdoch, J. (1997) Towards a geography of heterogeneous associations, *Progress in Human Geography.* 21, 3: 321–37.

Musterd, S., Murie, A., and Kesteloot, C. (2006) *Neighborhoods of Poverty: Urban Social Exclusion and Integration in Europe.* London: Palgrave Macmillan.

Nagar, R., Lawson, V., McDowell, L., and Hanson, S. (2002) Locating globalization: Feminist (re)readings of the subjects and spaces of globalization, *Economic Geography.* 78, 3: 257–84.

Nam, Y.J., and Jung, H.J. (2008) Welfare reform and older immigrants: Food stamp program participation and food insecurity, *Gerontologist.* 48, 1: 42–50.

Nam, Y., and Kim, W. (2012) Welfare reform and elderly immigrants' naturalization: Access to public benefits as an incentive for naturalization in the United States, *International Migration Review.* 46, 3: 656–79.

Nansen Initiative (2005) Available at: https://www.nanseninitiative.org/

Naples, N.A. (2009) Crossing borders: Community activism, globalization and social justice, *Social Problems.* 56, 1: 2–20.

Nassari, J. (2009) Refugees and forced migrants at the crossroads: Forced migration in a changing world, *Journal of Refugee Studies.* 22, 1: 1–10.

Nevins, J. (2002) *Operation Gatekeeper: The Rise of the 'Illegal Alien' and the Making of the U.S.–Mexico Boundary.* New York and London: Routledge.

Nevins, J. (2008) *Dying to Live: A Story of US Immigration in an Age of Global Apartheid.* San Francisco: Open Media Books.

Newbold, K.B. (2007) *Six Billion Plus: World Population in the Twenty-first Century,* 2nd edition. Lanham: Rowman and Littlefield.

New Straits Times (2015) 'Govt won't offer jobs to Rohingya migrants in Malaysia', 24 June.

New Yorker (2008) 'The Countertraffickers: rescuing the victims of the global sex trade', 5 May.

New York Times (2006) 'Leaving New York, with bodega in tow', 29 October.

New York Times (2007) 'Judge strikes down town's immigration law', 26 July.

New York Times (2007) 'Challenge in Connecticut over immigrants arrest', 26 September.

New York Times (2008a) 'After Iowa raid, immigrants fuel labor inquiries, 27 July'.

New York Times (2008b) 'Iowa rally protests raid and conditions at plant', 28 July.

New York Times (2009) 'A family divided by 2 words, legal and illegal', 26 April.

New York Times (2013) 'Ailing Midwestern cities extend a welcoming hand to immigrants', 6 October.

New York Times (2016) 'Macedonian police use tear gas to stop migrants at border', 10 April.

New York Times Magazine (2007) 'All immigration politics is local (and complicated, nasty and personal)', 5 August.

Nicholls, W.J. (2011) Cities and the unevenness of social movement space: The case of France's immigrant rights movement, *Environment and Planning A.* 43, 7: 1655–73.

Nicholls, W.J. (2013) Making undocumented immigrants into a legitimate political subject: Theoretical observations from the United States and France, *Theory Culture and Society.* 30, 3: 82–107.

Ní Laoire, C.N. (2000) Conceptualising Irish rural youth migration: A biographical approach, *International Journal of Population Geography*. 6: 229–43.

Ní Laoire, C.N. (2007) To name or not to name: Reflections on the use of anonymity in an oral archive of migrant life narratives, *Social and Cultural Geography*. 8, 3: 373–90.

Noiriel, G. (1984) *Longwy: Immigrés et prolétaires 1880–1980*. Paris: PUF.

Nyberg-Sorensen, N., Van Hear, N., and Engberg-Pedersen, P. (2002) The migration-development nexus evidence and policy options state-of-the-art overview, *International Migration*. 40, 5: 3–47.

Nyers, P., and Rygiel, K. (2012) Introduction: Citizenship, migrant activism and the politics of movement, in Nyers, P., and Rygiel, K. (eds) *Citizenship, Migrant Activism and the Politics of Movement*. London: Routledge.

OECD (2000) OECD (ed.) *Combating the Illegal Employment of Foreign Workers*. Paris: OECD.

OECD (2006) *International Migration Outlook 2006*. Paris: OECD.

OECD (2014) *Education at a Glance 2014*. Paris: OECD.

OECD/SOPEMI (2008) *International Migration Outlook Annual Report 2008*. Paris: OECD.

OECD/SOPEMI (2012) *International Migration Outlook 2012*. Paris: OECD.

OECD/SOPEMI (2013) *International Migration Outlook 2013*. Paris: OECD.

OECD/SOPEMI (2014) *International Migration Outlook 2014*. Paris: OECD.

OECD/SOPEMI (2015) *International Migration Outlook 2015*. Paris: OECD.

Ondiak, N. (2007) Refugees in a global era, *Journal of Refugee Studies*. 20, 3: 542–6.

Ong, A. (1999) *Flexible Citizenship: The Cultural Logics of Transnationality*. Durham and London: Duke University Press.

Ong, A. (2006) *Neoliberalism as Exception: Mutations in Citizenship and Sovereignty*. Durham and London: Duke University Press.

Open Doors (2008) Open Doors on-line. Available at: http://opendoors.iienetwork.org

Orrenius, P.M., and Zavodny, M. (2003) Do Amnesty programs reduce undocumented immigration? Evidence from IRCA, *Demography*. 40, 3: 437–50.

Orrenius, P.M., Zavodny, M., and Kerr, E. (2012) Chinese immigrants in the U.S. labor market: Effects of post-Tiananmen immigration policy, *International Migration Review*. 46, 2: 456–82.

Orozco, M. (2002) Globalization and migration: The impact of family remittances in Latin America, *Latin American Politics and Society*. 44, 2: 41–66.

Orozco, M., and Rouse, R. (2007) *Migrant hometown associations and opportunities for development: A global perspective*. Washington, DC: Migration Policy Institute.

Ozuekren, A.S., and Van Kempen, R. (2002) Housing careers of minority ethnic groups: Experiences, explanations and prospects, *Housing Studies*. 17, 3: 365–79.

Pai, H.H. (2004) The invisibles – migrant cleaners at Canary Wharf, *Feminist Review*. 78: 164–74.

Palidda, S. (ed.) (2011) *Racial Criminalization of Migrants in the 21st Century*. Farnham: Ashgate.

Palmgren, P.A. (2014) Irregular networks: Bangkok refugees in the city and region, *Journal of Refugee Studies*. 27, 1: 21–41.

Pande, A. (2013) 'The paper that you have in your hand is my freedom': Migrant domestic work and the sponsorship (Kafala) System in Lebanon, *International Migration Review*. 47, 2: 414–41.

Papadametriou, D.G. (2007) *The Age of Mobility: How to get more out of migration in the 21st century*. Washington, DC: Migration Policy Institute.

Papadopoulos, D., and Tsianos, V.S. (2013) After citizenship: Autonomy of migration, organizational ontology and mobile commons, *Citizenship Studies*. 17, 2: 178–96.

Papadoupoulou-Kourkoula, A. (2009) *Transit Migration: The Missing Link Between Emigration and Settlement*. Basingstoke: Palgrave.

Papastergiadis, N. (2006) The invasion complex: The abject other and spaces of violence, *Geografiskar Annaler*. 88 B, 4: 429–42.

Parekh, B. (2006) *Rethinking Multiculturalism: Cultural Diversity and Political Theory*, 2nd edition. Basingstoke: Macmillan.

Parnini, S.N., Othman, M.R., and Ghazali, A.S. (2013) The Rohingya refugee crisis and Bangladesh-Myanmar relations, *Asian and Pacific Migration Review*. 22, 1: 133–46.

Parreñas, R.S. (2001) *Servants of Globalization*. Stanford: Stanford University Press.

Paul, A.M. (2011) Stepwise international migration: A multistage migration pattern for the aspiring migrant, *American Journal of Sociology* 116, 6: 1842–86.

Pearlman, W. (2014) Competing for Lebanon's diaspora: Transnationalism and domestic struggles in a weak state, *International Migration Review*. 48, 1: 34–75.

Peck, J. (1996) *Workplace: The Social Regulation of Labour Markets*. New York: Guilford Press.

Peck, J. (2001) *Workfare States*. New York: Guilford Press

Peck, J., and Tickell, A. (2002) Neoliberalizing space, *Antipode*. 34, 3: 380–404.

Peck, J., and Tickell, A. (2006) Conceptualizing neoliberalism, thinking Thatcherism, in Leitner, H., Peck, J., and Sheppard, E.S. (eds) (2007) *Contesting Neoliberalism: Urban Frontiers*. New York: Guilford Press.

Peck, J., Theodore, N., and Brenner, N. (2010) Postneoliberalism and its malcontents, *Antipode*. 41, 1: 94–116.

Pecoud, A. (2015) *Depoliticising Migration: Global Governance and International Migration Narratives*. Basingstoke: Palgrave Macmillan.

Pedersen, M.H. (2012) Going on a class journey: The inclusion and exclusion of Iraqi refugees in Denmark, *Journal of Ethnic and Migration Studies*. 38, 7: 1101–17.

Pellerin, H. (2010) Governing labour migration in the era of GATS: The growing influence of *lex mercatoria*, in Gabriel, C., and Pellerin, H. (eds) *Governing International Labour Migration*. London: Routledge.

Peng, I. (2016) Testing the limits of welfare state changes: The slow-moving immigration policy reform in Japan, *Social Policy and Administration*. 50, 2: 278–95.

Penrose, J. (2013) Multiple multiculturalisms: Insights from the Edinburgh Mela, *Social and Cultural Geography*. 14, 7: 829–51.

Perchinig, B. (2006) EU citizenship and the status of third country nationals, in Bauböck, R. (ed.) (2006) *Migration and Citizenship: Legal Status, Rights and Political Participation*. Amsterdam: Amsterdam University Press.

Perrin, M.E., Hagopian, A., Sales, A., and Huang, B. (2007) Nurse migration and its implications for Philippine hospitals, *International Nursing Review*. 54, 3: 219–26.

Pessar, P.R. and Mahler, S. (2003) Transnational migration: Bringing gender in, *International Migration Review*. 37: 812–46.

Peutz, N. (2006) Embarking on an anthropology of removal, *Current Anthropology*. 47, 2: 217–41.

Phillimore, J., and Goodson, L. (2006) Problem or opportunity? Asylum-seekers, refugees, employment and social exclusion in deprived urban areas, *Urban Studies*. 43, 10: 1715–36.

Pierce, S. (2015) *Unaccompanied Child Migrants in U.S. Communities, Immigration Court, and Schools*, Policy Brief. Washington, DC: Migration Policy Institute.

Piore, M. (1979) *Birds of Passage: Migrant Labor and Industrial Societies*. New York: Cambridge University Press.

Piper, N. (2004) Gender and migration policies in Southeast and East Asia: Legal protection and socio-cultural empowerment of unskilled migrant women, *Singapore Journal of Tropical Geography*. 25, 2: 216–31.

Piper, N. (2006) Gendering the politics of migration, *International Migration Review*. 40, 1: 133–64.

Piper, N., and Roces, M. (2003) Introduction: Marriage and migration in an age of globalization, in Piper, N., and Roces, M. (eds) *Wife or Worker? Asian Women and Migration*. Lanham: Rowman and Littlefield.

Popke, J. (2007) Geography and ethics: Spaces of cosmopolitan responsibility, *Progress in Human Geography*. 31, 4: 509–18.

Portes, A. (1978) Migration and underdevelopment, *Politics and Society*. 8, 1: 1–48.

Portes, A. (1997) Immigration theory for a new century: Some problems and opportunities, *International Migration Review*. 31: 799–825.

Portes, A. (1999) Conclusion: Towards a new world – the origins and effects of transnational activities, *Ethnic and Racial Studies*. 22, 2: 463–77.

Portes, A. (2000) Book review of Massey, D. et al. 'Worlds of Motion', *International Migration Review*. 34, 3: 976–78.

Portes, A., and Bach, R.L. (1985) *Latin Journey: Cuban and Mexican Immigrants in the United States*. Berkeley: University of California Press.

Portes, A., Castells, M., and Benton, L. (1989) *The Informal Economy: Studies in Advanced and Less Developed Countries*. Baltimore: Johns Hopkins University Press.

Portes, A., and DeWind, J. (2004a) A cross-Atlantic dialogue: The progress of research and theory in the study of international migration, *International Migration Review*. 38, 3: 828–51.

Portes, A., and DeWind, J. (2004b) Conceptual and methodological developments in the study of international migration, *International Migration Review*. 38. Special issue.

Portes, A., and DeWind, J. (2007) *Rethinking Migration: New Theoretical and Empirical Perspectives*. New York and Oxford: Berghahn Books.

Portes, A., and Fernandez-Kelly, P. (2008) No margin for error: Educational and occupational achievement among disadvantaged children of immigrants, *Annals of the American Academy of the Political and Social Science*. 620: 12–36.

Portes, A., and Guarnizo, L., and Landolt, P. (1999) The study of transnationalism: Pitfalls and promise of an emergent research field, *Ethnic and Racial Studies*. 22, 2: 217–37.

Portes, A., and Rumbaut, R.G. (2006) *Immigrant America: A Portrait*, 3rd edition. Berkeley: University of California Press.

Portes, A., and Rumbaut, R.G. (2014) *Immigrant America: A Portrait*, 4th edition. Berkeley: University of California Press.

Portes, A., and Sensenbrenner, J. (1993) Embeddedness and immigration: Notes on the social determinants of economic action, *American Journal of Sociology*. 98, 6: 1320–50.

Portes, A., and Walton, J. (1981) *Labor, Class and the International System*. New York: Academic Press.

Portes, A., and Zhou, M. (1993) The new second generation: Segmented assimilation and its variants, *The Annals of the American Academy of Political and Social Science*. 530: 74–96.

Pratt, G. (1999) From registered nurse to registered nanny: Discursive geographies of Filipina domestic workers in Vancouver, BC, *Economic Geography*. 75, 3: 215–36.

Pred, A. (1977) The choreography of existence: Comments on Hägerstrand's time geography, *Economic Geography*. 53: 207–21.

Predelli, L.N. (2008) Religion, citizenship and participation – A case study of immigrant Muslim women in Norwegian Mosques, *European Journal of Women's Studies*. 15, 3: 241–60.

Preibisch, K. (2010) Pick-Your-Own Labor: Migrant workers and flexibility in Canadian agriculture, *International Migration Review*. 44, 2: 404–41.

Preston, V., Kobayashi, A., and Man, G. (2006) Transnationalism, gender, and civic participation: Canadian case studies of Hong Kong immigrants, *Environment and Planning A*. 38: 1633–51.

Price, M., and Benton-Short, L. (2008) *Migrants to the Metropolis*. Syracuse: Syracuse University Press.

Prothero, M. (1990) Labor recruiting organizations in the developing world: Introduction, *International Migration Review*. 24, 2: 221–8.

Purcell, M. (2003) Islands of practice and the Marston/Brenner debate: Toward a more synthetic critical human geography, *Progress in Human Geography*. 27, 3: 317–32.

Putnam, R.D. (2000) *Bowling Alone: The Collapse and Revival of American Community*. New York: Simon and Schuster.

Quassoli, F. (1998) Migrants in the Italian underground economy, *International Journal of Urban and Regional Research*. 23, 2: 212–31.

Radu, D. (2008) Social interactions in economic models of migration: A review and appraisal, *Journal of Ethnic and Migration Studies*. 34, 4: 531–48.

Raes, S., Rath, J., Dreef, M., Kumcu, A., Reil, F., and Zorlu, A. (2002) Amsterdam: Stitched up, in Rath, J. (ed.) *Unravelling the Rag Trade*. Oxford: Berg Press.

Raghuram, P. (2004) The difference that skills make: Gender, family migration strategies and regulated labour markets, *Journal of Ethnic and Migration Studies*. 30, 2: 303–21.

Raghuram, P. (2007) Which migration, what development: Unsettling the edifice of migration and development, University of Bielefeld, Center on Migration, Citizenship and Development, Working Paper 28.

Raghuram, P. (2008) Migrant women in male-dominated sectors of the labour market: A research agenda, *Population Space and Place*. 14, 1: 43–57.

Raghuram, P. (2012) Global care, local configurations: Challenges to conceptualizations of care, *Global Networks*. 12, 2: 155–74.

Raghuram, P., and Kofman, E. (2002) The state, skilled labour markets, and immigration: The case of doctors in England, *Environment and Planning A*. 34, 11: 2071–89.

Rajas, J. (2015) Disciplining the human rights of immigrants: Market veridiction and the echoes of eugenics in contemporary EU immigration policies, *Third World Quarterly*. 36, 6: 1129–44.

Rajkumar, D., Berkowitz, L., Vosko, L.F., Preston, V., and Latham, R. (2012) At the temporary–permanent divide: How Canada produces temporariness and makes citizens through its security, work, and settlement policies, *Citizenship Studies*. 16, 3–4: 483–510

Ram, M. *et al.* (2003a) The dynamics of informality: Employment relation in small firms and the effects of regulatory change, *Work, Employment and Society*. 15, 4: 845–61.

Ram, M., Edwards, P.K., Gilman, M.W., and Arrowsmith, J. (2003b) Once more into the sunset? Asian clothing firms after the national minimum wage, *Environment and Planning C: Government and Policy*. 21, 1: 71–88.

Ranis, G., and Fei, J.C.H. (1961) A theory of economic development, *The American Economic Review*. LI, 4: 533–58.

Rath, J. (ed.) (2002) *Unravelling the Rag Trade*. Oxford: Berg Press.

Rath, J. and Kloosterman, R. (2000) A critical review of research on immigrant entrepreneurship, *International Migration Review*. XXXIV, 3: 657–81.

Ravenstein, E.G. (1885) The Laws of Migration, *Journal of the Statistical Society*. 48, 2: 167–245.

Ravenstein, E.G. (1889) The Laws of Migration, *Journal of the Royal Statistical Society*. 52: 241–301.

Reed, H. Ludlow, and Baslow (2015) Forced migration, in White, M.J. (ed.) *International Handbook of Migration and Population Distribution.* New York and London: Springer.

Reich, M., Gordon, D., and Edwards, R. (1973) A theory of labor segmentation, *American Economic Review*, 63: 359–65.

Reniers, G. (1999) On the history and selectivity of Turkish and Moroccan migration to Belgium, *International Migration.* 37, 4: 679–713.

Reyneri, E. (2001) Migrants' involvement in the underground economy in the Mediterranean countries of the European Union. ILO – International Migration Working Paper no. 41.

Reyneri, E. (2004) Immigrants in a segmented and often undeclared labour market, *Journal of Modern Italian Studies* 9, 1: 71–93.

Reyneri, E., and Fullin, G. (2011) Labour market penalties of new immigrants in new and old receiving West European countries, *International Migration.* 49, 1: 31–57.

Riano, Y., and Baghdadi, N. (2007) I thought I could have a more egalitarian relationship with a European – the role of gender and geographical imaginaries in women's migration, *Nouvelles Questions Feministes.* 26, 1: 38–53.

Riccio, B. (2008) West African transnationalisms compared: Ghanaians and Senegalese in Italy, *Journal of Ethnic and Migration Studies.* 34, 2: 217–34.

Richmond, A.H. (2002) Globalization: Implications for immigrants and refugees, *Ethnic and Racial Studies.* 25, 5: 707–27.

Ridgley, J. (2008) Cities of refuge: Immigration enforcement, police, and the insurgent genealogies of citizenship in US sanctuary cities, *Urban Geography.* 29, 1: 53–77.

Rigby, J., and Schlembach, R. (2013) Impossible protest: No borders in Calais, *Citizenship Studies.* 17, 2: 157–72.

Robins, K., and Aksoy, A. (2001) From spaces of identity to mental spaces: Lessons from Turkish-Cypriot cultural experience in Britain, *Journal of Ethnic and Migration Studies.* 27, 4: 685–711.

Robinson, J. (2002) Global and world cities: A view from off the map, *International Journal of Urban and Regional Research.* 26, 3: 531–54.

Rodriguez, R.M. (2010) *Migrants for Export: How the Philippine State Brokers Labor to the World.* Minneapolis: University of Minnesota Press.

Rogaly, B. (2008) Intensification of workplace regimes in British horticulture: The role of migrant workers, *Population, Space and Place.* 14, 6: 497–510.

Rogaly, B. (2009) Spaces of work and everyday life: Labour geographies and the agency of unorganised temporary migrant workers, *Geography Compass.* 3, 6: 1975–87.

Room, G. (1995) Poverty and social exclusion: The new European agenda for policy and research, in Room, G. (ed.) *Beyond the Threshold: The Measurement and Analysis of Social Exclusion.* Bristol: The Policy Press.

Rosenblum, M.R. (2011) *E-verify: Strengths, weaknesses, and proposals for reform.* Washington, DC: Migration Policy Institute.

Ross, A. (2003) *No-collar: The Humane Workplace and its Hidden Costs.* Basic Books, New York.

Rotte, R. (2000) Immigration control in United Germany: Toward a broader scope of national policies, *International Migration Review*. 34, 3: 357–89.

Rouse, R. (1992) Making sense of settlement: Class transformation, cultural struggle and transnationalism among Mexican migrants in the United States, in Glick-Schiller, N., Basch, L., and Blanc-Szanton, C. (eds) *Towards a Transnational Perspective on Migration*. New York: New York Academy of Sciences.

Rudnyckyj, D. (2004) Technologies of servitude: Governmentality and Indonesian transnational labor migration, *Anthropological Quarterly*. 77, 3: 407–34.

Ruggie, J. G. (1982) International regimes, transactions, and change: Embedded liberalism in the postwar economic order, *International Organization*. 36, 2: 379–415.

Ruhs, M., and Anderson, B. (2009) Semi-compliance and illegality in migrant labour markets: An analysis of migrants, employers, and the state in the UK, *Population, Space, and Place*. 16, 3: 195–211.

Ruhs, M., and Anderson, B. (2013) Migrant workers: Inevitability or policy choice?, in Willem Holtslag, J., Kremer, M., and Schrijvers, E. (eds) *Making Migration Work*. Amsterdam: Amsterdam University Press.

Ruiz, N.G., Wilson, J.H., and Choudhury, S. (2012) The Search for Skills: Demand for H-1B Immigrant Workers in U.S. Metropolitan Areas, Brookings Institution paper.

Ryan, L. (2008) 'I had a sister in England': Family-led migration, social networks and Irish Nurses, *Journal of Ethnic and Migration Studies*. 34, 3: 453–70.

Sack, R. (1986) *Human Territoriality: Its Theory and History*. Cambridge: Cambridge University Press.

Safran, W. (1991) Diasporas in modern societies: Myths of homeland and return. *Diaspora*, 1: 83–99.

Said, E. (1978) *Orientalism*. London: Penguin Books.

Sainsbury, D. (2006) Immigrants' social rights in comparative perspective: Welfare regimes, forms of immigration and immigration policy regimes, *Journal of European Social Policy*. 16, 3: 229–44.

Sainsbury, D. (2012) *Welfare states and Immigrant Rights: The politics of inclusion and exclusion*. Oxford: Oxford University Press.

Saint-Blancat, C., and Cancellieri, A. (2014) From invisibility to visibility? The appropriation of public space through a religious ritual: The Filipino procession of Santacruzan in Padua, Italy, *Social and Cultural Geography*. 15, 6: 645–63.

Salehyan, I. (2009) U.S. asylum and refugee policy towards Muslim nations since 9/11, in Givens, T., Freeman, G., and Leal, D. (eds) *Immigration Policy and Security*. New York and London: Routledge.

Salih, R. (2001) Moroccan migrant women: Transnationalism, nation-states and gender, *Journal of Ethnic and Migration Studies*. 27, 4: 655–71.

Salt, J. (2000) Trafficking and human smuggling: A European perspective, *International Migration*. 38, 3: 31–56.

Salt, J., and Stein, J. (1997) Migration as a business: The case of trafficking, *International Migration*. 35, 4: 467–94.

Salzinger, L. (2003) *Genders in Production: Making Workers in Mexico's Global Factories.* Berkeley: University of California Press.

Samers, M. (1997a) The production and regulation of North African immigrants in the Paris automobile industry, 1970–1990, unpublished D.Phil thesis, Oxford University, Oxford, UK.

Samers, M. (1997b) The production of diaspora: Algerian emigration from colonialism to neo-colonialism (1840–1970), *Antipode.* 29, 1: 32–64.

Samers, M. (1998a) Immigration, 'ethnic minorities' and 'social exclusion' in the European Union: A critical perspective, *Geoforum.* 29, 2: 119–21.

Samers, M. (1998b) 'Structured coherence': Immigration, racism and production in the Paris car industry, *European Planning Studies.* 6, 1: 49–72.

Samers, M. (1999) 'Globalization', migration, and the geo-political economy of the 'spatial vent', *Review of International Political Economy.* 6, 2: 166–99.

Samers, M. (2001) 'Here to work': Undocumented immigration in the United States and Europe, *SAIS Review.* XXI, 1: 131–45.

Samers, M. (2002) Immigration and the Global City Hypothesis: Towards an alternative research agenda, *International Journal of Urban and Regional Research.* 26, 2: 389–402.

Samers, M. (2003a) Invisible capitalism: Political economy and the regulation of undocumented immigration in France, *Economy and Society.* 32, 4: 555–83.

Samers, M. (2003b) Immigration and the spectre of Hobbes: Some comments for the quixotic Dr. Bauder, *ACME: An International E-journal of Critical Geographies.* 2, 2: 210–17.

Samers, M. (2003c) Diaspora unbound: Muslim identity and the erratic regulation of Islam in France, *International Journal of Population Geography.* 9: 351–64.

Samers, M. (2004a) An emerging geopolitics of 'illegal' immigration in the European Union, *European Journal of Migration and Law.* 6: 27–45.

Samers, M. (2004b) Do welfare systems matter? A preliminary analysis of welfare retrenchment and the employment of young people in France, *Kolor.* 4, 2: 75–96.

Samers, M. (2005) The 'underground economy', immigration and economic development in the European Union: An agnostic-skeptic perspective, *International Journal of Economic Development.* 6, 3: 199–272.

Samers, M. (2010a) *Migration* (Key Ideas in Geography Series). London: Routledge.

Samers, M. (2010b) The 'socio-territoriality' of cities: An urban framework for analysis, in Glick-Schiller, N., and Caglar, A. (eds) *Locating Migration: Re-scaling Cities and Migrants.* Ithaca: Cornell University Press.

Samers, M. (2010c) Strange castles and courtyards: Explaining the political economy of undocumented immigration and undeclared employment, in Menz, G., and Caviedes, A. (eds) *Labour Migration in Europe.* London: Palgrave-Macmillan.

Samers, M. (2010d) At the heart of migration management: Immigration and labour markets in the European Union, in Gabriel, C., and Pellerin, H. (eds) *Governing International Labour Migration.* London: Routledge.

Samers, M. (2014) How to understand the incorporation of immigrants in European labour markets, in Martiniello, M., and Rath, J. (eds) *An Introduction to Immigration Studies: European Perspectives*. Amsterdam: University of Amsterdam Press.

Samers, M. (2015a) Migration, in Agnew, J., Mamadouh, V., Secor, A., and Sharp, J. (eds) *The Wiley Blackwell Companion to Political Geography*. Chichester: Wiley Blackwell.

Samers, M. (2015b) Migration policies, migration and regional integration in North America, in Talani, L.S., and McMahon, S. (eds) *Handbook of the International Political Economy of Migration*. Cheltenham and Northampton: Edward Elgar.

Samers, M. (2016) New Guestworker Regimes?, in Amelina, A., Horvath, K., and Meeus, B. (eds) *Anthology of Migration and Social Transformation: European Perspectives*. New York: Springer.

Samers, M. (2017) The economic lives of young people of immigrant origin in France's 'poorest city', *Journal of Ethnic and Migration Studies*.

Samers, M., and Snider, M. (2015) Finding work: The experience of immigrants in North America, in Bauder, H., and Shields, J. (eds) *Immigrant Experiences in North America*. Toronto: Canadian Scholar's Press, Inc.

Sandercock, L. (2003) *Cosmopolis II: Mongrel Cities of the 21st Century*. London: Continuum Books.

Sanders, J.M., and Nee, V. (1987) Limits of ethnic solidarity in the enclave economy, *American Sociological Review*. 52: 745–73.

Sanders, J.M., and Nee, V. (1992) Problems in resolving the enclave economy debate, *American Sociological Review*. 57, 3: 415–18.

Sanderson, M.R., Derudder, B., Timberlake, M., and Witlox, F. (2015) Are world cities also world immigrant cities? An international, cross-city analysis of global centrality and immigration, *International Journal of Comparative Sociology*. 56, 3–4: 173–97.

Saraiva, C. (2008) Transnational migrants and transnational spirits: An African religion in Lisbon, *Journal of Ethnic and Migration Studies*. 34, 2: 253–69.

Sassen, S. (1988) *The Mobility of Labor and Capital*. Cambridge: Cambridge University Press.

Sassen, S. ([1991] 2001, 2nd ed.) *The Global City: New York, London, Tokyo*. Princeton and Oxford: Princeton University Press.

Sassen, S. (1996a) *Losing Control: Sovereignty in an Age of globalization*. New York: Columbia University Press.

Sassen, S. (1996b) New Employment Regimes in cities: The impact on immigrant workers, *New Community*. 22, 4: 579–94.

Sassen, S. (1998) *Globalization and its Discontents*. New York: New Press.

Sassen, S. (1999) *Guests and Aliens*. New York: New Press.

Sassen, S. (2006a) *Cities in a World Economy*, 3rd edition. Thousand Oaks: Pine Forge Press.

Sassen, S. (2006b) *Territory, Authority, Rights: From Medieval to Global Assemblages*. Princeton: Princeton University Press.

Sassen-Koob, S. (1984) Notes on the incorporation of Third World women into wage-labor through immigration and off-shore production, *International Migration Review*. 118: 1114–67.

Saxenian, A.L. (2005) From brain drain to brain circulation: Transnational communities and regional upgrading in India and China, *Studies in Comparative International Development*. 40, 2: 35–61.

Sayad, A. (1977) Les trois âges de l'émigration Algérienne en France, *Actes de la recherche en Sciences Sociales*. 15: 59–80.

Sayad, A. (1991) *L'immigration ou les Paradoxes de l'Alterité*. Bruxelles: Editions Universitaires/De Boeck Université.

Sayer, A. (1984) *Method in Social Science*. London: Hutchinson Press.

Sayyid, S. (2000) Beyond Westphalia: Nations and diaspora – the case of the Muslim *Umma*, in Hesse, B. (ed.) *Un/settled Multiculturalisms: Diaspora, Entanglements, Transruptions*. London: Zed Books.

Schapendonk, J. (2011) Turbulent Trajectories: Sub-Saharan African Migrants Heading North. PhD dissertation, Radboud University, Nijmegen, Netherlands.

Schierup, C.-U., Hansen, P., and Castles, S. (2006) Understanding the dual crisis, in Schierup, C.-U., Hansen, P., and Castles, S. (eds) *Migration, Citizenship, and the European Welfare State*. Oxford: Oxford University Press.

Schneider, F. (2012) The shadow economy and work in the shadow: What do we (not) know? IZA Discussion Paper No. 6423, March.

Schon, J. (2015) Focus on the forest, not the trees: A changepoint model of forced displacement, *Journal of Refugee Studies*. 28, 4: 437–67.

Schuck, P. (2002) Plural citizenships, in Hansen, R., and Weil, P. (eds) *Dual Nationality, Social Rights and Federal Citizenship in the U.S. and Europe*. New York and Oxford: Berghahn Books.

Schuster, L. (2000) A comparative analysis of the asylum policy of seven European governments, *Journal of Refugee Studies*. 13, 1: 118–31.

Schuster, L. (2003) Sangatte: Smoke and mirrors, *Global Dialogue*. 4, 4: 57–68.

Schuster, L. (2005) A sledgehammer to crack a nut: Deportation, detention and dispersal in Europe, *Social Policy and Administration*. 39, 6: 606–21.

Schuster, L., and Majidi, N. (2015) Deportation stigma and re-migration, *Journal of Ethnic and Migration Studies*. 41, 4: 625–52.

Sciortino, G. (2000) Toward a political sociology of entry policies: Conceptual problems and theoretical proposals, *Journal of Ethnic and Migration Studies*. 26, 2: 213–38.

Sciortino, G., and Finotelli, C. (2015) Closed memberships in a mobile world? Welfare states, welfare regimes and international migration, in Talani, L.S., and McMahon, S. (eds) *Handbook of the International Political Economy of Migration*. Cheltenham and Northampton: Edward Elgar.

Scott, S. (2016) Venues and filters in managed migration policy: The case of the United Kingdom, *International Migration Review*, forthcoming.

Sejersen, T.B. (2008) 'I Vow to Thee My Countries': The expansion of dual citizenship in the 21st century, *International Migration Review*. 42, 3: 523–49.

Seol, D.-H., and Skrentny, J.D. (2004) South Korea: Importing undocumented workers, in Cornelius, W., Tsuda, T., Martin, P., and Hollifield, J. (eds) *Controlling Immigration: A Global Perspective*. Stanford: Stanford University Press.

Seol, D.-H., and Skrentny, J.D. (2009) Why is there so little migrant settlement in Asian countries? *International Migration Review*. 43, 3: 578–620.

Sheffer, (1986) A new field of study: Modern diasporas in international politics, in Sheffer, G. (ed.) *Modern Diasporas in International Politics*, London: Croom Helm.

Sheffer, G. (2003) *At Home Abroad: Diaspora Politics*. Cambridge: Cambridge University Press.

Sheller, M., and Urry, J. (2006) The new mobilities paradigm, *Environment and Planning A*. 38: 207–26.

Shoeb, M., Weinstein, H.M., and Halpern, J. (2007) Living in religious time and space: Iraqi refugees in Dearborn, Michigan, *Journal of Refugee Studies*. 20, 3: 441–59.

Sibley, D. (1995) *Geographies of Exclusion*. London: Routledge.

Siemiatycki, M. (2015) Continuity and change in Canadian immigration policy, in Bauder, H., and Shields, J. (eds) *Immigrant Experiences in North America: Understanding Settlement and Integration*. Toronto: Canadian Scholars Press.

Sigona, N. (2016) Everyday statelessness in Italy: Status, rights, and camps, *Ethnic and Racial Studies*. 39, 2: 263–79.

Silverman, M. (1992) *Deconstructing the Nation: Immigration, Racism and Citizenship in Modern France*. London: Routledge.

Silverman, S.J., and Hajela, R. (2015) Briefing: Immigration detention in the UK, The Migration Observatory, University of Oxford, third revision, report available at: http://www.migrationobservatory.ox.ac.uk/sites/files/migobs/ Immigration%20Detention%20Briefing.pdf

Silvey, R. (2004a) Power, difference and mobility: Feminist advances in migration studies, *Progress in Human Geography*. 28, 4: 490–506.

Silvey, R. (2004b) Transnational domestification: State power and Indonesian migrant women in Saudi Arabia, *Political Geography*. 23: 245–64.

Silvey, R. (2006) Geographies of gender and migration: Spatializing social difference, *International Migration Review*. 40, 1: 64–81.

Silvey, R. (2009) Development and geography: Anxious times, anemic geographies, and migration, *Progress in Human Geography*. 33, 4: 507–15.

Silvey, R., and Lawson, V. (1999) Placing the migrant, *Annals of the Association of American Geographers*. 89, 1: 121–32.

Singer, A. (2004) *The Rise of New Immigrant Gateways*. Brookings Institution Report.

Singer, A. and Massey, D. (1998) The social process of undocumented border crossing among Mexican migrants, *International Migration Review*. 32, 3: 561–92.

Sirkeci, I., Cohen, J.H., and Ratha, D. (eds) (2012) *Migration and Remittances During the Global Financial Crisis and Beyond*. Washington, DC: World Bank.

Sivanandan, A. (2001) Poverty is the new black, *Race and Class*. 43, 2: 1–5.

Sjaastad, L.A. (1962) The costs and returns of human migration, *Journal of Political Economy*. 70, 5: 80–93.

Skeldon, R. (1997 [2014]) *Migration and Development: A Global Perspective*. London: Routledge.

Skeldon, R. (2015) What's in a title? The fifth edition of the 'Age of Migration', *Ethnic and Racial Studies*. 38, 13: 2356–61.

Sklair, L. (2001) *The Transnational Capitalist Class*. Oxford: Blackwell.

Smith, A., and Stenning, A. (2006) Beyond household economies: Articulation and spaces of economic practice in postsocialism, *Progress in Human Geography*. 30: 190–213.

Smith, B.E., and Winders, J. (2008) 'We're here to stay': Economic restructuring, Latino migration and place-making in the US South, *Transactions of the Institute of British Geographers*. 33: 60–70.

Smith, M.P. (2001) *Transnational Urbanism: Locating Globalization*. Oxford: Basil Blackwell.

Smith, M.P. (2005) Transnational urbanism revisited, *Journal of Ethnic and Migration Studies*. 31, 2: 235–44.

Smith, M.P., and Guarnizo, L (1998) (eds) *Transnationalism from Below*. New Brunswick: Transaction Publishers.

Smith, N. (1984) *Uneven Development: Nature, Capital and the Production of Space*. Oxford: Basil Blackwell.

Social and Cultural Geography, Public religion and urban space in Europe. Special Issue, Volume 15, Number 6.

Soguk, N. (1999) *States and Strangers*. Minneapolis: University of Minnesota Press.

Soja, E. (1989) *Postmodern Geographies*. London: Verso.

Song, J. (2016) The politics of immigrant incorporation policies in Korea and Japan, *Asian Perspective*. 40, 1: 1–26.

Sørensen, N.N. (2012) Revisiting the migration-development nexus: From social networks and remittances to markets for migration control, *International Migration*. 50, 3: 61–76.

Sørensen, N.N., Van Hear, N., and Engberg-Pedersen, P. (2002) The migration-development nexus evidence and policy options state-of-the-art overview, *International Migration*. 40, 5: 3–47.

Sousa-Poza, A. (2004) Is the Swiss labour market segmented? An analysis using alternative approaches, *Labour*. 18, 1: 131–61.

Soysal, Y.N. (1994) *Limits of Citizenship: Migrants and Postnational Membership in Europe*. Chicago and London: University of Chicago Press.

Soysal, Y.N. (1997) Changing parameters of citizenship and claims-making: Organized Islam in European public spheres, *Theory and Society*. 26, 4: 509–27.

Spaan, E., Van Naerssan, T., and Kohl, G. (2002) Re-imagining borders: Malay identity and Indonesian migrants in Malaysia, *Tijdschrift Voor Economische en Sociale Geografie*. 93, 2: 160–72.

Sparke, M. (2006) A neo-liberal nexus: Economy, security and the biopolitics of citizenship on the border, *Political Geography*. 25, 2: 151–80.

Spence, L. (2005) Country of birth and labour market outcomes in London: An analysis of labour force survey and census data. London: GLA.

Spoonley, P., and Bedford, R. (2012) *Welcome to Our World? Immigration and the reshaping of New Zealand*. Auckland: Dunmore Publishing.

Staeheli, L.A., and Nagel, C.R. (2006) Topographies of home and citizenship: Arab-American activists in the United States, *Environment and Planning A*. 38: 1599–614.

Staeheli, L.A., Ehrkamp, P., Leitner, H., and Nagel, C. (2012) Dreaming the ordinary: Daily life and the complex geographies of citizenship, *Progress in Human Geography*. 36, 5: 628–44.

Stalker, P. (2000) *Workers without Frontiers: The Impact of Globalization on International Migration*. Geneva: ILO.

Standing, G. (2011) *The Precariat: The New Dangerous Class*. London: Bloomsbury Academic.

Stark, O. (1991) *The Migration of Labor*. Oxford: Basil Blackwell.

Stewart, E. (2008) Exploring the asylum-migration nexus in the context of health professional migration, *Geoforum*. 39, 1: 223–35.

Stewart, E., and Mulvey, G. (2014) Seeking safety beyond refuge: The impact of immigration and citizenship policy upon refugees in the UK, *Journal of Ethnic and Migration Studies*. 40, 7: 1023–39.

Storey, D. (2001) *Territory*. Harlow: Prentice Hall.

Stuesse, A., and Coleman, M. (2014) Automobility, immobility, altermobility: Surviving and resisting the intensification of immigrant policing, *Antipode*. 26, 1: 51–72.

Suárez-Orozco, M.M., and Páez, M.M. (eds) (2002) *Latinos: Remaking America*. Berkeley: University of California Press.

Suhrke, A. (1994) Environmental degradation and population flows, *Journal of International Affairs*. 47: 473–96.

Surak, K. (2008) Convergence in foreigners' rights and citizenship policies? A look at Japan, *International Migration Review*. 42, 3: 550–75.

Surak, K. (2015) Guestworker regimes globally: A historical comparison, in Talani, L.S., and McMahon, S. (eds) *Handbook of the International Political Economy of Migration*. Cheltenham and Northampton: Edward Elgar.

Swyngedouw, E. (1997) Excluding the other: The production of scale and scaled politics, in Lee, R., and Wills, J. (eds) *Geographies of Economies*. London: Arnold.

Taylor, P.J. (1996). Embedded statism and the social sciences: Opening up to new spaces. *Environment and Planning A*. 28: 1917–27.

Taylor, P.J. (2004) *World City Network: A Global Urban Analysis*. London: Routledge.

Taylor, S. (2005) From border control to migration management: The case of paradigm change in the western response to transborder population movement, *Social Policy and Administration*. 39, 6: 563–86.

Teitelbaum, M.S. (2008) Demographic analyses of international migration, in Brettell, C., and Hollifield, J.F. (eds) *Migration Theory: Talking Across Disciplines*, 2nd edition. New York: Taylor and Francis.

Theodore, N. (2003) Political economies of day labour: Regulation and restructuring of Chicago's contingent labour markets, *Urban Studies*. 40, 9: 1811–28.

Theodore, N., Blaauw, D., Schenck, C., Valenzuela, A., Schoeman, C., and Meléndez, E. (2015) Day labor, informality and vulnerability in South Africa and the United States, *International Journal of Manpower*. 36, 6: 807–23.

Thielemann, E. (2004) Why asylum policy harmonisation undermines refugee burden-sharing, *European Journal of Migration and Law*. 6, 1: 47–65.

Tichenor, D. (2015) Symposium Article. The political dynamics of unauthorized immigration: Conflict, change, and agency in time, *Polity*. 47, 283–301.

Tiilikainen, M. (2003) Somali women and daily Islam in the diaspora, *Social Compass*. 50, 1: 59–69.

Tilly, C. (2007) Trust networks in transnational migration, *Sociological Forum*. 22, 1: 3–24.

Tirman, J. (2004) Introduction: The movement of people and the security of states, in Tirman, J. (ed.) *The Maze of Fear: Security and Migration after 9/11*. New York: New Press.

Tobler, W. (1995) Migration, Ravenstein, Thornwaite, and beyond, *Urban Geography*. 16, 4: 327–43.

Todaro, M.P. (1969) A model of labor migration and urban unemployment in less developed countries, *The American Economic Review*. 59, 1: 138–48.

Todaro, M.P. (1976) *International Migration in Developing Countries*. Geneva: ILO.

Tollefsen, A., and Lindgren, U. (2006) Transnational citizens or circulating semi-proletarians? A study of migration circulation between Sweden and Asia, Latin America and Africa between 1968 and 2002, *Population, Space and Place*. 12: 517–27.

Topak, O. (2014) The biopolitical border in practice: Surveillance and death at the Greece-Turkey borderzones, *Environment and Planning D: Society and Space*. 32, 5: 815–33.

Torpey, J. (2000) *The Invention of the Passport: Surveillance, Citizenship and the State*. Cambridge: Cambridge University Press.

Torres, R., Heyman, R., Munoz, S., Apgar, L., Timm, E., Tzintzun, C., Hale, C.R., Mckiernan-Gonzalez, J., Speed, S., and Tang, E. (2013) Building Austin, building justice: Immigrant construction workers, precarious labor regimes and social citizenship, *Geoforum*. 45: 147–57.

Toyota, M., Yeoh, B.S.A., and Nguyen, L. (2007) Editorial introduction: Bringing the 'left behind' back into view in Asia: A framework for understanding the 'migration left-behind nexus', *Population, Space and Place*. 13: 157–61.

Tripier, M. (1990) *Immigration Dans la Class Ouvrière en France*. Paris: L'Harmattan.

Tsuda, T., and Cornelius, W.A. (2004) Japan: government policy, immigrant reality in Cornelius, W., Tsuda, T., Martin, P.L., and Hollifield, J. (eds) *Controlling Immigration: A Global Perspective*. Stanford: Stanford University Press.

Turner, S. (2008) Studying the tensions of transnational engagement: From the nuclear family to world-wide web, *Journal of Ethnic and Migration*. 34, 7: 1049–56.

Tyner, J. (2004) *Made in the Philippines: Gendered Discourses and the Making of Migrants*. London: Routledge Curzon.

UN (2000) Protocol to prevent, suppress and punish trafficking in persons, especially women and children, supplementing the United Nations Convention against Transnational Organized Crime, United Nations.

UN (2012) Tool kit on international migration, Department of Economic and Social Affairs Population Division Migration Section, June.

UN (2016) UN International Migration Report, advance copy, available at: http://www.un.org/en/development/desa/population/migration/publications/migrationreport/docs/MigrationReport2015_Highlights.pdf

UNDESA (2016) *International Migration 2015*. New York: UNDESA.

UNHCR (2008) Convention and Protocol Relating to the Status of Refugees, available at: http://www.unhcr.org/cgi-bin/texis/vtx/home.

UNHCR (2015a) States Parties to the 1951 Convention relating to the Status of Refugees and the 1967 Protocol.

UNHCR (2015b) *Statistical Yearbook 2014*. Geneva: UNHCR.

UNHCR (2016a) UNHCR Refugee Resettlement Trends 2015.

UNHCR (2016b) UNHCR website, available at: http://www.unhcr.org/pages/49c3646c11.html

UNRWA (2015) *About UNRWA*, Jerusalem: UNRWA.

U.S. Committee for Refugees and Immigrants (2008) *World Refugee Survey: The Worst Places for Refugees*.

Valenzuela, A. (2001) Day labourers as entrepreneurs, *Journal of Ethnic and Migration Studies*. 27, 2: 335–52.

Van Amersfoort, H. (1996) Migration: The limits of governmental control, *New Community*, 22, 2: 243–57.

Van Hear, N. (1998) *New Diasporas*. London: UCL Press.

Van Hear, N. (2014) Reconsidering migration and class, *International Migration Review*. 48, S1: 100–121.

Van Hear, N., and Sørensen, N.N. (eds) (2003) *The Migration-Development Nexus*, Geneva: IOM.

Van Hook, K., and Balistreri, J.S. (2006) Ineligible parents, eligible children: Food stamps receipt, allotments, and food insecurity among children of immigrants, *Social Science Research*. 35, 1: 228–51.

van Houtum, H. (2010) Human blacklisting: The global apartheid of the EU's external border regime, *Environment and Planning D: Society and Space*. 28, 6: 957–76.

van Houtum, H., and Pijpers, R. (2007) The European Union as a gated community: The two-faced border and immigration regime of the EU, *Antipode*. 39, 2: 291–309.

Van Liempt, I., and Doomernik, J. (2006) Migrant's agency in the smuggling process: The perspectives of smuggled migrants in the Netherlands, *International Migration*. 44, 4: 165–89.

Van Liempt, I., and Sersli, S. (2013) State responses and migrant experiences with human smuggling: A reality check, *Antipode*. 45, 4: 1029–46.

Van Parijs, P. (1992) Commentary: Citizenship exploitation, unequal exchange and the breakdown of popular sovereignty, in Barry, B., and Goodin, R.E. (eds) *Free Movement: Ethical Issues in the Transnational Migration of People and Money*. New Jersey: Harvester Wheatsheaf.

Van Tubergen, F. (2006) Religious affiliation and attendance among immigrants in eight western countries: Individual and contextual effects, *Journal of the Scientific Study of Religion*. 45, 1: 1–22.

Van Wijk, J. (2010) Luanda – Holanda: Irregular Migration from Angola to the Netherlands, *International Migration*. 48, 2: 1–30.

Vansemb, B. (1995) The place of narrative in the study of Third World migration: The case of spontaneous rural migration in Sri Lanka, *Professional Geographer*. 47, 4: 411–25.

Varsanyi, M. (2008) Immigration policing through the backdoor: City ordinances, the 'right to the city', and the exclusion of undocumented day laborers, *Urban Geography*. 29, 1: 29–52.

Varsanyi, M., Lewis, P.G., Provine, D.M., and Decker, S. (2012) A multilayered jurisdictional patchwork: Immigration federalism in the United States, *Law and Policy*. 34, 2: 138–58.

Vasta, E., and Kandilige, L. (2010) 'London the Leveller': Ghanaian Work Strategies and Community Solidarity, *Journal of Ethnic and Migration Studies*. 36, 4: 581–98.

Vaughan-Williams, N. (2015) 'We are not animals!' Humanitarian border security and zoopolitical spaces in Europe, *Political Geography*. 45: 1–10.

Veiga, U.M. (1999) Immigrants in the Spanish labour market, in Baldwin-Edwards, M., and Arango, J. (eds) *Immigrants and the Informal Economy in Southern Europe*. London: Frank Cass.

Veronis, L. (2006) The Canadian Hispanic Day parade, or how Latin American immigrants practice (sub)urban citizenship in Toronto, *Environment and Planning A*. 38: 1653–71.

Vertovec, S. (1999) Conceiving and researching transnationalism, *Ethnic and Racial Studies*. 22, 2: 447–62.

Vertovec, S. (2001) Transnationalism and identity, *Journal of Ethnic and Migration Studies*. 27, 4: 573–82.

Vertovec, S. (2004) Migrant transnationalism and modes of transformation, *International Migration Review*. 38, 3: 970–1001.

Vertovec, S. (2007) Super-diversity and its implications, *Ethnic and Racial Studies*. 30, 6: 1024–54.

Viladrich, A. (2011) Beyond welfare reform: Reframing undocumented immigrants' entitlement to health care in the United States, a critical review, *Social Science and Medicine*. 74: 822–9.

Vink, M.P. (2007) Dutch 'multiculturalism' beyond the pillarisation myth, *Political Studies Review*. 5: 337–50.

Vink, M.P., and Bauböck, R. (2013) Citizenship configurations: Analysing the multiple purposes of citizenship regimes in Europe, *Comparative European Politics*. 11, 5: 621–48.

Vink, M.P., and de Groot, G-R (2010) Citizenship attribution in Western Europe: International framework and domestic trends, *Journal of Ethnic and Migration Studies*. 36, 5: 713–34.

Voigt-Graf, C. (2004) Towards a geography of transnational spaces: Indian transnational communities in Australia, *Global Networks*. 4, 1: 25–49.

Waever, O., Buzan, B., Kelstrup, M., and Lemaitre, P. (1993) *Identity, Migration and the New Security Agenda in Europe*. New York: St. Martin's Press.

Wahlbeck, O. (2002) The concept of diaspora as an analytical tool in the study of refugee communities, *Journal of Ethnic and Migration Studies*. 28: 221–38.

Wahlbeck, O. (2007) Work in the kebab economy – a study of the ethnic economy of Turkish immigrants in Finland, *Ethnicities*. 7, 4: 543–63.

Waldinger, R. (2008) The border within: Citizenship facilitated and impeded, a review of 'Becoming a Citizen: Incorporating Immigrants and Refugees in the United States and Canada', by Irene Bloemraad, *Contemporary Sociology– A Journal of Reviews*. 37, 4: 306–8.

Waldinger, R. (2015) *The Cross-Border Connection: Immigrants, Emigrants, and their Homelands*. Cambridge: Harvard University Press.

Waldinger, R., and Fitzgerald, D. (2004) Transnationalism in question, *American Journal of Sociology*. 109, 5: 1177–95.

Waldinger, R., and Lichter, M. (2003) *How the Other Half Works: Immigration and the Social Organization of Labor*. Berkeley: University of California Press.

Waldinger, R., and Luthra, R.R. (2010) Into the Mainstream? Labor market outcomes of Mexican-origin workers, *International Migration Review*. 44, 4: 830–68.

Wallerstein, I. (1974) *The Modern World-System, vol. I: Capitalist Agriculture and the Origins of the European World-Economy in the Sixteenth Century*. New York/ London: Academic Press.

Wallerstein, I. (1979) *The Capitalist World-Economy*. Cambridge: Cambridge University Press.

Wall Street Journal (2015a) 'Migrants make Sahara outpost a boomtown', 16 July.

Wall Street Journal (2015b) 'In Greece, debt and migration crises collide', 31 August.

Wall Street Journal (2015c) 'At sea's edge, a migrant family hesitates', 19–20 September.

Wall Street Journal (2015d) 'Berlin gets tough on economic migrants', 25 September.

Walsh, J. (2007) Navigating Globalization: Immigration Policy in Canada and Australia, 1945–2007, *Sociological Forum*. 23, 4: 786–813.

Walters, W. (2004) Secure borders, safe haven, domopolitics, *Citizenship Studies*. 8, 3: 237–60.

Walters, W. (2006) Rethinking borders beyond the state, *Comparative European Politics*. 4: 141–59.

Walters, W. (2008) Anti-political economy: Cartographies of 'illegal immigration' and the displacement of the economy, in Best, J., and Paterson, M. (eds) *Cultural Political Economy*. London and New York: Routledge.

Walters, W. (2010) Anti-illegal immigration policy: The case of the European Union, in Gabriel, C., and Pellerin, H. (eds) *Governing International Labour Migration*. London and New York: Routledge.

Walters, W. (2015) Migration, vehicles, and politics: Three theses on viapolitics, *European Journal of Social Theory*. 18, 4: 469–88.

Walton-Roberts, M. (2004) Rescaling citizenship: Gendering Canadian immigration policy, *Political Geography*. 23, 3: 265–81.

Walzer, M. (1983) *Spheres of Justice*. New York: Basic Books.

Wang, H.Z. (2007) Hidden Spaces of Resistance of the Subordinated: Case Studies from Vietnamese Female Migrant Partners in Taiwan, *International Migration Review*. 41: 706–27.

Wang, W.W., and Fan, C.C. (2012) Migrant Workers' Integration in Urban China: Experiences in Employment, Social Adaptation, and Self-Identity, *Eurasian Geography and Economics*. 53, 6: 731–49.

Ward, K., and England, K. (2007) Introduction: Reading neoliberalization, in Ward, K. and England, K. (eds) *Neoliberalization: States, Networks, Peoples*. Oxford: Blackwell Publishing.

Washington Post (2015a) 'People in Europe are full of fear' over refugee influx', 3 September.

Washington Post (2015b) 'A smugglers' haven in the Sahara', 20 July.

Waslin, M. (2009) Immigration policy and the Latino community since 9/11, in Givens, T., Freeman, G., and Leal, D. (eds) *Immigration Policy and Security*. New York and London: Routledge.

Waters, J. (2003) Flexible citizens? Transnationalism and citizenship amongst economic immigrants in Vancouver, BC, *The Canadian Geographer*. 47, 3: 219–34.

Waters, M.C., and Jiminez, T.R. (2005) Assessing immigrant assimilation: New empirical and theoretical challenges, *Annual Review of Sociology*. 31: 105–25.

Watson, I. (2010) Multiculturalism in South Korea: A critical assessment, *Journal of Contemporary Asia*. 40, 2: 337–46.

Weber, F. (2016) Labour Market Access for Asylum Seekers and Refugees under the Common European Asylum System, *European Journal of Migration and Law*. 18, 1: 34–64.

Weiner, M. (1995) *Global Migration Crisis: Challenges to States and Human Rights*. New York: Harper Collins.

Wells, M. (1996) *Strawberry Fields: Politics, Class and Work in California Agriculture*. Ithaca: Cornell University Press.

White, A. (2011) Polish Migration in the UK – local experiences and effects, *AHRC Connected Communities symposium: Understanding Local Experiences and Effects of New Migration*, 2011-09-26, Sheffield Hallam University, Sheffield.

White, J. (1998) Old wine, cracked bottle? Tokyo, Paris and the global city hypothesis, *Urban Affairs Review*. 33: 451–77.

White, P., Winchester, H., and Guillon, M. (1987) South-east Asian refugees in Paris: The evolution of a minority community, *Ethnic and Racial Studies*. 10, 1: 48–61.

Wilkinson, M., and Craig, C. (2012) Wilful negligence: Migration policy, migrants' work and the absence of social protection in the UK, in Carmel, E., Cerami, A., and Papadopoulos, T. (eds) *Migration and Welfare in the New Europe*. Bristol: The Policy Press.

Willen, S.S. (2010) Citizens, 'real' others, and 'other' others: Governmentality, biopolitics, and the deportation of undocumented migrants from Tel Aviv, in DeGenova, N., and Peutz, N. (eds) *The Deportation Regime: Sovereignty, Space and the Freedom of Movement.* Durham: Duke University Press.

Williams, A.M., and Balaz, V. (2012) Migration, Risk, and Uncertainty: Theoretical Perspectives, *Population, Space and Place.* 18, 2: 167–80.

Williams, A.M., Baláz, V., and Wallace, C. (2004) International labour mobility and uneven regional development in Europe: Human capital, knowledge and entrepreneurship, *European Urban and Regional Studies.* 11, 1: 27–46.

Williams, C.C. (2010), The changing conceptualisation of informal work in developed economies, in Marcelli, E., Williams, C.C., and Joassart, P. (eds), *Informal Work in Developed Nations.* Routledge, London, 11–33.

Williams, C.C., and Lansky, M.A. (2013) Informal employment in developed and developing economies: Perspectives and policy responses, *International Labor Review.* 152, 3–4: 355–80.

Williams, C.C., and Nadin, S. (2014) Evaluating the participation of the unemployed in undeclared work: Evidence from a 27-nation European survey, *European Societies.* 16, 1: 68–89.

Williams, C.C., and Windebank, J. (1998) *Informal Employment in the Advanced Economies.* London: Routledge.

Willis, K., and Yeoh, B. (eds) (2000) *Gender and Migration.* Cheltenham and Northampton: Edward Elgar.

Wills, J., May, J., Datta, K., Evans, Y., Herbert, J., and McIlwaine, C. (2008) *London's Changing Migrant Division of Labour.* London: Queen Mary University, Department of Geography.

Wilpert, C. (1998) Migration and informal work in the new Berlin: New forms of work or new sources of labour, *Journal of Ethnic and Migration Studies.* 24, 2: 269–94.

Wilson, J.H., and Habecker, S. (2008) The lure of the capital city: An anthro-geographical analysis of recent African immigration to Washington, DC, *Population, Space and Place.* 14: 433–48.

Wilson, K.L., and Portes, A. (1980) Immigrant enclaves: An analysis of the labor market experiences of Cubans in Miami, *American Journal of Sociology.* 86: 305–19.

Wilson, T.D. (1993) Theoretical approaches to Mexican wage labor migration, *Latin American Perspectives.* 20, 3: 98–129.

Wilson, T.D. (1998) Weak ties, strong ties: Network principles in Mexican migration, *Human Organization.* 57, 4: 394–403.

Wimmer, A. (2015) Race-centrism: A critique and research agenda, *Ethnic and Racial Studies.* 38, 13: 2186–205.

Wimmer, A., and Glick-Schiller, N.G. (2002) Methodological nationalism and the study of migration, *Archives Européene de Sociologie.* 43, 2: 217–40.

Wimmer, A., and Glick-Schiller, N.G. (2003) Methodological nationalism, the social sciences, and the study of migration: An essay in historical epistemology, *International Migration Review.* 37, 3: 576–610.

Winant, H. (2015) Race, ethnicity and social science, *Ethnic and Racial Studies.* 38, 13: 2176–85.

Winders, J. (2012) Seeing immigrants: Institutional visibility and immigrant incorporation in new immigrant destinations, *The ANNALS of the American Academy of Political and Social Science*. 641, 1: 58–78.

Witteborn, S. (2015). Becoming (Im)perceptible: Forced migrants and virtual practice, *Journal of Refugee Studies*. 28, 3: 350–67.

Wolpe, H. (1980) *The Articulation of Modes of Production*. London: Routledge.

Wolpert, J. (1965) Behavioural aspects of the decision to migrate, *Papers of the Regional Science Association*. 15: 159–69.

Wolpert, J. (1966) Migration as an adjustment to environmental stress, *Journal of Social Issues*. 22, 4: 92–102.

World Bank (2012) *World Development Indicators 2012*. Washington: World Bank.

World Bank (2016) Migration and remittances: Recent developments and outlook, *Migration and Development Brief*. No.26, April.

Wright, R., and Ellis, M. (1996) Immigrants and the changing ethnic-racial division of labor in New York City, 1970–1990, *Urban Geography*. 17: 317–53.

Wright, R., and Ellis, M. (2000a) The ethnic and gender division of labor compared among immigrants to Los Angeles, *International Journal of Urban and Regional Research*. 24, 3: 583–600.

Wright, R., and Ellis, M. (2000b) Race, region and the territorial politics of immigration in the US, *International Journal of Population Geography*. 6: 197–211.

Yanasmayan, Z. (2015) Citizenship on paper or at heart? A closer look into the dual citizenship debate in Europe, *Citizenship Studies*. 19, 6–7: 785–801.

Yeates, N. (2004) A dialogue with 'global care chain' analysis: Nurse migration in the Irish context, *Feminist Review*. 77: 79–95.

Yeates, N. (2009) *Globalising Care Economies and Migrant Workers: Explorations in Global Care Chains*. Palgrave: Basingstoke.

Yeates, N. (2010) The globalization of nurse migration: Policy issues and responses, *International Labour Review*. 149, 4: 423–40.

Yeates, N. (2012) Global care chains: A state-of-the-art review and future directions in care transnationalization research, *Global Networks*. 12, 2: 135–54.

Yeoh, B.S.A., Chee, H.L., and Baey, G.H.Y. (2013) The place of Vietnamese marriage migrants in Singapore: Social reproduction, social 'problems' and social protection, *Third World Quarterly*. 34, 10: 1927–41.

Yeoh, B.S.A., Chee, H.L., and Vu, T.K.D. (2013) Global householding and the negotiation of intimate labour in commercially-matched international marriages between Vietnamese women and Singaporean Men, *Geoforum*. 51: 284–93.

Yeoh, B.S.A, and Huang, S. (1998) Negotiating public space: Strategies and styles of migrant female domestic workers in Singapore, *Urban Studies*. 35, 3: 583–602.

Yeoh, B.S.A., and Huang, S. (2010) Transnational domestic workers and the negotiation of mobility and work practices in Singapore's home-spaces, *Mobilities*. 5, 2: 219–36.

Yeoh, B.S.A., and Lin, W. (2013) Chinese migration to Singapore: Discourses and discontents in a globalizing nation-state, *Asian and Pacific Migration Journal*. 22, 1: 31–54.

Yeoh, B.S.A., and Soco, M.A. (2014) The cosmopolis and the migrant domestic worker, *Cultural Geographies*. 21, 2: 171–87.

Yeoh, B.S.A., and Willis, K. (1999) 'Heart' and 'wing', nation and diaspora: Gendered dimensions in Singapore's regionalisation process, *Gender, Place and Culture*. 64, 4: 355–72.

Zelinsky, W., and Lee. B. (1998) Heterolocalism: An alternative model of the socio-spatial behaviour of immigrant ethnic communities, *International Journal of Population Geography*. 4, 4: 281–98.

Zetter, R. (1988) Refugees, repatriation, and root causes, *Journal of Refugee Studies*. 1, 2: 99–106

Zetter, R. (2007) More labels, fewer refugees: Remaking the refugee label in an era of globalization, *Journal of Refugee Studies*. 20, 2: 172–92.

Zetter, R., and Morrissey, J. (2014) The environment-mobility nexus: Reconceptualizing the links between environmental stress, (im)mobility and power, in Fiddian-Qasmiyeh, E., Loescher, G., Long, K., and Sigona, N. (eds) *The Oxford Handbook of Forced Migration*. Oxford: Oxford University Press.

Zhang, H.X., Kelly, P.M., Locke, C., Winkels, A., and Adger, W.N. (2007) Migration in a transitional economy: beyond the planned and spontaneous dichotomy in Vietnam, *Geoforum*. 37: 1066–81.

Zhou, M. (1992) *Chinatown: The Socioeconomic Potential of an Urban Enclave*. Philadelphia: Temple University Press.

Zhou, M. (1997) Segmented assimilation: Issues, controversies, and recent research on the new second generation, *International Migration Review*. 31, 4: 825–58.

Zhou, M., Lee, J., Vallejo, J.A., Tafoya-Estrada, R., and Xiong, Y.S. (2008) Success attained, deterred, and denied: Divergent pathways to social mobility in Los Angeles's new second generation, *Annals of the American Academy of Political and Social Science*. 620: 37–61.

Zimmerman, S.E. (2009) Irregular secondary movements to Europe: Seeking asylum beyond refuge, *Journal of Refugee Studies*. 22, 1: 74–96.

Zimmerman, W., and Tumlin, K.C. (1999) *Patchwork Policies: State Assistance for Immigrants under Welfare Reform*. The Urban Institute, Occasional Paper 24.

Zlotnick, H. (1998) International migration 1965–1996: An overview, *Population and Development Review*. 24, 3: 429–68.

Zolberg, A. (2002) Guarding the gates, on-line paper, available at: http://www.newschool.edu/icmec/guardingthegates.html

Zolberg, A. (2006) Managing a world on the move, *Population and Development Review*. 32, Supplement S: 222–53.

Zuberi, D., and Ptashnick, M. (2012) In search of a better life: The experiences of working poor immigrants in Vancouver, Canada, *International Migration*. 50, Issue Supplement s1: e60–e93.

INDEX

Note: 'f' after a page number indicates a figure; 'm' a map; 't' a table; and 'n' a note

Taylor & Francis eBooks

Helping you to choose the right eBooks for your Library

Add Routledge titles to your library's digital collection today. Taylor and Francis ebooks contains over 50,000 titles in the Humanities, Social Sciences, Behavioural Sciences, Built Environment and Law.

Choose from a range of subject packages or create your own!

Benefits for you

» Free MARC records
» COUNTER-compliant usage statistics
» Flexible purchase and pricing options
» All titles DRM-free.

Benefits for your user

» Off-site, anytime access via Athens or referring URL
» Print or copy pages or chapters
» Full content search
» Bookmark, highlight and annotate text
» Access to thousands of pages of quality research at the click of a button.

REQUEST YOUR **FREE** INSTITUTIONAL TRIAL TODAY | **Free Trials Available** We offer free trials to qualifying academic, corporate and government customers.

eCollections – Choose from over 30 subject eCollections, including:

Archaeology	Language Learning
Architecture	Law
Asian Studies	Literature
Business & Management	Media & Communication
Classical Studies	Middle East Studies
Construction	Music
Creative & Media Arts	Philosophy
Criminology & Criminal Justice	Planning
Economics	Politics
Education	Psychology & Mental Health
Energy	Religion
Engineering	Security
English Language & Linguistics	Social Work
Environment & Sustainability	Sociology
Geography	Sport
Health Studies	Theatre & Performance
History	Tourism, Hospitality & Events

For more information, pricing enquiries or to order a free trial, please contact your local sales team:
www.tandfebooks.com/page/sales

 Routledge
Taylor & Francis Group | The home of Routledge books | **www.tandfebooks.com**

Printed in the United States
by Baker & Taylor Publisher Services